U0931058

中国科学院生物标本馆

主编　乔格侠

科学出版社
北　京

内 容 简 介

中国科学院生物标本馆包括隶属于中国科学院19个科研机构的20家生物标本馆（博物馆），是我国动物、植物、菌物及化石等标本保藏、研究和科学教育的主要实体场馆，是国家科技资源两个共享服务平台——国家动物标本资源库和国家植物标本资源库的牵头单位。中国科学院生物标本馆的标本藏量、管理和利用，代表着我国战略生物资源的整体水平，在国家生物资源的保护与可持续利用中具有不可替代的重要作用，对国家生物安全发挥了重要的支撑与保障作用。本书阐述了中国科学院生物标本馆的发展历史，以及各收藏机构的基本概况，总结了近年来其在战略生物资源收集和保藏、服务科研支撑和国家需求等方面所取得的各项成果，并分析了生物标本馆的未来发展趋势。

本书可供涉及生物学领域的研究院所和高校的科研工作者，涉及生物标本收藏的各标本馆、博物馆及其他收藏机构和农、林、环保、自然保护区工作人员，以及对生物标本收藏感兴趣的个人阅读参考。

图书在版编目（CIP）数据

中国科学院生物标本馆=Biological Collections of Chinese Academy of Sciences: 汉英对照 / 乔格侠主编. —北京：科学出版社，2021.9

ISBN 978-7-03-068688-6

Ⅰ. ①中…　Ⅱ. ①乔…　Ⅲ. ① 生物－标本－博物馆－介绍－中国－汉、英　Ⅳ. ①Q-34

中国版本图书馆CIP数据核字（2021）第075480号

责任编辑：马　俊　李　迪　高璐佳 / 责任校对：郑金红

责任印制：肖　兴 / 封面设计：北京图阅盛世文化传媒有限公司

科学出版社 出版

北京东黄城根北街16号

邮政编码：100717

http://www.sciencep.com

北京九天鸿程印刷有限责任公司 印刷

科学出版社发行　各地新华书店经销

*

2021年9月第　一　版　开本：889 × 1194　1/16

2021年9月第一次印刷　印张：16 3/4

字数：534 000

定价：298.00元

（如有印装质量问题，我社负责调换）

Biological Collections of Chinese Academy of Sciences

Chief Editor QIAO Gexia

Science Press
Beijing

QIAO Gexia
Biological Collections of Chinese Academy of Sciences

Biological collections of Chinese Academy of Sciences includes 20 biological collections or museums belonging to 19 institutions of Chinese Academy of Sciences. They are the main entity venues for preservation, research and scientific education of specimens such as animals, plants, fungi and fossils in China. They are also the leading unit of the two sharing service platforms of national scientific and technological resources: the National Animal Collection Resource Center and the National Plant Specimen Resource Center. The quantity, management and utilization of the specimens of the biological collections of CAS represents the overall level of China's strategic biological resources, play an irreplaceable role in the protection and sustainable utilization of national biological resources, and play an important supporting and guarantee role in the national biological safety. This book describes the history of the development of the biological collections of CAS, as well as the basic situation of every collections, summarizes the achievements in strategic biological resource collection and preservation, servicing support for scientific research and national demand in recent years, and analyzes the future development trend of the biological collections.

This book can be used as a reference for the researchers of institutes and universities in the fields of biology, the collections, herbariums, museums and other institutions involved in the collection of biological specimens, the staff of agriculture, forestry, environmental protection and nature reserves, as well as the individuals interested in the collection of biological specimens.

ISBN 978-7-03-068688-6

生态文明：共建地球生命共同体

谨以此书献给《生物多样性公约》第十五次缔约方大会

Ecological Civilization—Building a Shared Future for All Life on Earth

This book is dedicated to the 15th Meeting of the Conference of the Parties to the *Convention on Biological Diversity*

丛书编委会

Editorial Committee of the Series

《中国科学院生物标本馆》编委会

主　任

孙　命　　中国科学院科技促进发展局

副主任

周　桔　　中国科学院科技促进发展局

曾　艳　　中国科学院科技促进发展局

主　编

乔格侠　　中国科学院动物研究所

编写者（按姓氏汉语拼音排序）

蔡　磊　　中国科学院微生物研究所

陈　军　　中国科学院动物研究所

陈世龙　　中国科学院西北高原生物研究所

冯　缨　　中国科学院新疆生态与地理研究所

高信芬　　中国科学院成都生物研究所

贺　鹏　　中国科学院动物研究所

胡光万　　中国科学院武汉植物园

孔宏智　　中国科学院植物研究所

李　淳　　中国科学院古脊椎动物与古人类研究所

李家堂　　中国科学院成都生物研究所

李剑武　　中国科学院西双版纳热带植物园

李维薇　　中国科学院昆明动物研究所

彭焱松　　中国科学院庐山植物园

乔格侠　　中国科学院动物研究所

谭烨辉　　中国科学院南海海洋研究所

王少青　　中国科学院海洋研究所

王永栋　　中国科学院南京地质古生物研究所

杨祝良　　中国科学院昆明植物研究所

殷海生　　中国科学院分子植物科学卓越创新中心

袁海生　　中国科学院沈阳应用生态研究所

张奠湘　　中国科学院华南植物园

张先锋　　中国科学院水生生物研究所

Editorial Committee of Biological Collections of Chinese Academy of Sciences

序

生物标本是人类了解自然、探究生命奥秘的直接材料，也是一种重要的战略生物资源。生物标本馆则是肩负着妥善长久保藏这一重要生物资源责任的场所。

中国的生物标本事业起步较晚，但中国科学院历来重视生物标本的收集和保藏工作，并始终坚持将其作为生命科学领域的重点内容之一进行培育和建设。在各个以生物为研究对象的研究所成立之时，往往会同时建立用于集中保藏标本的部门。生物标本馆的历史，同时也是这些研究所的发展史。在追溯这些研究所的历史时，我们甚至会发现其标本馆在研究所还未成立之时就已经存在，而其保藏的一些标本甚至已经存在了一两百年。中华人民共和国成立后，在对以中国科学院馆藏为主体的生物标本的大量研究基础上，经过几代生物分类学工作者的努力，诞生了我国生物领域的一项重大科学成果：《中国植物志》、《中国动物志》、《中国孢子植物志》的编研和出版。这是我们对自己国土上的生物的一次全面普查，也为中国其他生命科学领域，特别是生物多样性保护、生物安全及生物资源可持续利用等方面奠定了坚实的基础。所以无论从记录自然生命历史的角度，还是见证研究所、科学院，乃至整个学科发展的角度，或是保存生物资源的战略意义方面，生物标本都是一笔宝贵财富。

近年来，随着我国在生物领域的发展以及国家对生物资源的重视和投入力度的加大，生物标本馆在资源收集、保藏和管理、利用等方面都有了明显进步。尤其是中国科学院体系的生物标本馆，目前已有由 19 个研究所作为依托单位的 20 家标本馆（博物馆），收集保藏了全国 60% 的生物标本，并在生物多样性研究与保护、服务国家经济建设和开展公众科学教育方面取得了一系列成绩。

2018 年，中国科学院战略生物资源管理委员会组织全院生物标本馆开展了“生物标本馆（博物馆）综合评估”，对近年来所取得的成绩进行了全面梳理和总结，该书即对中国科学院生物标本馆在战略生物资源方面所做工作的集中展示。此外，“《生物多样性公约》第十五次缔约方大会”（COP15）即将在我国昆明召开，届时将通过该书向参会人员展示中国在生物多样性方面的工作。

然而与发达国家相比，我国在生物标本收藏、管理和可持续利用方面还有较大差距，在战略生物资源保护与发展方面仍然任重而道远。我们需要不断加强对战略生物资源的科学认知，深入开展生物资源的全面调查，进一步推进生物标本资源的保藏、利用与共享。同时坚持以生态文明建设为引领，发挥战略生物资源在服务“健康中国”和“美丽中国”建设方面的重要作用，致力于推进“后 2020 时代”的全球生物多样性保护和经济可持续发展进程。

地球本身是一个整体，其生态环境不会因国境线而改变。随着各国经济发展，国际贸易和协作逐步深入，而全球性生态环境问题却日益严峻，生物多样性变化趋势也不容乐观。为了建设人类命运共同体，维护良好的生存环境，各国需要也必须携手共进，共同应对和解决全球性的复杂生态问题，需要在生物多样性方面加强全球合作。

该书为中英文双语出版，目的就是向国际社会展示中国科学院在生物资源和生物多样性等领域所做出的努力与贡献，希望能与其他国家建立更多联系，创造更多合作机会。我们愿与世界各国携手，为人类了解自然和地球生态环境，为人类命运共同体的构建贡献一份力量。

中国科学院院士

2020 年 7 月 22 日

Foreword

Biological specimens are not only the direct material for human to understand nature and explore the mystery of life, but also important strategic biological resources. The biological collections are the places to take the responsibility of long-term preservation of these important biological resources.

The cause of biological specimens in China started late, but the Chinese Academy of Sciences has always attached great importance to the collections and preservation of biological specimens and has always insisted on its cultivation and construction as one of the key contents in the field of life science. At the time of the establishment of biological research institutes, departments for the centralized collection of specimens are often established at the same time. The history of the biological collections is also the development history of these institutes. When we trace back to the history of these institutes, we can even find that the collections existed before the establishment of the institutes, and some specimens have existed for one or two hundred years. After the founding of the People's Republic of China, on the basis of a large number of studies on biological specimens collected by the Chinese Academy of Sciences and the efforts of several generations of taxonomists, an important scientific achievement in the field of biology in China was born: the compilation, research and publication of the *Flora Reipublicae Popularis Sinicae* (*Flora of China*), *Fauna Sinica* (*Fauna of China*) and *Flora Cryptogamarum Sinicarum* (*Cryptogamic Flora of China*). This is a comprehensive survey of the living things in our own land and has laid a solid foundation for other fields of life sciences in China, especially in the areas of biodiversity protection, biological security and sustainable utilization of biological resources. Therefore, no matter from the perspective of recording the history of natural life, witnessing the development of institutes, the Chinese Academy of Sciences and even the whole discipline of biology in China, or the strategic significance of preserving biological resources, biological specimens are valuable assets.

In recent years, with the development of biological research in China and the increase of national attention and investment in biological resources, biological collections have made significant progress in the collection, preservation, management and utilization of resources. The Chinese Academy of Sciences, in particular, has 20 collections (museums) affiliated to 19 research institutes, which collect and preserve 60% biological specimens of the country, and have made a series of achievements in biodiversity research and protection, servicing national economic construction and public science education.

In 2018, the Strategic Biological Resources Management Committee of the Chinese Academy of Sciences organized all the biological collections of the Chinese Academy of Sciences to carry out the "Evaluation of the Biological Collections (Museums)", comprehensively summarizing the achievements made in recent years. This book is a concentrated display of the work of the biological collections of the Chinese Academy of Sciences in the aspect of strategic biological resources. On the other hand, the 15th Meeting of the Conference of the Parties to the *Convention on Biological Diversity* (COP15) will be held in Kunming, China. At that time, this book will show China's work on biodiversity to all participants.

However, compared with developed countries, there is still a big gap in the collection, management and sustainable utilization of biological specimens and a long way to go in the protection and development of strategic biological resources in China. We need to constantly strengthen the scientific understanding of strategic biological resources, carry out comprehensive investigations of them, and further promote the preservation, utilization and sharing of them. At the same time, we should adhere to the guidance of ecological civilization construction, give play to the important role of strategic biological resources in serving the construction of "Healthy China" and "Beautiful China", and strive to promote the process of global biodiversity protection and economic sustainable development in the "Post 2020 Era".

The earth itself is a whole, and its ecological environment will not be changed by the boundary line. With the economic development of various countries, international trade and cooperation are gradually deepening, but the global ecological environment problems are increasingly serious, and the trend of biodiversity change is not optimistic. In order to build a community of shared future for mankind and maintain a good living environment, all countries need and must work together to deal with and solve global complex ecological problems and strengthen global cooperation in biodiversity.

This book is published in both Chinese and English. The purpose of it is to show the efforts and contributions made by the Chinese Academy of Sciences in the fields of biological resources and biodiversity to the international community. It is hoped that more contacts can be established with other countries and more opportunities for cooperation will be generated. We are willing to join hands with other countries in the world to help mankind understand the nature and the ecological environment of the earth and contribute to the construction of a community of shared future for mankind.

CHEN Yiyu
Academician of the Chinese Academy of Sciences
July 22, 2020

目　录

Contents

国家动物博物馆

國家动物博物館
National Zoological Museum of China

1

引　言

Introduction

生物标本是自然界各种生物的最真实、最直接的表现形式和实物记录，是生物学研究领域的重要素材，在分类学、系统学、生态学等各个分支学科中都有着十分重要的地位。同时，生物标本在许多领域都为社会带来利益，包括生物安全、药物研发、公共卫生和安全、环境变化监测以及传统分类学和系统学研究等。

生物标本馆则是长久妥善保存这些记录的场馆，其建立的目的就是让生物标本的收藏范围不断扩大、保存时间不断延长，从而为科研提供多种多样的研究材料，为社会大众展示千姿百态的自然生命，为国民经济建设和国家生物安全提供资源保障。

中国的生物标本收藏与生物标本馆孕育于 19 世纪，起步于 20 世纪初，发展于中华人民共和国成立之后。其中最重要、保藏标本数量最多的是隶属于中国科学院系统的标本馆。

中国科学院生物标本馆是国家动物、植物、菌物、化石等标本保藏、研究和科学教育的重要实体，是全国生物标本集中保藏的最主要场所，在国家生物资源的保护与可持续利用中具有不可替代的重要作用。截至 2019 年底，中国科学院共有 20 个生物标本馆（博物馆），隶属于 19 个研究所，保藏各类生物标本共计 2141.5 万号，其中定名标本占一半以上，近一半标本完成了数字化工作。

自 2014 年以来，中国科学院生物标本馆取得了一系列可喜的成绩：在战略生物资源收集与保藏方面，配合国家重大研究计划，各馆开展针对性考察和采集，共组织国内或国际合作考察 1400 余次，采集标本 317.3 万号；在服务国家战略方面，标本馆参与国家重大战略、重大研究计划和国门生物安全建设，“十三五”以来各馆主持或者支撑的院内外项目共计 491 项，经费达 5.7 亿元，对科研工作发挥了巨大的支撑作用；在经典分类及生物物种资源研究方面，共鉴定标本 103 万余号，显著提升了馆藏水平，还支撑了 5600 余篇论文和 230 余部专著的发表，对中国生物物种资源研究发挥了重要作用；在公众教育与科普服务方面，共举办 2000 余场各类科普活动，受众达 446 万余人次，提供社会咨询服务 1.8 万余人次，发表科普著作 80 余部、科普文章 700 余篇，得到了社会各界的高度认可。

但与发达国家相比，我国在标本收藏、管理及可持续利用方面还有差距。未来将在中国科学院的领导下，建立中国科学院生物标本馆分类管理体系；依托国家科技资源共享服务平台，建设集生物标本收集、保藏、研究、支撑与科普为一体的国家生物标本资源共享服务平台；依托“国家动物标本资源库”和“国家植物标本资源库”，以标本收集、保存、鉴定、研究为主要内容，继续服务国家重大项目；针对“一带一路”国家、全球生物多样性热点地区，开展联合考察与研究、资源共享、人才培养等国际合作；加强标本馆自身能力建设，提升支撑科技创新能力，构建合理的人才队伍，确保体系的可持续发展。

The biological specimen is the most real and direct manifestation and physical record of all kinds of creatures in nature. It is an important material in the field of biological research, and plays an important role in taxonomy, systematics, ecology and other branches. At the same time, biological specimens bring benefits to society in many fields, including biosafety, drug research and development, public health and safety, environmental change monitoring and traditional taxonomic and systematic research.

The biological collection is a long-term and proper venue for keeping these records. The purposes of its establishment are to make the collection of biological specimens continue to expand and the preservation time continue to extend, so as to provide a variety of research materials for scientific research, show a variety of natural lives for the public, and provide resource guarantee for national economic construction and national biosafety.

China's biological specimens and collections were conceived in the 19th century, started in the early 20th century, and developed after the founding of the People's Republic of China. Among them, the most important and the largest number of specimens are the collections affiliated to the Chinese Academy of Sciences.

As an important entity of national animal, plant, fungus, fossil and other specimen preservation, research and scientific education, the biological collections of the Chinese Academy of Sciences are the most important place for centralized preservation of national biological specimens, which plays an irreplaceable role in the protection and sustainable utilization of national biological resources. By the end of 2019, there are 20 biological collections / museums under the Chinese Academy of Sciences, belonging to 19 institutes, and a total of 21.415 million biological specimens were preserved, more than half of which had been identified and nearly half of which had been digitized.

Since 2014, the biological collections of the Chinese Academy of Sciences has made a series of gratifying achievements: In terms of strategic biological resources collection and preservation, in cooperation with the national major research plan, each collection has carried out targeted investigation and collection, organized more than 1,400 domestic and international cooperative investigations, collected 3.173 million specimens. In terms of serving the national strategy, the collections have participated in the major national strategy and research plan, and national biosafety construction, and since the 13th Five-Year Plan, there have been 491 internal and external projects presided over or supported by collections / museums, with a total funding of 570 million yuan, which has played a huge supporting role in scientific research. In terms of classical taxonomy and biological species resource research, more than 1.03 million specimens have been identified, which has significantly improved the collection level, supported more than 5,600 papers and more than 230 monographs and played an important role in China's biological species resources research. In terms of public education and popular science services, more than 2,000 various popular science activities have been held, with an audience of more than 4.46 million people, providing social advisory services for more than 18,000 people, and more than 80 popular science works and 700 popular science articles were published, which have been highly recognized by the society.

However, compared with developed countries, there is still a gap in specimen collection, management and sustainable utilization in China. In the future, under the leadership of the Chinese Academy of Sciences, the classification management system will be established. Relying on the national science and technology resource sharing service platform, a national biological specimen resource sharing service platform integrating biological specimen collection, preservation, research, support and science popularization will be built. Relying on the "National Animal Collection Resource Center" and the "National Plant Specimen Resource Center", with specimen collection, preservation, identification and research as the main content, continue to serve major national projects. Focus on "the Belt and Road (B&R)" countries and the global biodiversity hotspot, international cooperation of joint investigation and research, resource sharing, and personnel training will be carried out. The capacity building of the collections should be strengthened, the ability to support scientific and technological innovation should be enhanced, and a reasonable talent team should be built. In this way, the sustainable development of the system can be ensured.

2

经典分类与生物标本馆

Classical taxonomy and biological collections

生物标本最初由私人收藏发展而来，最后支撑起了一个学科——生物分类学。传统的生物分类或称经典分类，十分依赖于生物标本，因为这是其最主要的研究材料。而生物标本馆则是集中保存生物标本的场所，也是分类学家开展研究工作的地方。生物标本馆有时也被称为自然历史博物馆、自然博物馆、博物馆、标本馆或标本室，本书统称生物标本馆。

2.1　现代生命科学中的经典分类

生物分类学是一门传统学科，也是现代生物学的重要组成部分，它使研究者能够以精确的方式将各种生物区分开来，并表明其各自之间有着怎样的进化关系。早期的分类学研究主要是本草学的研究，如李时珍的《本草纲目》(成书于 1578 年)，把 1195 种药用植物进行了系统的分类整理，并附有 1100 多幅插图。《本草纲目》迄今仍被大量研究和使用，为人类利用野生植物资源战胜疾病做出了重要贡献；林奈 1753 年发表的《植物种志》一书，采用拉丁文双名法记载了当时世界上已知植物 7700 种，奠定了分类学的基础，标志着生物分类学达到了成熟阶段；自达尔文 1859 年发表《物种起源》，现代分类学研究在进化论思想指导下开始发展。

分类学是人类认识自然界生物多样性的金钥匙，为形形色色的物种提供科学的分类系统。在科学的分类系统下，对物种给予正确的分类和命名，提供详细的形态学、生态学和地理分布资料。否则错误的分类和名称必将引起其他科学研究成果的错误与混乱。生命的起源和进化的研究一直是生命科学研究的重要命题，分类学研究是其研究的根基。分类学研究物种的个体变异、种群变异、生态和生殖隔离，这些基础知识是遗传学和发育生物学研究必不可少的。分类系统的建立和系统发育分析与重建，为研究进化提供了框架，并展示了进化的格局，也为生态学、生物地理学和古生物学研究探讨生命的起源与各大类群生物的演化，以及地质历史时期生态环境的变化和全球气候变化提供生物学依据。生物多样性保护更是离不开分类学的调查和评估。从其研究理论和方法，以及研究对象和内容上来说，分类学本身就是研究生物多样性的一门学科。

现代生物分类学研究在中国起步很晚，直到 20 世纪初，中国学者才逐渐开始采集、研究本土的动植物标本，比欧美学者在世界各地大规模采集晚了近 2 个世纪。在此之前，中国的生物标本被欧、美、俄、日等所派人员大量采集，运输回国进行研究并发表研究成果。新中国成立之后，中国政府高度重视自然资

At first, biological specimens developed from private collections, and finally supported a discipline—biological taxonomy. The traditional biological taxonomy, or classic taxonomy, is very dependent on biological specimens, because it is the most important research material for taxonomy. The biological collection is not only the place where the biological specimens are concentrated, but also the place where taxonomists carry out research work. The biological collection is sometimes called the museum of natural history, the museum of nature, the museum or the herbarium ("herbarium" is only for plant specimens). In this book, it will be called the biological collection.

2.1 Classical taxonomy in modern life science

Taxonomy is not only a traditional subject, but also an important part of modern biology. It enables researchers to distinguish all kinds of organisms in an accurate way and show what kind of evolutionary relationship they have. The early taxonomic research mainly focused on the study of Materia Medica, such as LI Shizhen's *Compendium of Materia Medica* (completed in 1578). 1,195 kinds of medicinal plants were systematically classified and sorted, with more than 1,100 illustrations attached. *Compendium of Materia Medica* has been widely studied and used so far, which has made an important contribution to the use of wild plant resources to fight diseases. Linnaeus published *Species Plantarum* in 1753, which recorded 7,700 known plants in the world at that time by binomial nomenclature, laying the foundation of taxonomy, marking the maturity of biological taxonomy. Since *On the Origin of Species* (Darwin) was published in 1859, modern taxonomic research was under the guidance of evolutionism, and began to develop.

Taxonomy is the golden key for human beings to understand the biodiversity of nature and provides a scientific classification system for all kinds of species. Under the scientific classification system, species are classified and named correctly, and detailed morphological, ecological and geographical distribution data are provided. Otherwise, the wrong classification and name will lead to the mistakes and confusion of other scientific research results. The study of the origin and evolution of life has always been an important topic in the study of life science, and the research of taxonomy is the foundation of its research. Taxonomy focuses on studies of individual variation, population variation, ecological and reproductive isolation of species, which are essential for genetics and developmental biology. The establishment of a taxonomic system and the analysis and reconstruction of phylogeny provide a framework for the study of evolution, show the pattern of evolution, and provide a biological basis for the study of ecology, biogeography and paleontology to explore the origin of life and the evolution of major groups of organisms, as well as the changes of ecological environment and global climate change in geological history. Biodiversity protection is inseparable from the investigation and evaluation of taxonomy. In terms of its research theories and methods, as well as its research objects and contents, taxonomy itself is a subject of biodiversity research.

Modern taxonomic research started very late in China. Until the beginning of the 20th century, Chinese scholars began to collect and study local animal and plant specimens, nearly two centuries later than European and American scholars in large-scale collections around the world. Before that, China's biological specimens were collected and published by European, American, Russian, Japanese and other countries in large quantities, and transported back to their own country for research and publishing research results. After the founding of the People's Republic of China, the government attached great importance to the exploitation and utilization of natural resources and carried out many large-scale comprehensive investigations, which led to a rapid increase in the collection of biological specimens and a growing number of taxonomic practitioners in China. On this basis, with the efforts of the whole country, the compilation and research of taxonomic works such as *Flora Reipublicae Popularis Sinicae*, *Fauna Sinica*, *Economic Insect Fauna of China*, *Economic Fauna of China*, *Animal Atlas of China*, *Flora Cryptogamarum Sinicarum*, *Flora Fungorum Sinicorum* were carried out and published successfully. Among them, the *Economic Insect Fauna of China* won the second prize of National Natural Science in 2002, and *Flora Reipublicae Popularis Sinicae* won the first prize of National Natural Science in 2009. The publication of these taxonomic works indicates that the collection, research, protection and utilization of biological resources dominated by China have been developed and expanded.

The basis of the protection and utilization of biological resources is taxonomy. Without accurate identification of species, there is no reasonable use and protection of resources. But China is vast in territory and rich in resources. There are quite a lot of biological groups that nobody or few people have carried out basic work such as field investigation, specimen collection, classification and identification so far, and many groups have important value and economic significance. There are thousands of nature reserves in China. What species and what to protect in these reserves need taxonomists to conduct field investigation and identification. Agriculture, forestry, horticulture, customs and other departments are also inseparable from the technical support of taxonomists - for example, urban greening and introduction and domestication can not be completed without the knowledge of classical classification. In addition, with the growth of population and the improvement of people's quality of life, people's demand for healthy food and medicine is constantly increasing. Scientists have been trying to find resources (including genetic resources) and inspiration from wildlife, which needs the close cooperation of

源的挖掘利用，开展了多次大规模的综合考察，使中国的生物标本收集量迅速增加，分类学从业人数也不断增加。在此基础上，集全国之力，开展《中国植物志》、《中国动物志》、《中国经济昆虫志》、《中国经济动物志》、《中国动物图谱》、《中国孢子植物志》、《中国真菌志》等志书的编研工作，并顺利出版，其中《中国经济昆虫志》于2002年荣获国家自然科学奖二等奖，《中国植物志》于2009年荣获国家自然科学奖一等奖。这些志书的出版，标志着由中国自己主导的生物资源收集、研究、保护与利用已经得到发展和壮大。

生物资源保护与利用的基础就是分类学，不对物种做出准确的鉴定，就谈不上合理利用和保护资源。但我国幅员辽阔，资源丰富，有不少生物类群至今无人或很少有人开展野外调查、标本采集、分类鉴定等基础工作，而且这些类群中不乏具有重要的利用价值和经济意义的物种。中国有几千个自然保护区，这些保护区内有什么物种，要保护什么，需要分类学家进行野外考察、鉴定来确定。农业、林业、园艺、海关等部门也离不开分类学家的技术支持——如城市的绿化以及引种驯化没有经典分类的知识是无法完成的。另外，随着人口的增长和人们生活质量的提高，人们对健康食品和药物的需求不断增加，科学家一直试图从野生生物中发掘资源（包括遗传资源）和灵感，这需要分类学家的密切配合方能完成。特别是当进入21世纪后，各种源自野生动物的传染病在人类群体中大肆传播，如严重急性呼吸综合征（SARS）、禽流感、埃博拉出血热，以及目前在全球肆虐的新型冠状病毒肺炎，给人类健康和社会进步带来了极大危害。这说明人类对野生动物的了解还远远不够，科学家还有许多工作要做。

经典分类是阐明一个国家或地区物种组成的基础学科，对生物多样性研究和资源的可持续利用，以及对其他学科的发展具有不可取代的作用。当前，科学技术高速发展，生命科学的研究呈多样化和微观化的趋势，宏观的分类学研究在分子生物学研究的大潮下，正在采用新的研究手段和获取新的数据。虽然经典分类学面临着研究人员缺乏、经费困难等诸多问题，但经典分类学研究并没有失去其科学价值和社会意义，特别是在中国这样一个生物多样性极其丰富的大国。经典分类学研究始终是国家生物资源保护和利用的基础。

2.2 生物标本和标本馆

作为自然界生命的历史记录，生物标本最初是少数有钱人的私人珍品收藏，用来满足其猎奇和炫耀的心理。随着数量的积累，标本越来越多地引发人们的思考，从而逐渐转变为科学研究的材料。标本馆或者博物馆的功能也从以往的仅仅用于展示，转变为研究和展示并重。

生物标本是自然界各种生物的最真实、最直接的表现形式和实物记录，是生物多样性的载体和表现形式，是人类认识自然的历史见证和档案，是人类研究、分析、监测生物多样性动态变化和探索物种起源演化的重要科学依据。生物标本是生物学研究领域的重要素材，在分类学、系统学、生态学等各个分支学科中都有着十分重要的地位。

早在1831-1836年，达尔文跟随英国皇家海军贝格尔舰进行环球资源调查，开展了科学采集，对收集的大量标本进行分类观察研究，形成了物种进化论认识，和另外一名英国博物学家华莱士共同创立了进化论，彻底改变了过去西方世界关于物种神创论的主流观点，也彻底改变了人类的世界观和对自然世界的哲学认识，奠定了现代生物科学的基础。如果分类学是其他生命科学的基础，那么标本就是基础中的基础。

生物标本在许多领域都为社会带来利益，包括生物安全、药物研发、公共卫生和安全、环境变化监测以及传统分类学和系统学研究等。分类学及其主要研究材料（标本）中蕴含的信息量巨大，包括大量馆藏标本和文献信息，从形态到分子，为数以万计的不同物种建立户口簿和档案馆。提供物种形态资料和标本，甚至DNA条形码，为物种的快速识别与鉴定服务。大量标本的分类信息，包括文字与图像资料，使我们能

taxonomists. Especially after entering the 21st century, all kinds of infectious diseases from wild animals spread in human groups, such as SARS, Avian influenza, Ebola hemorrhagic fever, and the present Coronavirus Disease 2019 (COVID-19) that has been raging around the world. They have caused great harm to human health and social progress. This shows that human beings don't know enough about wild animals, and scientists still have a lot of work to do.

Classic taxonomy is a basic discipline to clarify the species composition of a country or region, which plays an irreplaceable role in biodiversity research, sustainable utilization of resources and the development of other disciplines. At present, with the rapid development of science and technology, the research of life science is in the trend of diversification and microcosmic. Under the tide of molecular biology research, macro taxonomic research is adopting new research methods and obtaining new data. Although classical taxonomy is faced with many problems, such as lack of researchers, financial difficulties and so on, the research of classical taxonomy has not lost its scientific value and social significance, especially in China, which is a big country with extremely rich biodiversity. Classic taxonomy research is always the foundation of national biological resources protection and utilization.

2.2 Biological specimens and collections

As a historical record of natural life, biological specimens were originally collected by rich people as private treasures to satisfy their curiosity and show off psychology. With the accumulation of the quantity, the specimens are more and more arousing people's thinking, thus gradually turning into the materials of scientific research. The function of the collection or herbarium or museum has also changed from being only used for exhibition in the past to paying equal attention to research and exhibition.

Biological specimens are the most real and direct manifestation and physical record of all kinds of creatures in nature, the carrier and manifestation of biological diversity, the historical witness and archive of human understanding of nature, and the important scientific basis for human research, analysis and monitoring of the dynamic changes of biological diversity and exploring the origin and evolution of species. The biological specimen is an important material in the field of biological research, which plays an important role in taxonomy, systematics, ecology and other branches.

As early as 1831-1836, Darwin followed the British Royal Navy's "Beagle" survey of global resources, carried out scientific collection and classified observation and research on a large number of collected specimens, and formed the understanding of species evolution theory. He co-founded evolution theory with another British naturalist Wallace, completely changed the mainstream view of species creationism in the western world in the past and completely changed human's world view and philosophical understanding of the natural world, and laid the foundation of modern biological science. If taxonomy is the foundation of other life sciences, then specimens are the foundation of foundation.

The biological specimens can bring benefits to society in many fields, including biosafety, drug research and development, public health and safety, environmental change monitoring, and traditional taxonomy and systematic research. Taxonomy and specimen information is huge, including collected specimens and literature information, from morphology to molecules, to establish household registers and archives for tens of thousands of different species. Provide species morphological data and specimens, even DNA bar-coding for the rapid identification and identification of species. A large number of specimen classification information, including text and image information, enables us to quickly grasp species-related information through network services. International projects such as databases of names, specimens and photos, online Flora or Fauna and online encyclopedias are all new manifestations of the research results of the specimen and classical taxonomy in the information age. In the era of network information, big data is playing an unexpected role. The taxonomy itself is also an information science. Each species is an information database, and each specimen is one of the information. Species depend on each other in the ecosystem, including us, a large number of specimens and species information comprehensive analysis, mining can explore such as a rare and endangered species distribution changes and population dynamics, global climate change impact on human society and other important topics.

Contribution of biological specimens to agricultural and forestry production. In modern agricultural and forestry production, using natural enemies to control diseases and pests is one of the only ways to improve the quantity and quality of agricultural products. It depends on taxonomists' accurate identification of diseases, pests and natural enemies. On this basis, we can formulate practical control measures, so as to achieve twice the result with half the effort. For example, as early as the 1920s, fern weevil *Syagrius intrudens* was very harmful to ferns in the Hawaiian forest. However, it is only known from the literature that this insect has been found in greenhouses in Australia and Ireland, and there is no other record of its distribution. In 1921, an old fern weevil specimen collected in 1857 was found in a private collection in Sydney, Australia, with specific information about the place where it was collected. A small population of the fern weevil and a small cocoon wasp parasitizing the weevil larvae were found in the forest recorded on the label. This kind of small

够通过网络服务快速掌握物种的有关信息。国际上正在开展的名称、标本、图像数据库和在线生物志与网络大百科全书等项目，均是标本和经典分类学研究成果在信息时代新的体现。在网络信息时代，大数据正在发挥意想不到的作用。分类学本身也是一门信息科学，每个物种都是一个信息数据库，每个标本都是其中的一条信息。物种在生态系统中相互依存，包括我们人类，通过对大量的标本和物种信息的综合分析、挖掘，可以探讨诸如某个珍稀濒危物种的分布区变化和种群动态、全球气候变化对人类社会的影响等重大命题。

生物标本对农林业生产的贡献 在现代农林业生产中，利用天敌防控病虫害是提升农产品数量和质量的必由之路之一。这有赖于分类学家对病害、虫害以及天敌生物物种的准确鉴定，在此基础上，才能制定切合实际的防控措施，从而达到事半功倍的效果。例如，早在20世纪20年代，一种蕨象甲（*Syagrius intrudens*）在夏威夷森林中对蕨类危害严重。但从文献中仅能得知此虫在澳大利亚和爱尔兰的温室中被发现过，无其他地方分布记载。1921年有人在澳大利亚悉尼的一个私人收藏中看到一号1857年采集的蕨象甲标本，到标签记载的森林中进行调查后，发现了该甲虫的一个小种群和寄生这种甲虫幼虫的一种小茧蜂。然后采集这种小茧蜂送到夏威夷进行繁殖用于防治蕨象甲，取得良好效果。看似默默无闻的标本，在沉默了65年后发挥了重要的作用。

生物标本对作物病害流行研究的贡献 19世纪40年代，在爱尔兰发生了马铃薯晚疫病，这是一个世界闻名的事件。因为病害发生，作为人们主食的马铃薯严重减产，导致巨大的饥荒，使爱尔兰人口急剧下降。为了澄清这一作物病害大发生的原因，近年来许多研究人员利用经过形态研究保藏在标本馆里的当年爱尔兰马铃薯晚疫病大流行时的标本，以及不同时期的标本，采用各种分子生物学的方法和手段进行研究，阐明了致病疫霉引致的马铃薯晚疫病流行机制，为防治类似的作物大病害提供了不可多得的基础资料。

生物标本能够服务于国家经济发展，催生新的产业 在长期菌物分类学研究的基础上，1957年中国科学院微生物研究所邓叔群院士从野外采集的标本上分离获得了灵芝菌种，并于1959年实现了灵芝的人工栽培（菌物标本馆至今保藏有当时的凭证标本）。1970年中国科学院微生物研究所真菌室研究人员进一步完善了灵芝栽培技术，获得世界上第一瓶人工培养收集的灵芝孢子粉，并将栽培技术无偿向社会进行推广，催生了灵芝的巨大产业。目前中国灵芝年产量超过11万吨，形成了人工栽培、产品开发利用的完整产业链，年产值达数十亿美元。

生物标本对全球生态学研究的贡献 利用馆藏标本能够揭示全球气候变化。近年来中国科学院植物研究所利用馆藏的1931-2009年采集的香蒲属腊叶标本，建立植物气孔指数与大气CO_2浓度回归方程，然后运用气孔指数法，以化石叶片气孔指数作为大气CO_2浓度的代用指标，重建了第三纪与第四纪之交温室向冷室转型期的大气二氧化碳浓度，并整合了前人数据，构建出首条500万年以来陆地生态系统中全球大气CO_2浓度变化曲线，探讨了二氧化碳对全球气候变化的影响，有助于政府和社会各界共同理解、思考、制定对策来应对当前和未来的气候变暖，兼具理论和现实双重意义。

生物标本对生物多样性调查和评估的基础作用 《中国生物多样性保护战略与行动计划（2011—2030年）》在其规定的十个优先领域中，其中一个是"开展生物多样性调查、评估与监测"。生物资源的调查、收集和保护是生物学研究与生物资源利用的基础，关系到国家的资源安全、社会发展和植物学相关研究，具有重要的战略意义。而开展生物多样性调查、评估、收集和保护等工作都需要经典分类学及标本的支撑。

所以，生物标本是一类重要的战略生物资源，随着国家生物与生态安全，以及外来入侵害虫的预警、预报、检测与监测的日益重要，以及重大动物疫病的频繁暴发，急需从馆藏生物标本资源中获取关键的基础信息和研发原材料。

生物标本有如此重要的作用，为了更好地为人类服务，就必须将其集中保存。生物标本馆则是长久妥善保存这些生物标本的场馆，其建立的目的就是让生物标本的收藏范围不断扩大、保存时间不断延长，从

cocoon wasp was collected and sent to Hawaii for propagation to control fern weevil, and good results were achieved. The seemingly unknown specimen played an important role after 65 years of silence.

Contribution of biological specimens to the study of crop diseases. The potato late blight in Ireland in 1840s is a world-famous event. Because of the disease, the production of potatoes, which is the main food for people, has been seriously reduced, resulting in a huge famine and a sharp decline in the population of Ireland. In order to clarify the cause of the occurrence of the disease, in recent years, many researchers have used the specimens collected in the herbarium through morphological research during the epidemic period of potato late blight in Ireland, as well as the specimens collected in different periods, to study the epidemic mechanism of potato late blight caused by Phytophthora by using various molecular biological methods and hands similar crop diseases provide rare basic data.

Biological specimens can serve the development of the national economy and promote new industries. On the basis of long-term mycological research, academician Teng Shu-Chün of the Institute of Microbiology of CAS isolated *Ganoderma lucidum* strain from the samples collected in the field in 1957, and realized the artificial cultivation of *Ganoderma lucidum* in 1959 (Up to now, the Fungarium has preserved the certificate specimens at that time). In 1970, researchers from the Department of Fungi, Institute of Microbiology, CAS further improved the cultivation technology of *Ganoderma lucidum*, obtained the first bottle of spore powder collected by artificial cultivation in the world, and promoted the cultivation technology to the society free of charge, giving birth to the huge industry of *Ganoderma lucidum*. At present, the annual output of *Ganoderma lucidum* in China is more than 110,000 tons, forming a complete industrial chain of artificial cultivation and product development and utilization, with an annual output value of billions of dollars.

Contribution of biological specimens to global ecological research. Using collected specimens, we can reveal global climate change. In recent years, the Institute of Botany, CAS has collected dehydrated specimens (1931-2009) from *Typha* in its collection, established the regression equation between the stomatal index of plants and the atmospheric CO_2 concentration, and then reconstructed the atmospheric CO_2 concentration during the transition period from the greenhouse to the cold chamber at the turn of the Tertiary to Quaternary Period by using the stomatal index method and taking the stomatal index of fossil leaves as the substitute index of atmospheric CO_2. Based on the previous data, this study constructed the first global atmospheric CO_2 concentration curve in the terrestrial ecosystem since 5 million years ago, and discussed the impact of CO_2 on global climate change. It is helpful for the government and all walks of life to understand, think about and formulate countermeasures to deal with the current and future climate warming, which has both theoretical and practical significance.

The basic role of biological specimens in biodiversity survey and assessment. Among the ten priority areas stipulated in *China National biodiversity conservation strategy and action plan (2011-2030)*, one of them is to carry out biodiversity survey, assessment and monitoring. The investigation, collection and protection of biological resources are the basis of biological research and utilization of biological resources, which are related to national resource security, social development and botany-related research, and have important strategic significance. The work of biodiversity investigation, assessment, collection and protection needs the support of classical taxonomy and specimens.

Therefore, the biological specimen is an important strategic biological resource. With the increasing importance of national biological and ecological security, early warning, prediction, detection and monitoring of alien invasive pests, and frequent outbreaks of major animal diseases, it is urgent to obtain basic information and research and development raw materials from biological specimen resources.

Biological specimens play such an important role. In order to serve human better, they must be preserved in a centralized way. The biological collection is a long-term and proper venue for keeping these biological specimens. The purposes of its establishment are to make the collection of biological specimens continue to expand and the preservation time continue to extend, so as to provide a variety of research materials for scientific research, show a variety of natural lives for the public, and provide resource guarantee for national economic construction and national biosafety.

Compared with other museums used to show the development of human civilization, the biological collections are not only the display of rare treasures and not just to meet the public's curiosity psychology, but also to systematically present the rich and colorful life of nature, which is more conducive to people's research and exploration, and to produce new knowledge. So that people's understanding of nature is more in-depth and comprehensive. In this way, the biological collection will become a research base for taxonomists and a cradle for cultivating taxonomic talents.

On the other hand, in order to give back to society, popularize scientific knowledge and spread the idea of loving nature and loving life, most of the biological collection will sort out the important and interesting knowledge on this basis, together with some beautiful specimens, to show to the public. In this way, the biological collection also undertakes the educational function that some schools cannot provide in the classroom. Moreover, the knowledge is displayed through the specimen, which has the advantage that ordinary books cannot provide. The most cutting-edge scientific knowledge from human

而为科研提供多种多样的研究材料，为社会大众展示千姿百态的自然生命，为国民经济建设和国家生物安全提供资源保障。

相比其他用于展示人类文明发展历程的博物馆，生物标本馆不仅仅是奇珍异宝的展示，也不单单是为了满足大众的猎奇心理，而是有目的、系统地将大自然丰富多彩的生命进行集中呈现，更有利于人们去研究和探索，并产生新的知识，让人类对自然的理解更加深入和全面。这样，标本馆就成为分类学工作者进行研究的基地，也是培养生物分类学人才的摇篮。

此外，为了回馈社会，普及科学知识，传播热爱自然、热爱生命的理念，大部分生物标本馆都会整理已经研究得出的重要、有趣的知识，连同一些美观的标本，一起展示给社会大众。这样生物标本馆也就承担了一部分学校课堂所无法提供的教育功能，而且这些知识是通过标本实物展现的，有着普通书本所不具有的优势，来自人类探索自然的最前沿的科学知识，更直观也更容易被接受。

我们可以把生物标本馆看作大自然的“图书馆”，那么其中的每一个物种便是一本“书”，每一号标本便是储存生命信息的一页“纸”。书的价值体现在人们不断的翻阅中，而这一本本“自然之书”也等待着每一位充满好奇和使命感的人们去“阅读”，去思考，去发现。

2.3 中国生物标本馆的发展历程

近代生物标本的采集和生物标本馆的建立均始于欧洲，伴随着殖民扩张，其在世界范围内大量收集标本。自诞生以来，生物标本馆的使命之一就是要尽可能多地收集标本，扩大收藏范围。发达国家本土的标本收集与保存在 100 多年以前就已经完成，并基本摸清了本土的生物资源家底。而至少在 150 多年前，他们就已将标本的搜集范围延伸到南美、东南亚、非洲、东亚和中亚等生物多样性丰富与特殊的地区。

中国是全球物种多样性最丰富的 12 个国家之一，这些宝贵的生物资源使发达国家垂涎不已。中国的生物标本收藏与生物标本馆建设孕育于 19 世纪外国传教士在中国开展的生物资源调查和采集，当时许多外国传教士和分类学家借传教深入到中国的广大地区，大量采集各类生物标本，并且基于这些标本描述并发现了大量新的物种，如大熊猫、珙桐等。他们将大部分的标本带回本国，保存在本国一些大型的博物馆中，只有一部分留在中国。随着标本采集力度的加大，为了保存这些珍贵的动植物标本，有不少传教士相继在中国建立博物馆，其中最早的就是法国传教士韩伯禄（P. M. Heude，1836-1902）于 1868 年在上海创立的徐家汇博物院(Musée de Zi-Ka-Wei)，以收藏动植物标本为主。该博物馆在 1930 年划归私立的震旦大学管理，后被称为震旦博物院（Musée de Heude）。

从 20 世纪 20 年代开始，陆续有留学回国的生物分类学家开始对中国生物物种多样性进行研究。自此中国开始建立自己的生物标本馆，如 1928 年成立的静生生物调查所、1929 年成立的北平研究院动物研究所、1930 年成立的中央研究院自然历史博物馆。

中华人民共和国成立后，在国家的大力支持和领导下，在中国科学院的主持下，组织全国有关科研院所及高等院校的生物分类学家，于 1958 年和 1959 年分别启动了《中国植物志》和《中国动物志》编研工作，尤其是在 1973 年召开了包括《中国孢子植物志》在内的“三志”工作会议，全面启动和加强了中国生物资源的研究。目前已完成了《中国植物志》的中英文版本，《中国动物志》和《中国孢子植物志》也已出版了多卷。

伴随着“三志”的编研，中国在生命科学领域的研究不断深入和扩大，生物标本馆事业也得到了发展和壮大。特别是在 20 世纪 80 年代，各个综合性大学以及农林、医学类大专院校和地区性师范类专科学校陆续建立起了生物系，在全国掀起了生物分类学热潮，配套的用于教学的大量生物标本馆或标本室也相继建立。至 20 世纪 90 年代初期，我国仅植物标本馆就有 300 余家，而生物标本馆总量据估计超过 500 家。

exploration of nature is more intuitive and easier to accept.

We can regard the biological collection as the "library" of nature, so each species is a "book", and each specimen is a page "paper" storing life information. The value of the book is reflected in people's continuous reading, and this "book of nature" is waiting for everyone who is full of curiosity and sense of mission to "read", to think, and to find.

2.3 The development course of Chinese biological collections

震旦博物院动物标本室
Animal Collection of Musée de Heude

The collection of modern biological specimens and the establishment of biological collections began in Europe. With their colonial expansion, a large number of specimens were collected around the world. Since its birth, one of the missions of the biological collection is to collect as many specimens as possible and expand the collection range. The collection and preservation of local specimens in developed countries have been completed more than 100 years ago, and the background of local biological resources has been basically found out. Since at least 150 years ago, they have extended the collection of specimens to South America, Southeast Asia, Africa, East Asia and Central Asia and other regions with rich and special biodiversity.

China is one of the 12 countries with the richest species diversity in the world. These precious biological resources are coveted by the developed countries. The collection of biological specimens and the construction of biological collection in China were born in the 19th century when foreign missionaries carried out the investigation and collection of biological resources in China. At that time, many foreign missionaries and taxonomists went deep into the vast areas of China for collecting biological specimens with the purpose of missionary work. Based on these specimens, a large number of new species, such as giant panda (*Ailuropoda melanoleuca*) and dove tree (*Davidia involucrata*), were described and found. They brought most of the specimens back to their country and kept them in some large museums in their country and only a part of them remained in China. With the increasing efforts of specimen collection, in order to preserve these precious animal and plant specimens, many missionaries have set up museums in China before and after, among which the earliest is the Musée de Zi-Ka-Wei, which was founded in Shanghai in 1868 by the French missionary P. M. Heude (1836-1902), mainly collecting animal and plant specimens. In 1930, the museum was under the management of the private Aurora University, which was later called Musée de Heude.

Since the 1920s, there have been many taxonomists who have returned to China to study the biodiversity of China. Since then, China has begun to establish its own biological collection, such as the Fan Memorial Institute of Biology established in 1928, the Institute of Zoology of the National Academy of Peiping established in 1929, and the Natural History Museum, Academia Sinica established in 1930.

After the founding of P. R. China, with the strong support and leadership of the country, and under the auspices of CAS, it organized the taxonomists of relevant scientific research institutes and colleges and universities across the country to launch the compilation and research of *Flora Reipublicae Popularis Sinicae* and *Fauna Sinica* in 1958 and 1959 respectively. The "Three Flora / Fauna" (*Flora Reipublicae Popularis Sinicae, Fauna Sinica* and *Flora Cryptogamarum Sinicarum*) working conference was held in 1973, then research on biological resources in China was comprehensively launched and strengthened. At present, the English versions of *Flora Reipublicae Popularis Sinicae* have been completed, and many volumes of *Fauna Sinica* and *Flora Cryptogamarum Sinicarum* have been published.

静生生物调查所所址
Building of Fan Memorial Institute of Biology

北平研究院动物研究所标本制作室
Specimen preparation room of the Institute of Zoology of the National Academy of Peiping

随着科学技术的进步和生物学自身的发展，分类学热潮逐渐退去，部分体量很小的标本馆也随之消失。据相关调查和统计，至2016年，全国运转正常的生物标本保藏机构还有250余家，收藏总量为4000万-4500万号/份。而随着国民经济的不断发展，已有的生物标本馆的场馆建设和保藏条件也得到了极大改善，其中大多数拥有专门的保藏场馆和保藏器具，特别是一些大的科研院校的生物标本馆近几年重新修建或翻建了场馆，购置了新的保藏器具，提升了保藏环境条件，接近或达到了国际先进标本馆的水平，标本也得到了安全保藏。

近年来，各生物标本馆积极参与国家及地方自然资源考察任务，发挥了重要作用，并为获得的生物标本提供了长久保藏场所。全国标本馆平均每年新增加标本超过120万号。一些科研院校还积极开展国际合作，对周边国家和地区以及一些生物多样性研究热点地区开展标本收集，极大地增加了我国生物标本藏量和覆盖地区面积。

然而，高校和地方生物标本馆虽数量众多，但具有一定藏量规模（动物标本40万号以上，植物标本5万号以上，菌物标本1万号以上）的标本馆不足30家，大多数馆藏量较小，其主要功能多集中在教学辅助和科普教育上，除某些专类馆或特色馆以外，大都难以形成规模，无法确保相应的保藏、管理、研究等工作很好地开展。

中国科学院作为国家战略科技力量的重要组成，自1949年成立以来，一直将生物标本的采集、保藏、研究作为一项重要工作，给予了高度重视。其收藏主要包括震旦博物院、静生生物调查所、北平研究院等中国最早收藏的动植物标本，以及新中国成立之后，国家启动的青藏高原、横断山区、西双版纳、十万大山等地区的十几项大型综合科学考察所采集的大部分标本。

同时，在中国科学院二期“创新工程”建设中，中国科学院生物标本馆得到财政部改扩建专项资金项目的支持，对当时全院（包括院地共建）26个生物标本馆的硬件环境进行了很大的改善。所有标本馆得到了改建或扩建，总面积约88 000平方米，从根本上解决了以前存在的保藏空间不足、保藏环境简陋等问题；标本柜大多采用新购置的可移动式密集柜，不但节省了空间，便于标本管理和取放，而且加强了防虫、防尘、避光等有利于标本安全保藏的举措；在温湿度控制、消防预警方面也有所提升。通过这一系列的重要建设，中国科学院生物标本馆的硬件条件已达到国际同类馆的中等水平。

中国科学院目前拥有20家标本馆（博物馆），包括亚洲最大的动物、植物和菌物标本馆。截至2019年底，全院共收藏各类生物标本（含化石）2126.9万号，占中国生物标本总量的60%以上。经过几代人的努力，中国科学院生物标本馆在场馆建设、保藏条件、藏量及类群、涵盖范围、研究水平、运行管理、数字化建设、人才队伍、科研支撑、社会服务等方面得到了极大提升，发挥了重要作用，做出了突出贡献，赢得了社会赞誉。

2.4 中国科学院生物标本馆的贡献

中国科学院生物标本馆体系虽然场馆数量较少，但目前其馆藏标本数量占全国的60%以上，生物门类齐全，采集范围广泛，收藏历史悠久，能够代表整个中国生物标本馆事业的发展历程。在近70年的发展过程中，中国科学院生物标本馆始终注重国家需求，服务国家经济发展和生态文明建设，为生物资源可持续

With the compilation and research of the “Three Flora/Fauna”, the research in the field of life science in China has been continuously deepened and expanded, and the cause of biological collection has also been developed and expanded. Especially in the 1980s, various comprehensive universities, as well as agricultural and forestry, medical colleges and local normal colleges have successively established departments of biology, which has set off a wave of biological taxonomy in the country, and a large number of biological collections or herbariums for teaching have also been established. By the early 1990s, there were more than 300 herbariums in China, and the total number of biological collections is estimated to be more than 500.

With the progress of science and technology and the development of biology itself, the upsurge of taxonomy gradually receded, and some small collections also disappeared. According to relevant surveys and statistics, by 2016, there were more than 250 well-functioning biological specimen preservation institutions in China, with a total collection of 40-45 million. With the continuous development of the national economy, the construction and preservation conditions of the existing biological collections have also been greatly improved. Most of them have special preservation venues and instruments. In recent years, the biological collections of some large scientific research institutions have rebuilt the venues, purchased new preservation instruments, and improved the environmental conditions for preservation, which are close to or reach the international advanced level, and the specimens have also been safely preserved.

In recent years, the biological collections actively participated in the national and local natural resources investigation tasks, played an important role, and provided a long-term preservation place for the obtained biological specimens. There are more than 1.2 million new specimens added to the collections every year. Some scientific research institutions also actively carry out international cooperation to collect specimens from surrounding countries and regions and some biodiversity research hotspots, greatly enriching China’s biological specimen reserves and coverage areas.

However, although there are a large number of University and local biological collections, there are less than 30 collections with a certain scale of storage (more than 400,000 animal specimens, more than 50,000 plant specimens, more than 10,000 fungus specimens). Most of the collection is small, and their main functions are mainly focused on teaching assistance and popular science education. Many of them are difficult to form a scale except for some special collections. So it is impossible to ensure the corresponding work of preservation, management and research has been well carried out.

As an important part of the national strategic scientific and technological forces, CAS has always taken the collection, preservation and research of biological specimens as an important work since its establishment in 1949 and attached great importance to it. Its collection mainly includes the earliest animal and plant specimens collected in China, such as the Musée de Heude, the Fan Memorial Institute of Biology and the National Academy of Peiping, as well as most of the specimens collected by more than ten large-scale comprehensive scientific expedition in Qinghai-Tibet Plateau, Hengduan Mountains Region, Xishuangbanna, Mt. Shiwanda and other areas launched by the state after the founding of P. R. China.

At the same time, in the second phase of the “Innovation Project” of the CAS, the biological collection was supported by the special fund project of reconstruction and expansion of the Ministry of Finance, which greatly improved the hardware environment of 26 biological collections of CAS (including the collections jointly established by CAS and local government). All the collections have been rebuilt or expanded, with a total area of about 88,000 m^2, which fundamentally solves the problems of insufficient storage space and poor storage environment; most of the specimen cabinets adopt the newly purchased movable dense cabinets, which not only saves space, facilitates specimen management and collection, but also strengthens measures such as insect prevention, dust prevention, light avoidance, etc., which are conducive to specimen safe storage; in terms of temperature and humidity control and fire alarm, it also proposes a promotion. Through this series of important construction, the hardware condition of the biological collection of CAS has reached the medium level of the similar international museums.

At present, CAS has 20 collections (museums), including the largest animal, plant and fungus collections in Asia. By the end of 2019, CAS has collected 21.269 million biological specimens (including fossils), accounting for 60% of the total biological specimens in China. Through the efforts of generations, the biological collections of CAS have been greatly improved in the aspects of venue construction, preservation conditions, reserves and groups, coverage, research level, operation management, digital construction, talent team, scientific research support, social services, etc., and played an important role, made outstanding contributions and won social praise.

2.4 The contribution of the biological collections of CAS

Although the number of the biological collections of CAS is relatively small, at present, its collection of specimens accounts for more than 60% of the whole country, with complete biological categories, a wide range of collection, and a long collection history, which can represent the development process of the whole Chinese biological specimen collection. In nearly 70 years of development, the biological collections of CAS have always paid attention to national needs, served national economic development and ecological civilization construction, contributed to the sustainable utilization of biological resources, and did a

利用贡献力量，围绕生物多样性研究与保护、国家经济建设和公众科学教育等方面做了大量工作。

生物多样性研究与保护

当代中国生物学领域最重要的一项工作就是“三志”的编研。这是一项获得国家支持的重大工程。“三志”即《中国植物志》、《中国动物志》和《中国孢子植物志》。这三部志书以现代生命科学的研究方法对中国生物资源进行全面普查，极大地提高了中国对其自身生物资源的认知水平，成为中国开展各项生物学研究的最坚实的基础，促进了生物分类学以及其他生物学科的快速发展。在“三志”的编研过程中，各种有关生物资源保护与利用的研究和书籍不断涌现，研究成果层出不穷。而其中大部分工作由中国科学院的相关研究所承担，如动物研究所、植物研究所、微生物研究所等。同时，编研过程中所采集的大部分标本也保存在中国科学院生物标本馆中，各标本馆也为这些志书的编研提供了大量的标本查阅服务工作。

“三志”的编研以及其他相关研究使人们认识到野生动植物具有重要的生态价值和实用价值，而部分物种的分布范围却越来越小，数量越来越少，许多物种面临着生存危机。因此国家也开始了生物多样性的研究和保护之路。1956 年，全国人民代表大会通过一项议案，提出了建立自然保护区的问题。同年 10 月林业部草拟了《天然森林伐区（自然保护区）划定草案》，并在广东省建立了中国的第一个自然保护区——鼎湖山自然保护区。这也是唯一隶属于中国科学院的自然保护区。自 20 世纪 70 年代末、80 年代初以来，中国自然保护事业迅速发展。

与此同时，相关法律法规也开始诞生和完善。1950 年 5 月，《关于稀有动物保护办法》颁布，规定禁止任意捕猎四川松潘等地的大熊猫等稀有动物。1962 年 9 月，国务院颁布《关于积极保护和合理利用野生动物资源的指示》，强调“野生动物资源是国家的自然财富”，要“切实保护”、“合理利用”。1988 年 11 月，中国第一部为保护野生动物而订立的法律——《中华人民共和国野生动物保护法》颁布，次年正式发布了《国家重点保护野生动物名录》。之后，《中国国家重点保护野生植物名录(第一批)》于 1999 年 8 月由国务院批准。2019 年和 2020 年又对《国家重点保护野生动物名录》和《国家重点保护野生植物名录》进行了调整发布。

在中国生物多样性研究和保护的过程中，中国科学院承担了大量艰苦工作。在这一时期各个阶段，也积累了许多生物标本，它们被妥善、安全地保存在中国科学院生物标本馆中，是这一段历史的起始和发展过程的最直接的记录。

服务国家经济建设

生物资源是地球生物圈内丰富的生物多样性与人类智慧相结合的产物，是人类健康食品、药物、多种工业材料，以及农作物抗干旱、抗低温、抗辐射等抗逆境的种质资源和基因资源宝库，更是生态文明建设以及人类可持续发展的重要物质基础。以生物资源和生物技术为基础的绿色生物经济正在兴起，并将成为经济转型和社会可持续发展的重要出路。

生物标本作为一种重要的生物资源，可直接服务于园林绿化、药物研发以及工业、农业和林业等产业，增加经济产出。在这一过程中，中国科学院生物标本馆起到了非常重要的作用。例如，“仙草”灵芝在中国有两千多年的药用历史，具有非常重要的经济价值。1957 年中国科学院微生物研究所邓叔群院士从野外采集的标本上分离获得了灵芝菌种，并于 1959 年实现了灵芝的人工栽培。菌物标本馆至今保藏有当时的凭证标本。基于馆藏模式标本，研究人员进一步完善了灵芝栽培技术，如今其产业年产值已达约 200 亿元，是目前人工栽培规模最大的药用真菌。此外，中国橡胶和烟草等重要产业的建立及发展，离不开中国科学院科研人员的努力，西双版纳热带植物园标本馆至今保藏有中国科学院西双版纳热带植物园创始人蔡希陶采集的烟草标本以及中国第一份橡胶标本，见证着这段艰辛的历程。

lot of work around biodiversity research and protection, national economic construction and public science and education.

Biodiversity research and protection

The most important work in the field of biology in contemporary China is the compilation and research of "Three Flora/Fauna". This is a major project supported by the state. "Three Flora/Fauna" are *Flora Reipublicae Popularis Sinicae, Fauna Sinica* and *Flora Cryptogamarum Sinicarum*. These three works have made a comprehensive survey of China's biological resources with the research methods of modern life science, greatly improving China's cognitive level of its own biological resources, becoming the most solid foundation for China's biological research, and promoting the rapid development of biological taxonomy and other biological disciplines. In the process of compilation and research, a variety of research and books about the protection and utilization of biological resources are emerging, and the research results are endless. Most of the work is undertaken by the relevant research institutes of CAS, such as the Institute of Zoology, the Institute of Botany, the Institute of Microbiology, etc. At the same time, most of the specimens collected in the compilation and research process are also preserved in the biological collections of CAS, which also provides a large number of specimen reference services for the compilation and research of these works.

The compilation and research of "Three Flora/Fauna" and other related researches make people realize that wild animals and plants have important ecological value and practical value, while the distribution range of some species is getting smaller and smaller, the number is getting less and less, many species are facing survival crisis. Therefore, the country has also started the road of biodiversity research and protection. In 1956, the National People's Congress of China adopted a proposal to raise the issue of establishing Nature Reserves. In October of the same year, the Ministry of Forestry drafted the *Draft of Delimitation of Natural Forest Cutting Area (Nature Reserve)*, and established Dinghushan Nature Reserve, China's first nature reserve, in Guangdong Province. This is also the only natural reserve under CAS. Since the late 1970s and the early 1980s, China's natural protection industry has developed rapidly.

At the same time, relevant laws and regulations began to be born and improved. In May 1950, the *Measures for the Protection of Rare Animals* was promulgated, stipulating that the hunting of giant pandas and other rare animals in Songpan, Sichuan Province and other places was prohibited. In September 1962, the State Council issued the *Directive on the Active Protection and Rational Use of Wildlife Resources*, emphasizing that "wildlife resources are the natural wealth of the country", and that "effective protection" and "rational use" should be carried out. In November 1988, China's first law for the protection of wildlife: *the Protection of Wild Animals of P. R. China* was promulgated. The following year, the *List of National Key Protected Wild Animals* was officially released. After that, the *List of National Key Protected Wild Plants* (the first batch) was released in August 1999. In 2019 and 2020, the *List of National Key Protected Wild Animals* and the *List of National Key Protected Wild Plants* were adjusted and released.

In the process of biodiversity research and conservation in China, CAS has undertaken a lot of hard work. During this period, many biological specimens were also accumulated, which were properly and safely preserved in the biological collections of CAS, which is the most direct record of the beginning and development of this history.

Serving national economic construction

Biological resources are the product of the combination of rich biodiversity and human wisdom in the earth's biosphere. They are the treasure house of human healthy food, medicine, various industrial materials, crop germplasm resources and gene resources resistant to drought, low temperature, radiation and other adversity. They are also the important material basis for the construction of ecological civilization and the sustainable development of human beings. The green bioeconomy based on biological resources and biotechnology is on the rise and will become an important way for economic transformation and social sustainable development.

As an important biological resource, the biological specimens can directly serve for landscaping, drug research and development, industry, agriculture, forestry and other industries, and increase economic output. In this process, the biological collections of CAS play a very important role. For example, "Magic Herb" the Lingzhi has a medical history of more than 2000 years in China, which has a very important economic value. In 1957, academician TENG Shu-Chün of the Institute of Microbiology, CAS isolated Lingzhi strains from the samples collected in the field, and in 1959, the artificial cultivation of Lingzhi was realized. Up to now, the Fungarium has preserved the certificate specimens of that time. Based on the collection of type specimens, the researchers further improved the cultivation technology of Lingzhi. Now the annual output value of the industry has reached about 20 billion yuan, which is the largest scale of artificial cultivation of medicinal fungi. In addition, the establishment and development of China's rubber and tobacco and other important industries are inseparable from the efforts of researchers of CAS. The Herbarium of Xishuangbanna Tropical Botanical

生物标本作为一类重要的国家战略资源，虽然不像土地资源、能源资源、水资源等可以导致国与国之间的直接冲突，但通过对丰富的生物标本资源进行研究，可以提供对有害和入侵生物进行有效防治的方法，以及对生态环境进行有效保护的策略，为国家生物安全提供重要的信息，为重大决策提供准确的科学依据，从而最大限度地降低生物因素所造成的经济损失。例如，成都生物研究所两栖爬行动物标本馆在草原毒蛇的治理、国家动物博物馆在马铃薯甲虫防控体系的建设、昆明动物博物馆在烟草害虫和草地贪夜蛾的防控等一系列生物防治工程中起到了重要的辅助和支撑作用。

公众科学教育

标本馆除了为中国农林害虫防控、入侵生物防治做出贡献外，还为动物有关的重大事件解释提供有力的专家支持、准确的标本比对和正确的舆论导向，担负着向广大公众普及生物科学知识的责任。

与西方国家的自然类博物馆的建立历史不同，最初中国生物标本馆建立的主要目的是为科学研究服务，公众教育只占其工作的一小部分。随着国民教育的普及和科学认知水平的提高，公众对获取科学知识的需求也逐步增加。近几年国家更是提高了对科普工作的重视程度，将其放在与科技创新同等重要的位置。中国各生物标本馆也适时调整工作重心，将科普教育列为标本馆的一项重要工作。特别是部分中国科学院生物标本馆专门设立负责公众教育的部门和场馆，向全社会开放，几乎每个标本馆或博物馆都已成为国家或地方的科普教育基地，成为全国或当地最有影响力的自然类教育机构。

例如，国家动物博物馆前身为中国科学院动物研究所标本馆，于 2007 年新建标本展示馆，同时标本馆与展示馆合并为“国家动物博物馆”。展示馆于 2009 年 5 月 17 日正式向社会公众开放，成为中国科学院动物研究所面向社会服务的一个窗口，生态文明教育和科学普及的基地，拉近科研工作者与社会大众距离的平台。展示馆建筑格局仿法国巴黎自然历史博物馆，总建筑面积 7500 平方米，其中展览面积 5500 平方米，共分为 3 层 8 个固定展厅，包括动物多样性与进化展厅、无脊椎动物展厅、国门生物安全展厅、濒危动物展厅、鸟类展厅、动物与人展厅、昆虫展厅和蝴蝶展厅。特别是国门生物安全展厅，是中国科学院动物研究所、国家动物博物馆与国家质量监督检验检疫总局（现为国家市场监督管理总局）动植物检疫监管司联合设立的，内容涵盖了与国门生物安全有关的法律法规、截获总体情况和相关案例等。该展厅于 2015 年 7 月 17 日开展并对外开放，已接待观众超过 30 万人次，并配合国家质量监督检验检疫总局相关部门开展科普专题活动 5 次，举行新闻发布会 3 次，成为国家向社会大众宣传国门生物安全知识的一个重要窗口。因此，国家质量监督检验检疫总局将国家动物博物馆指定为“国门生物安全宣传教育基地”。经过十年的发展，展示馆已经成为中国重要的科普宣传基地和国内最大的普及动物科学知识的专业博物馆。

此外，还有以某个生物细分类群为主的科普场馆，如成都生物研究所两栖爬行动物标本馆，也有以某个地域为展示对象的科普场馆，如西北高原生物研究所青藏高原生物标本馆。这些场馆无不以展示地球生命及其多样性为使命，向公众宣传生态环境保护的重要性，传播着人与自然和谐共处的理念。

Garden has preserved tobacco specimens collected by TSAI Hse-Tao, who is the founder of Xishuangbanna Tropical Botanical Garden of CAS, and the first rubber specimen in China, witnessing this arduous process.

As a kind of important national strategic resources, biological specimen, unlike land resources, energy resources, water resources, can lead to direct conflicts between countries, but through the research of abundant biological specimen resources, it can provide effective methods for the prevention and control of harmful and invasive organisms, as well as strategies for the effective protection of the ecological environment, provide important information to improve the national biological security and accurate scientific basis for major decisions, so as to minimize the economic losses caused by biological factors. For example, the management of grassland vipers by the Herpetological Museum of Chengdu Institute of Biology, the construction of potato beetle control system by the National Zoological Museum of China, and the prevention and control of tobacco pests and Meadow moths (*Spodoptera frugiperda*) by Kunming Natural History Museum of Zoology, have played an important supporting role in a series of biological control projects.

Public science education

In addition to contributing to the prevention and control of agricultural and forestry pests and invasive biological control in China, the collections also provide strong expert support, accurate specimen comparison and correct public opinion guidance for the interpretation of animal-related major events, and is responsible for popularizing the knowledge of biological science to the general public.

Different from the history of the establishment of natural museums in western countries, the primary purpose of the establishment of China's biological collections was to serve scientific research, and public education only accounted for a small part of its work. With the popularization of national education and the improvement of the level of scientific cognition, the public's demand for acquiring scientific knowledge is gradually increasing. In recent years, the government has paid more attention to science popularization and put it in the same important position as scientific and technological innovation. China's biological collections also timely adjusted the focus of work, set science education as an important work. In particular, some biological collections of CAS specially set up departments and venues responsible for public education, which are open to the whole society. Almost every collection or museum has become a national or local popular science education base and the most influential natural education institution in the country or in the local area.

For example, the National Zoological Museum of China (NZMC), formerly the Collection of the Institute of Zoology, CAS, built a new specimen exhibition hall in 2007. At the same time, the collection and the exhibition hall were merged into the "National Zoological Museum of China". The exhibition hall was officially opened to the public on May 17, 2009. It has become a window of the Institute of Zoology for social services, a base for ecological civilization education and scientific popularization, and a platform for narrowing the distance between scientific researchers and the public. The structure of the exhibition hall is similar to the French National Museum of Natural History (Muséum National d'Histoire Naturelle) in Paris, with a total construction area of 7,500 m^2, including 5,500 m^2 of exhibition area, which is divided into three floors and eight fixed exhibition halls, including Animal Diversity and Evolution Exhibition Hall, Invertebrate Exhibition Hall, National Biosafety Exhibition Hall, Endangered Animal Exhibition Hall, Bird Exhibition Hall, Animal and Human Exhibition Hall, Insect Exhibition Hall and Butterfly Exhibition Hall. In particular, the National Biosafety Exhibition Hall is jointly set up by the Institute of Zoology of CAS, NZMC and the former Department of Animal and Plant Quarantine and Supervision of the General Administration of Quality Supervision, Inspection and Quarantine, covering laws and regulations related to national biosafety, the overall situation of interception and relevant cases. This exhibition hall was opened to the public on July 17, 2015, and has received more than 300,000 visitors. It has cooperated with the relevant departments of AQSIQ to carry out five science popularization activities, held three press conferences, and has become an important window for the country to publicize the national biosafety knowledge to the public. Therefore, AQSIQ designated NZMC as the "National Biosafety Publicity and Education Base". After ten years of development, the exhibition hall has become an important science popularization base in China and the largest professional museum for popularizing animal science knowledge in China.

In addition, there are also popular science venues based on a biological subdivision group, such as the Herpetological Museum of Chengdu Institute of Biology, and popular science venues based on a certain region, such as the Qinghai-Tibet Plateau Museum of Biology of Northwest Institute of Plateau Biology. All these museums or collections take displaying the life and diversity of the earth as their mission, publicize the importance of ecological environment protection to the public, and spread the concept of harmonious coexistence of human and nature.

3

中国科学院生物标本馆现状

Current situation of Biological Collections of Chinese Academy of Sciences

中国科学院生物标本馆是国家动物、植物、菌物、化石等标本保藏、研究和科学教育的重要实体，是全国生物标本集中保藏的最主要场所，在国家生物资源的保护与可持续利用中具有不可替代的重要作用。自20世纪50年代建立以来，中国科学院生物标本馆一直担负着中国战略生物资源的收集、整理、保藏、研究与可持续利用的国家和历史使命，代表着国家战略生物资源的整体水平，致力于研究生物多样性、揭示生物进化规律，为国民经济可持续发展服务。

2008年，中国科学院成立了中国科学院生物标本馆（博物馆）工作委员会（以下简称“工委会”），主要任务是在中国科学院科技促进发展局领导下，协助调查中国科学院生物标本馆（博物馆）的工作状况，提出改革、发展和建设的意见及建议，编制中长期发展战略规划报告，并进行检查评估；促进中国科学院生物标本馆（博物馆）之间的科技协作、学术交流和信息资源的共建共享，促进与国内外生物标本馆（博物馆）之间的合作，组织和参与与生物标本馆（博物馆）工作有关的学术活动，促进中国科学院生物标本馆（博物馆）的对外宣传与交流。

在工委会成立之后，逐渐形成了一套管理措施体系，如建立了标本馆年度考核办法，优化了考核指标体系，坚持每年度进行工作总结，并组织咨询专家组每年进行评估，之后根据评估结果对各馆的运行补助费进行动态调整，等等。

3.1 中国科学院生物标本馆概况

中国科学院目前拥有由19个研究所作为依托单位的20家生物标本馆（博物馆），收集保藏的生物标本资源涵盖了动物、植物、菌物、古生物等，拥有中国乃至亚洲最大的生物标本馆，以及一系列中国最大、最有特色的专类标本馆，是中国最大、国际上具有重要影响力的生物标本资源保藏体系与数字化数据网络，也是生物标本资源整合与共享利用的平台。全院20家生物标本馆（博物馆）分布于全国13个省（自治区、直辖市）所收藏生物标本的采集地基本覆盖全国所有地区和生境类型（包括海域）。

3.2 中国科学院生物标本收藏

截至2019年11月，全院标本馆（博物馆）共保藏各类生物标本共计2141.5万号，其中定名标本占一

As an important entity of national animal, plant, fungus, fossil and other specimen preservation, research and scientific education, the biological collections of CAS are the most important places for centralized preservation of national biological specimens, which plays an irreplaceable role in the protection and sustainable utilization of national biological resources. Since its establishment in the 1950s, it has been shouldering the national and historical mission of the collection, sorting, preservation, research and sustainable utilization of China's strategic biological resources, representing the overall level of national strategic biological resources, committed to the research of biological diversity, revealing the law of species evolution, and serving the sustainable development of national economy.

In 2008, CAS established the Working Committee of Biological Collection (Museum) of CAS (hereinafter referred to as the "Working Committee"), whose main task is to assist in investigating the working conditions of the biological collections (Museum) of CAS under the leadership of the Bureau of Science & Technology for Development CAS, put forward opinions and suggestions on reform, development and construction, and formulate the medium and long-term development strategy and to plan reports and conduct inspection and evaluation; to promote scientific and technological cooperation, academic exchange and information resources sharing between collections, to promote cooperation with domestic and foreign biological collections, to organize and participate in academic activities related to the work of biological collections, and to promote the publicity and exchange of the biological collection of CAS.

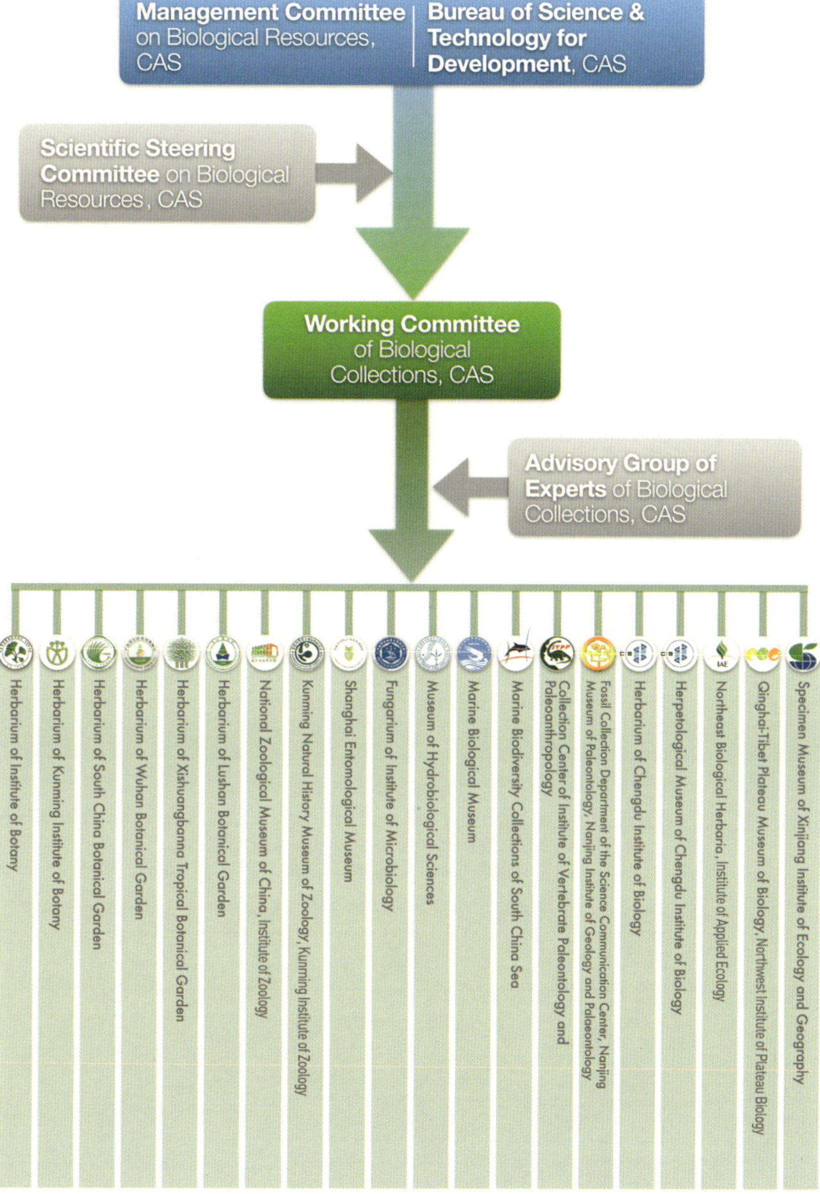

Organization chart of the Working Committee

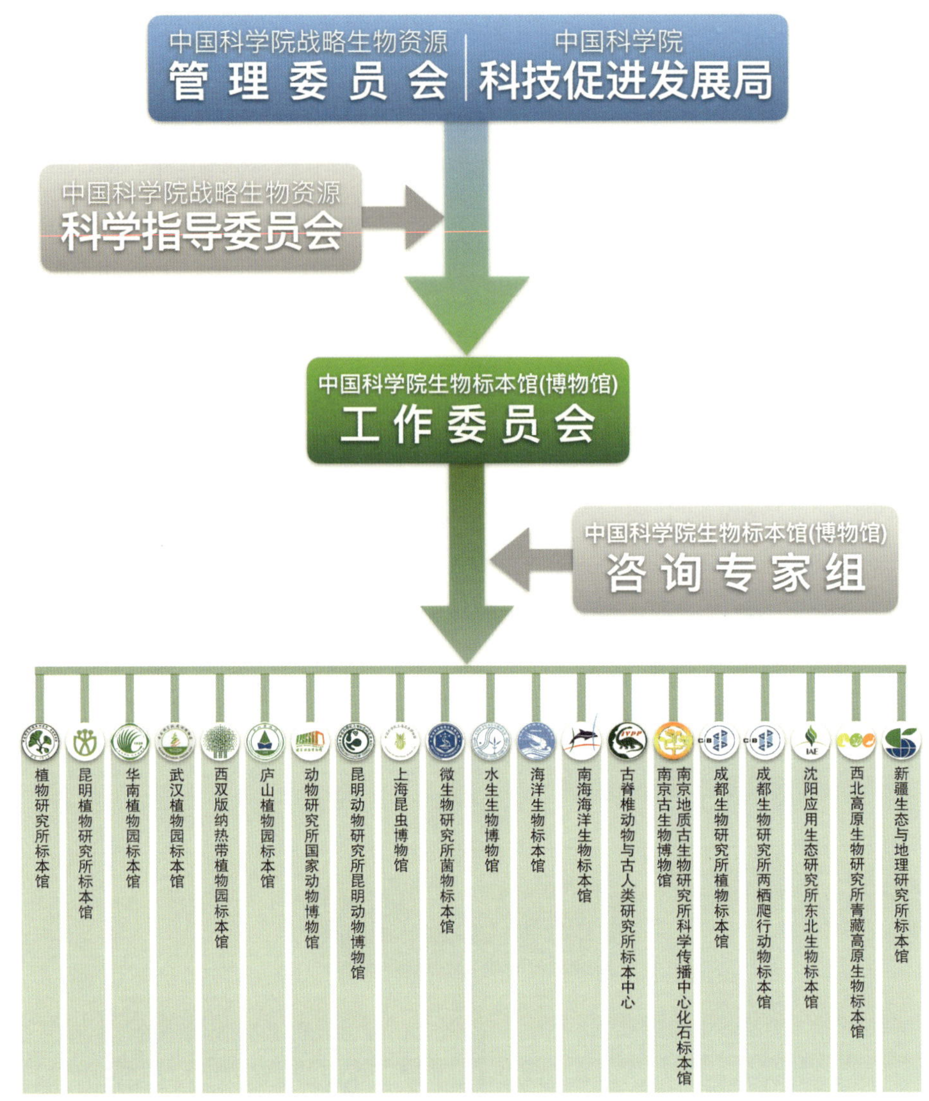

工委会组织结构图

生物标本资源收集保藏现状（单位：万号）

Collection and preservation status of biological specimen resources (×10k)

类别 Category	内容 Content	数量 Number
总体保藏现状 General preservation status	现有馆藏标本数 Number of specimens	2141.5
	现有定名标本数 Number of classified specimens	1167.9
	现有模式标本数 Number of type specimens	33.36
	已录入标本数字化信息数 Number of digitalized specimens	1011.9
各类群保藏状况 Preservation status of various groups	动物标本 Animal specimens	1239.0
	植物标本 Plant specimens	772.2
	菌物标本 Fungi specimens	60.6
	化石标本 Fossil specimens	69.7

工委会各成员单位及标本收藏现状（单位：万号）

Member units and specimen collection status of the Working Committee (×10k)

序号 Order number	名称 Name	隶属研究所 Institute affiliated to	收藏类群 Species groups of collections	标本主要产地 Main area where collections from	馆藏数量 Number of specimens
1	国家动物博物馆 National Zoological Museum of China	动物研究所 Institute of Zoology	动物 Animal	全国 China	856.0
2	植物研究所标本馆 Herbarium of Institute of Botany	植物研究所 Institute of Botany	植物 Plant	全国 China	282.9
3	昆明动物博物馆 Kunming Natural History Museum of Zoology	昆明动物研究所 Kunming Institute of Zoology	动物 Animal	西南 Southwest China	89.9
4	昆明植物研究所标本馆 Herbarium of Kunming Institute of Botany	昆明植物研究所 Kunming Institute of Botany	植物 Plant	西南 Southwest China	155.0
5	海洋生物标本馆 Marine Biological Museum	海洋研究所 Institute of Oceanology	海洋动植物 Marine animal and plant	中国沿海、西太平洋及南北极等海域 China's coast, Western Pacific and North and South poles	85.0
6	古脊椎动物与古人类研究所标本中心 Collection Center of Institute of Vertebrate Paleontology and Paleoanthropology	古脊椎动物与古人类研究所 Institute of Vertebrate Paleontology and Paleoanthropology	化石 Fossil	全国 China	23.5
7	菌物标本馆 Fungarium	微生物研究所 Institute of Microbiology	菌物 Fungi	全国 China	53.6
8	华南植物园标本馆 Herbarium of South China Botanical Garden	华南植物园 South China Botanical Garden	植物 Plant	华南 South China	111.1
9	上海昆虫博物馆 Shanghai Entomological Museum	中国科学院分子植物科学卓越创新中心 CAS Center for Excellence in Molecular Plant Sciences	动物（昆虫） Animal (insect)	华东 East China	127.0
10	成都生物研究所两栖爬行动物标本馆 Herpetological Museum of Chengdu Institute of Biology	成都生物研究所 Chengdu Institute of Biology	动物（两栖爬行） Animal (amphibian and reptile)	全国 China	11.7
11	南京地质古生物研究所科学传播中心化石标本馆、南京古生物博物馆 Fossil Collection Department of the Science Communication Center of Nanjing Institute of Geology and Palaeontology, Nanjing Museum of Palaeontology	南京地质古生物研究所 Nanjing Institute of Geology and Palaeontology	化石 Fossil	全国 China	46.2
12	青藏高原生物标本馆 Qinghai-Tibet Plateau Museum of Biology	西北高原生物研究所 Northwest Institute of Plateau Biology	动植物 Animal and plant	青藏高原 Qinghai-Tibet Plateau	55.5
13	水生生物博物馆 Museum of Hydrobiological Sciences	水生生物研究所 Institute of Hydrobiology	水生动植物 Aquatic animal and plant	全国 China	40.0
14	南海海洋生物标本馆 Marine Biodiversity Collections of South China Sea	南海海洋研究所 South China Sea Institute of Oceanology	海洋动物 Marine animal	南部海域 South China Sea	22.6
15	武汉植物园标本馆 Herbarium of Wuhan Botanical Garden	武汉植物园 Wuhan Botanical Garden	植物 Plant	华中 Central China	30.0
16	东北生物标本馆 Northeast Biological Herbaria	沈阳应用生态研究所 Institute of Applied Ecology	动物（水生甲虫）、植物、菌物 Animal (aquatic beetles), plant and fungi	东北 Northeast China	62.2
17	新疆生态与地理研究所标本馆 Specimen Museum of Xinjiang Institute of Ecology and Geography	新疆生态与地理研究所 Xinjiang Institute of Ecology and Geography	动植物 Animal and plant	新疆 Xinjiang	10.6
18	西双版纳热带植物园标本馆 Herbarium of Xishuangbanna Tropical Botanical Garden	西双版纳热带植物园 Xishuangbanna Tropical Botanical Garden	植物 Plant	西南 Southwest China	22.9
19	成都生物研究所植物标本馆 Herbarium of Chengdu Institute of Biology	成都生物研究所 Chengdu Institute of Biology	植物 Plant	西南 Southwest China	36.8
20	庐山植物园标本馆 Herbarium of Lushan Botanical Garden	庐山植物园 Lushan Botanical Garden	植物 Plant	华南 South China	19.0

工委会各成员单位分布

半以上，达到 1167.9 万号；模式标本 33.36 万号；近一半标本完成了数字化工作。部分场馆收藏范围遍及全国，其他则以所在地区为主，辐射周围区域。从收藏类群来看，其中 4 个以动物为主，如动物研究所国家动物博物馆；7 个以植物为主，如植物研究所标本馆；1 个以微生物为主，如微生物研究所菌物标本馆；3 个以水生生物为主，如水生生物研究所水生生物博物馆；2 个以化石为主，如古脊椎动物与古人类研究所标本中心；另外 3 个是地方性综合场馆，如西北高原生物研究所青藏高原生物标本馆。

中国科学院生物标本馆根据主要收藏类群的分类

Categories according to the main collection groups of CAS Biological Collections

标本馆分类 Category	主要收藏标本类群 Main groups of collected specimens	标本馆数量 Number of collections
动物类馆 Zoological Collection/Museum	动物标本 Animal specimens	4
植物类馆 Herbarium	植物标本 Plant specimens	7
微生物类馆 Fungarium	微生物标本（菌物等） Microbiological specimens (fungi, etc.)	1
水生生物类馆 Hydrobiont Collection/Museum	水生生物标本（水生动物、植物、微生物等） Aquatic specimens (aquatic animals, aquatic plants, microorganisms)	3
古生物类馆 Paleontology Collection/Museum	古生物标本（动植物化石） Palaeontological specimens (fossils of animals and plants)	2
地方性综合类馆 Local Comprehensive Collection/Museum	动物、植物、微生物等各类群生物标本 Specimens of various groups of animals, plants, microorganisms, etc.	3

Distribution of member units of the Working Committee

After the establishment of the Working Committee, a set of management measures system has been gradually formed, such as the establishment of the annual assessment method for the collections, the optimization of the assessment index system, the annual work summary, and the organization of the advisory expert group to assess each year, and then the dynamic adjustment of the operation subsidy of each collection according to the assessment results, etc.

3.1 A survey of CAS biological collections

At present, CAS has 20 biological collections with 19 research institutes as supporting units. The biological specimen resources collected and preserved include animals, plants, fungi, paleontology, etc. It has the largest collection in China and even Asia, as well as a series of the largest and most distinctive special collections in China. The collection system and digital data network of biological specimen resources are also the largest in China and internationally influential. And they are also platforms for the integration and sharing of biological specimen resources. The 20 biological collections are distributed in 13 provinces, cities and autonomous regions. The collection places of biological specimens basically cover all regions and habitat types (including sea areas) of the country.

3.2 Biological specimen collection of CAS

As of November 2019, CAS biological collection/museum has preserved a total of 21.415 million biological specimens, of which more than half are identified specimens, reaching 11.679 million; type specimens are 333,600; nearly half of the specimens have completed the digital work. Some of the collections are collected all over the country, while others are mainly in their own areas, radiating the surrounding areas. From the perspective of collection groups, four of them are mainly animals, such as the National Zoological Museum of China of Institute of Zoology; seven are mainly plants, such as the Herbarium of Institute of Botany; one is mainly microorganisms, such as the Fungarium of Institute of Microbiology; three are mainly aquatic organisms, such as the Museum of Hydrobiological Sciences of Institute of Hydrobiology; two are mainly fossils, such as the Collection Center of Institute of Vertebrate Paleontology and Paleoanthropology; the other three are local comprehensive venues, such as the Qinghai-Tibet Plateau Museum of Biology of Northwest Institute of Plateau Biology.

3.3 中国科学院生物标本馆介绍

中国科学院目前拥有 20 家生物标本馆（博物馆），按照其主要收藏类群，大致可以分为六类。其中每个标本馆的基本信息、简介、历史沿革、收藏概况、特色收藏等内容分述如下。

1）动物类馆

中国科学院动物研究所国家动物博物馆

基本信息

隶属于：中国科学院动物研究所
地址：中国北京市朝阳区北辰西路 1 号院
网址：http://museum.ioz.ac.cn/index.html
建馆年份：1950 年
收藏类群：无脊椎动物、鱼类、两栖类、爬行类、鸟类和哺乳类
标本藏量：856 万号
馆长：乔格侠，研究员
联系人：张莉莉
电子邮箱：zhll@ioz.ac.cn
联系电话：+86 10 64807267

标本馆简介

中国科学院动物研究所国家动物博物馆是中国科学院战略生物资源的重要保藏地和重要的科普场所，拥有亚洲最大的生物标本馆和国内最大的动物标本展示馆，是集动物分类研究、动物标本收藏和动物科学知识普及三位一体的国立学术机构。标本馆已有超过 150 年收藏历史，收藏有各类动物标本 856 万余号，占中国科学院生物标本总藏量的 1/3 多；展示馆展陈 5000 余种动物标本和展品，长年对外开放。

国家动物博物馆标本馆
Collection of National Zoological Museum of China

标本馆内景
Inside view of the collection

3.3 Introduction to the CAS biological collections

At present, CAS has 20 biological collections, which can be roughly divided into six categories according to its main collection groups. The basic information, brief introduction, history, collection overview and characteristic collection of each collections are described as follows.

1) Zoological Collection/Museum

National Zoological Museum of China (NZMC), Institute of Zoology, CAS

Basic Information

Affiliated to: Institute of Zoology, CAS
Address: No.1 Beichen West Road, Chaoyang District, Beijing, P. R. China
Website: http://museum.ioz.ac.cn/index.html
Established in: 1950
Species groups of collection: Invertebrates, fish, amphibians, reptiles, birds and mammals
Collection quantity: 8,560,000
Curator: Dr. QIAO Gexia, Professor
Contacts: Dr. ZHANG Lili
E-mail: zhll@ioz.ac.cn
Tel: +86 10 64807267

Introduction

The National Zoological Museum of China (NZMC) of the Institute of Zoology is an important place to preserve the strategic biological resources and science popularize of CAS. It has the largest biological collection in Asia and the largest animal specimen exhibition museum in China. It is a national academic institution integrating the research of animal classification, collection of animal specimens and popularization of animal science knowledge. With a history of more than 150 years, the museum has collected more than 8.56 million animal specimens of various kinds, accounting for more than 1/3 of the total collection of biological specimens of CAS. The exhibition hall has exhibited more than 5,000 kinds of animal specimens and exhibits, which have been open to the public for whole years.

History

The collection of NZMC has a collection history of more than 150 years, which can be traced back to the Musée de Zi-Ka-Wei founded in Shanghai in 1868 by a French missionary P. M. Heude (1836-1902), renamed as Musée de Heude in 1930. On the basis of the Musée de Heude, Fan Memorial Institute of Biology, Institute of Zoology of the National Academy of Peiping, Natural History Museum, Academia Sinica and the Palace Museum, Animal Collection Sorting Committee of Chinese Academy of Sciences was established in 1950. With the establishment of the Institute of Zoology of CAS in 1962, it is divided into "Insect collection", "Vertebrate collection" and "Invertebrate collection". Then they were integrated into the animal collection of the Institute of Zoology, CAS in 1996. In 2007, the animal collection and the newly built specimen exhibition hall were merged into the National Zoological Museum of China.

国家动物博物馆展示馆

Exhibition hall of National Zoological Museum of China

Overview of Collections

The animal collection includes invertebrate branch, insect branch, fish and amphibious reptile branch, bird branch, mammal branch and specimen digital branch. There are more than 8.56 million animal specimens in the collection, more than 1/3 of the total collection of biological specimens of CAS. The collection of specimens almost includes all the

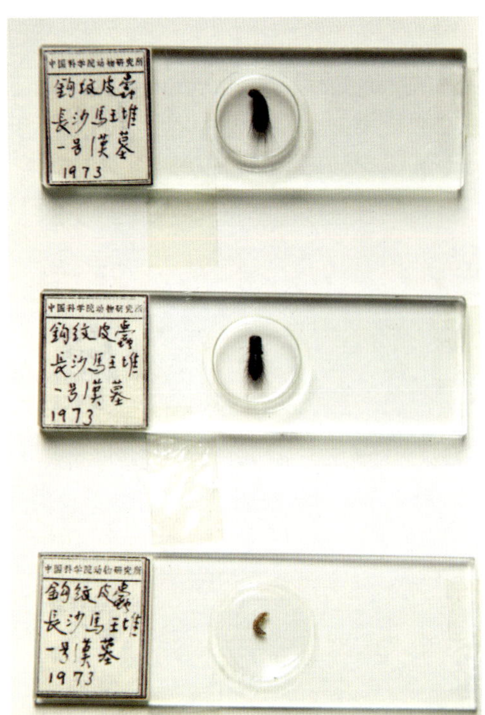

马王堆汉墓出土昆虫标本
Insect specimens unearthed from Mawangdui Han Tomb

历史沿革

国家动物博物馆标本馆已有超过 150 年收藏历史，最早可追溯至 1868 年由法国传教士韩伯禄（P. M. Heude, 1836-1902）在上海创建的徐家汇博物院（Musée de Zi-Ka-Wei），1930 年改称震旦博物院（Musée de Heude）。动物标本馆是在原震旦博物院、静生生物调查所、北平研究院动物研究所、中央研究院自然历史博物馆及故宫博物院等机构所属标本馆的基础上，于 1950 年成立"中国科学院动物标本整理委员会"，随着 1962 年中国科学院动物研究所成立，分为"昆虫标本馆"、"脊椎动物标本馆"和"无脊椎动物标本馆"三个馆，1996 年整合为中国科学院动物研究所动物标本馆。2007 年动物标本馆与新建成的标本展示馆合并为"国家动物博物馆"。

收藏概况

动物标本馆包括无脊椎动物标本分馆、昆虫标本分馆、鱼类及两栖爬行类标本分馆、鸟类标本分馆、兽类标本分馆和标本数字化分馆。标本馆现有各类动物标本 856 万余号，超过中国科学院生物标本收藏总量的 1/3。馆藏标本几乎包括了在中国分布的各主要类群和代表性种类，采集范围覆盖了包括香港、台湾在内的全国各地，以及三十余个国家。现有定名标本 65 000 余种约 350 万号，模式标本 1.04 万余种 10.8 万余号。借助计算机网络技术，开展了标本数字化工作，建成了国内最大动物数字化标本馆。目前，已标准化整理和数字化表达 255 万号动物标本并网上共享信息数据。

特色收藏

国家动物博物馆全面收集各类现生动物标本，不但在数量上在中国乃至亚洲名列前茅，而且经过多年的建设和发展，逐步形成了收藏历史悠久、类群较为齐全、覆盖区域广泛、模式标本收藏丰富、濒危保护物种标本较多等特色。

历史悠久的标本

马王堆汉墓出土昆虫标本 在长沙马王堆汉墓发掘过程中，发现有类似昆虫的出土物，送交国家动物博物馆。经专家鉴定，确认为昆虫，并将其妥善保藏。初步统计，数量约为 10 号。

中国最早的自然博物馆——震旦博物院收藏的动物标本 震旦博物院由法国天主教耶稣会神父韩伯禄（P. M. Heude）于清同治七年（1868 年）在上海徐家汇创建，当时名为"徐家汇博物院（Musée de Zi-Ka-Wei）"。民国十九年（1930 年）改名为"震旦博物院（Musée de Heude）"，归震旦大学管理。该馆收藏有大量采自中国华北和长江中下游，以及东南亚地区的动物标本。1953 年，该馆大部分动物标本移交中国科学院动物研究所标本馆收藏。据初步统计，共计 5000 余号，包括哺乳类、鸟类、爬行类、无脊椎动物、昆虫。

静生生物调查所动物标本 1928 年，为纪念前一年病逝的原民国政府教育总长范源濂（字静生）而在北平创立"静生生物调查所"，该所汇集了一大批后来成为中国生物学研究领域先驱的学者，如秉志、胡先骕、翁文灏、寿振黄、张春霖、刘崇乐、沈嘉瑞等。该所创立之后，对中国动植物资源开展了普查，采集了大

major groups and representative species in China, and range from all around China, including Hong Kong and Taiwan, and also from more than 30 other countries. There are more than 65,000 species and 3.5 million specimens of designated specimens, and more than 10,400 species and 108,000 type specimens. With the help of computer network technology, the collection has carried out the digitization of specimens and built the largest animal digital collection in China. At present, 2.55 million animal specimens have been standardized, collated and digitally expressed, and information data have been shared online.

Special Collections

NZMC collects all kinds of living animal specimens in an all-round way, which not only ranks among the best in quantity in China and even Asia, but also has gradually formed the characteristics of long collection history, complete groups, wide coverage area, rich collection of type specimens, and many specimens of endangered species after years of construction and development.

Specimens with a long history

Insect specimens unearthed from Mawangdui Han Tomb. During the excavation of Mawangdui Han Tomb in Changsha, unearthed objects similar to insects were found, which were sent to NZMC for expert identification and confirmed as insects. NZMC properly preserved them. According to preliminary statistics, the number is about 10.

Animal specimens collected by the earliest natural museum in China. The Musée de Heude: the Musée de Heude was founded in Shanghai in 1868 by P. M. Heude, a French Catholic Jesuit priest. At that time, it was called " Musée de Zi-Ka-Wei". In 1930, it was renamed "Musée de Heude" and was under the management of the Aurora University. The museum collects a lot of animal specimens from North China, the middle and lower reaches of the Yangtze River and Southeast Asia. In 1953, most of the animal specimens were handed over to the collection of the Institute of Zoology, CAS. According to preliminary statistics, there are about 5,000 species in total, including mammals, birds, reptiles, invertebrates.

Animal specimens from Fan Memorial Institute of Biology. In 1928, in memory of FAN Yuanlian, former Minister of Education of the Republic of China, who died one year ago, the "Fan Memorial Institute of Biology" was founded in Peiping. The institute gathered many pioneering scholars who later start the field of biological research in China, such as BING Zhi, HU Hsen-Hsu, WENG Wenhao, SHOU Zhenhuang, ZHANG Chunlin, LIU Chongle, SHEN Jiarui, etc. After its establishment, the institute has carried out a general survey of animal and plant resources in China and collected a lot of specimens. It is regarded as the starting institution of modern biological research in China. The Institute of Zoology, the Institute of Botany and other institutes are all started from it. At present, many animal specimens collected by Fan Memorial Institute of Biology are preserved in NZMC, with a number of tens of thousands.

Animal specimens from the imperial palace of the Qing Dynasty. In the 1950s, more than 20 animal skins of the former imperial palace of the Qing Dynasty were handed over from the Palace Museum to CAS and are now preserved in NZMC.

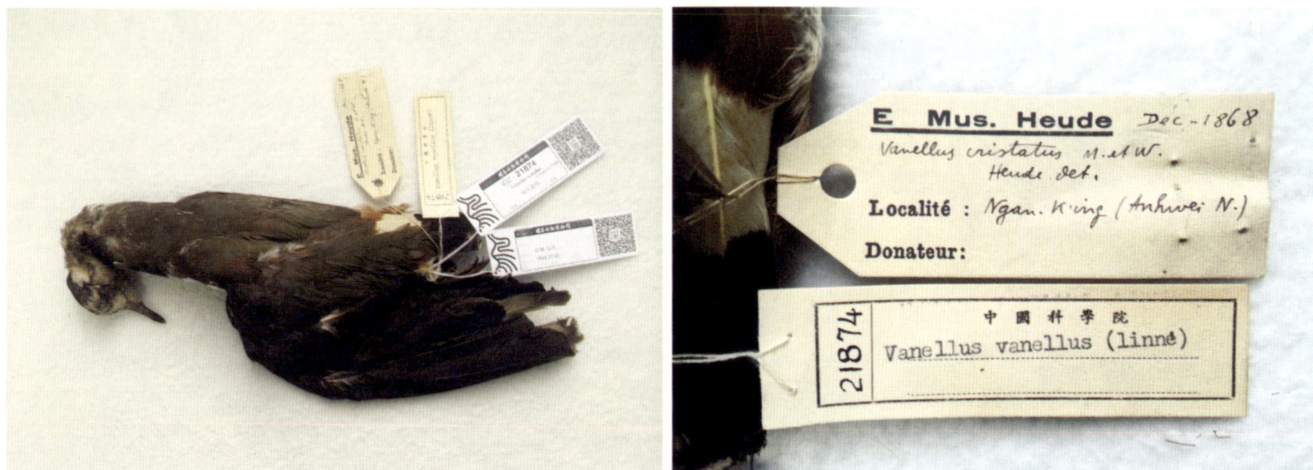

震旦博物院标本
Specimens from the Musée de Heude

静生生物调查所标本
Specimens from Fan Memorial Institute of Biology

清皇宫动物标本
Animal specimens from the Imperial Palace of the Qing Dynasty

量的标本。静生生物调查所被认为是中国现代生物学研究的起始机构，中国科学院动物研究所、中国科学院植物研究所等均以该所为各自机构的起源。静生生物调查所当初收藏的很多动物标本目前保藏在国家动物博物馆，数量达数万号。

清皇宫动物标本　20 世纪 50 年代，由故宫博物院向中国科学院移交一批原清皇宫的动物皮张，现保藏在国家动物博物馆，数量有 20 余号。

此外，标本馆尚保存有原京师大学堂、原中央研究院、原国立武汉大学、原国立厦门大学等民国时期的高等院校和科研机构收藏的动物标本数十号。

具有特殊历史意义的标本

国礼标本　中华人民共和国成立后，个别外国领导人赠送给中国领导人的动物标本礼品，或国外政府机构赠送给中国政府机构的动物标本礼品。初步统计有 10 余号。

具有独特学术价值的标本

标本馆还保藏着一些见证中国动物学发展历史的或具有特殊学术价值的标本，如由中国现代动物学创始人秉志先生亲手制作的鲤鱼头骨标本、极危物种云南闭壳龟标本（标本馆收藏 2 号，全球仅有 8 号标本）、白暨豚标本（3 号）等。

珍稀濒危动物标本

中华人民共和国成立之后，国家先后组织了多次大规模的科学考察，标本馆的科研工作者积极参与了这些考察，采集了大量的动物标本，其中很多物种现已被列入《国家重点保护野生动物名录》及《濒危野生动植物种国际贸易公约》（CITES）附录Ⅰ、附录Ⅱ和附录Ⅲ，已不能随意在野外采集。

蒙古国科学院赠送给中国科学院的标本
Specimens presented to CAS from the Mongolian Academy of Sciences

秉志先生亲手制作的鲤鱼头骨标本
Specimens of carp skull prepared by Mr. BING Zhi

In addition, there are dozens of animal specimens collected by Peking Normal University, Academia Sinica, National Wuhan University, National Xiamen University and other universities and scientific research institutions of the former government of the Republic of China.

Specimens with special historical significance

National gift specimen. After the founding of the People's Republic of China, some foreign leaders give animal specimen gifts to China's leaders, or foreign government agencies give animal specimen gifts to the government agencies. There are more than 10 specimens.

阳彩臂金龟
Cheirotonus jansoni

Specimen with unique academic value

There are also some specimens that have witnessed the development history of zoology in China or have special academic value, such as specimen of carp skull prepared by Mr. BING Zhi, the founder of modern zoology in China, specimens of critically endangered species Yunnan box turtle (*Cuora yunnanensis*, only 8 in the world, 2 in NZMC), 3 specimens of Chinese river dolphin (*Lipotes vexillifer*), etc.

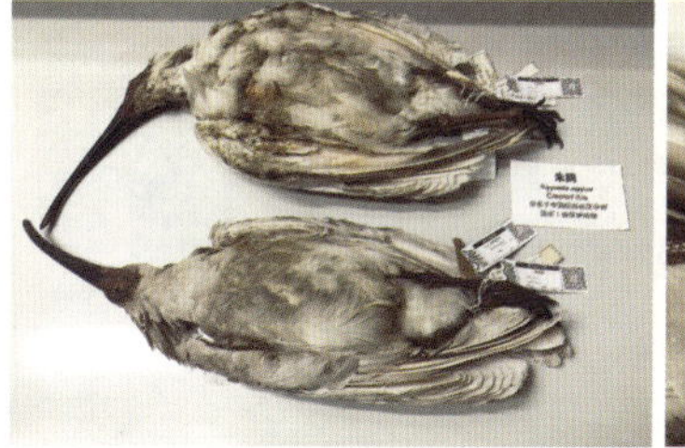

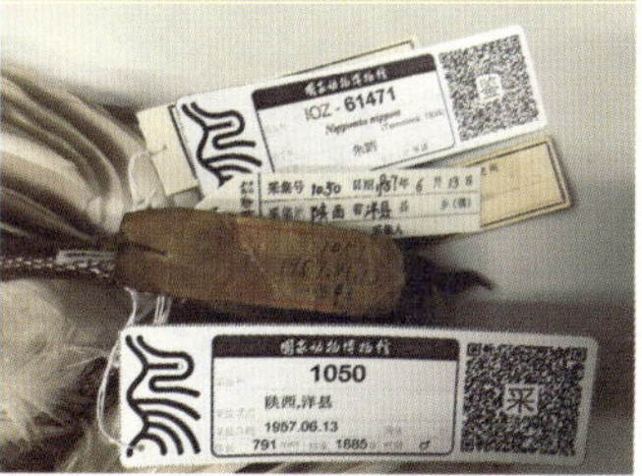

1957 年采自陕西洋县的朱鹮标本
Specimens of *Nipponia nippon* collected from Yangxian County, Shaanxi Province in 1957

Rare and endangered animal specimens

After the founding of the People's Republic of China, the country has organized many large-scale scientific investigations. The scientific researchers of NZMC actively participated in these investigations and collected a large number of animal specimens, many of which have been listed in the *List of National Key Protected Wild Animals*, Appendix I, Appendix II and Appendix III of the *Convention on International Trade in Endangered Species of Wild Fauna and Flora* (CITES). These specimens now can not be collected in the field at will.

中国科学院昆明动物研究所昆明动物博物馆

基本信息

隶属于：中国科学院昆明动物研究所
地址：中国云南省昆明市教场东路 32 号
网址：http://museum.kiz.cas.cn
建馆年份：1959 年
收藏类群：哺乳类、鸟类、两栖类、爬行类、鱼类、昆虫、软体动物等
标本藏量：89.9 万号
馆长 / 联系人：吴东东，研究员
电子邮箱：wudongdong@mail.kiz.ac.cn
联系电话：+86 871 65161258

标本馆简介

中国科学院昆明动物研究所昆明动物博物馆（以下简称“昆明动物博物馆”）建于 1959 年，是国家二级博物馆、国家科普教育基地，是中国西南地区规模最大、收藏量最为丰富的动物专题博物馆。昆明动物博物馆由标本馆和展示馆两部分组成。标本馆与昆明动物研究所同期建于 1959 年，是历史最悠久的支撑部门。60 年来，历经几代科学工作者的艰苦努力，昆明动物博物馆馆藏各类动物标本近 90 万号（其中模式标本 613 种 / 亚种 6518 号），涵盖了哺乳类、鸟类、两栖类、爬行类、鱼类、昆虫、软体动物等多个动物类群。展示馆由中国科学院与云南省合作建成，于 2006 年 11 月正式对公众开放。展示馆大楼建筑面积 7350 平方米（其中展示和公共服务面积约 6500 平方米）。展出各类动物标本，包括中国一、二级重点保护动物 90 余种，还有许多珍稀种、濒危种、观赏种、药用种、资源种及食用种类等，这些陈列标本充分展现了云南“动物王国”的风采和中国南方动物的概貌。

60 年来，广大科研人员依托馆藏标本资源，发表论文 1000 多篇，出版专著数十部，如：自然地理区域相关专著有《横断山区鸟类》、《横断山区鱼类》、《横断山区两栖爬行动物》以及《高黎贡山地区脊椎动物考察报告（第二册 鸟类）》、《珠江鱼类志》、《西南武陵山地区昆虫》等；以行政区划为依据，有《中国哺乳动物彩色图鉴》、《中国哺乳动物种和亚种分类名录与分布大全》、《中国动物志 鸟纲 第八卷 雀形目：阔嘴

昆明动物博物馆
Kunming Natural History Museum of Zoology

昆明动物博物馆昆虫展厅
Insect exhibition hall of KNHMZ

Kunming Natural History Museum of Zoology (KNHMZ), Kunming Institute of Zoology (KIZ), CAS

Basic Information

Affiliated to: Kunming Institute of Zoology, CAS
Address: No.32 Jiaochang Donglu, Kunming, Yunnan Province, P. R. China
Website: http://museum.kiz.cas.cn
Established in: 1959
Species groups of collection: Mammals, birds, amphibians, reptiles, fish, insects, mollusks, etc.
Collection quantity: 899,000
Curator/Contacts: Dr. WU Dongdong, Professor
E-mail: wudongdong@mail.kiz.ac.cn
Tel: +86 871 65161258

Introduction

Kunming Natural History Museum of Zoology (KNHMZ) of KIZ, CAS was established in 1959. It is a national second-level museum and a national science education base. It is the largest and most abundant animal thematic museum in Southwest China. The KNHMZ consists of a specimen museum and an exhibition hall. The specimen museum was established in 1959, at the same time as the KIZ, and is the oldest supporting department. Over the past 60 years, the KNHMZ has collected nearly 900,000 animal specimens (including 613 species / 6,518 subspecies model specimens) through the hard work of several generations of scientists, including such multiple animal groups as mammals, birds, amphibians, reptiles, fish, insects, mollusks, etc. The exhibition hall was jointly established by CAS and Yunnan Province, and was officially opened to the public in November 2006. The exhibition building covers 7,350 m^2 (including an exhibition and public service area of approximately 6,500 m^2). It displays various animal specimens, including more than 90 Chinese first- and second-class protected animal species, as well as many rare, endangered, ornamental, medicinal, resource, and edible species. These specimens on display highlight Yunnan as an "Animal Kingdom" and provides an excellent overview of animals from southern China.

Over the past 60 years, relying on the specimen resources of the museum, researchers from KIZ have published more than 1,000 papers and dozens of monographs. Monographs-related natural geography include *Birds of the Hengduan Mountains Region*, *Fishes of the Hengduan Mountains Region*, *Amphibians and Reptiles of the Hengduan Mountains Region*, *Report on Vertebrate Inspection in the Gaoligong Mountains* (Vol. 2: Aves) (name in Chinese only), *Fish Records in the Pearl River* (name in Chinese only), and *Insects of Wuling Mountains Area, Southwestern China*. Administrative division monographs include *A Field Guide to the Mammals of China*, *A Complete Checklist of Mammal Species and Subspecies in China: A Toxonomic and Geographic Reference*, and *Fauna Sinica Aves Vol. 8: Passeriformes (Eurylaimidae—Irenidae)*. Monographs on regional biological resources and pest control include *Scientific Investigation Report on Biological Resources in the Honghe Region of Southern Yunnan* (name in Chinese only), *Research on the*

昆明动物博物馆鸟类展厅
Bird exhibition hall of KNHMZ

昆明动物博物馆哺乳类展厅
Mammal exhibition hall of KNHMZ

昆明动物博物馆骨骼展厅
Skeleton exhibition hall of KNHMZ

鸟科—和平鸟科》等；地区生物资源及虫害防治方面相关专著有《云南南部红河地区生物资源科学考察报告》、《西南资源生物开发战略研究》和《云南粮食作物害虫及天敌昆虫图册》等；动物生物学和形态解剖学相关专著，有《树鼩生物学》、《叶猴生物学》、《金丝猴的解剖》和《中国雉类白腹锦鸡》等，并荣获国家级、省部级、地州级等多项科研成果奖。

收藏概况

昆明动物博物馆作为国家战略生物资源保藏机构，由兽类库、鸟类库、两栖爬行库、鱼类库、昆虫库、外来入侵物种库、生命条形码凭证标本库、青藏高原标本库、动物图片库、动物声像档案库等组成，标本馆建筑总面积为 2460 平方米（其中贮藏面积 2111 平方米）。现保藏各类动物标本近 90 万号，定名标本数量 43 万余号(约占馆藏的 48%)，涵盖约 6650 种，脊椎动物定名率高达 90%。馆藏标本中脊椎动物标本约占馆藏量的 1/3，其中鱼类约 22 万号，两栖爬行类约 4.7 万号，鸟类约 2.6 万号，哺乳类约 2.3 万号；另外，昆虫约 56.7 万号，软体动物约 1.6 万号。动物标本由假剥制标本、皮张、骨骼、浸泡标本、姿态标本、生态景观标本、动物巢穴、鸟卵等类型组成，是中国西南地区馆藏面积最大、收藏量最为丰富、种类最为齐全的动物标本库。

标本的采集范围以中国南方热带、亚热带地区为主，基本涵盖了中国西南地区所有生态环境各种类型的标本，具有浓郁的西南地区特色。同时，根据昆明动物研究所的特色动物研究类群需要，也有不少采自新疆、内蒙古、黑龙江等地的种类，覆盖了中国 33 个省份 1200 个县，还有部分采自老挝、缅甸、越南、泰国、肯尼亚、日本、美国等十多个国家的标本。昆明动物博物馆依托人才团队及馆藏标本优势，每年为数千名国内外学者提供标本查阅、咨询服务，支撑服务科研和科普事业。

特色收藏

云南动物种类数为中国之冠，素有“动物王国”之称。云南动物物种极为丰富，有脊椎动物 2242 种，占全国的 51.4%;云南特有动物物种 351 种，其中，鱼类特有种 270 种；有陆生国家重点保护野生动物 236 种，占全国重点保护野生动物的 55.6%。按照保护级别划分，国家Ⅰ级重点保护陆生野生动物 58 种，国家Ⅱ级重点保护陆生野生动物 178 种。昆明动物博物馆的馆藏资源立足云南动物特色，例如，云南记录有画眉亚科鸟类 30 属 118 种，占中国所记录属数的 94%，占所记录种数的 82%，昆明动物博物馆馆藏画眉亚科鸟类标本 29 属 117 种。

亚洲象姿态标本
Specimens of Asian elephant (*Elephas maximus*)

标本馆保存着 20 世纪 50 年代以来采集的数万号脊椎动物标本，具有重要的科研价值和历史意义。这些标本来自数次大规模、具有重大意义的野外考察，见证了昆明动物研究所第一批科学

Strategy of Biological Development of Southwestern Resources (name in Chinese only), and *Illustrative Plates of Food Crop Pests and Natural Enemy Insects in Yunnan* (name in Chinese only). In addition, monographs on animal biology, morphology, and anatomy include *Biology of Chinese Tree Shrews*, *Biology of the Leaf Monkey* (name in Chinese only), *The Anatomy of the Golden Monkey*, and *The Chinese Phasianids: Lady Amherst's Pheasant*. In addition, various national, provincial, and state-level scientific research achievement awards have been obtained.

Overview of Collections

As a national strategic biological resource preservation institution, the KNHMZ consists of an animal specimen collection room, bird specimen collection room, amphibian and reptile specimen collection room, fish collection room, insect collection room, alien invasive species collection room, life barcode voucher specimen bank, and Tibetan Plateau specimen bank. The KNHMZ also contains a picture library and animal audiovisual archive. The total construction area of the herbarium is 2,460 m^2 (including a storage area of 2,111 m^2). The KNHMZ has collected nearly 900,000 animal specimens, 430,000 of which are identified specimens (accounting for 48% of all the collection), covering 6,650 species, with 90% of the identified vertebrates. Nearly 1/3 of all vertebrate specimens in China are preserved, including about 220,000 fish, 47,000 amphibians, 26,000 birds, 23,000 mammals; Besides, there are 567,000 insects, and 16,000 mollusks. The types of animal specimens are composed of bare replica specimens, skins, skeletons, immersed specimens, postured specimens, ecological landscape specimens, animal nests, bird eggs, etc. Having the most abundant collection and most complete animal specimen types, KNHMZ is the largest animal collection in Southwest China.

The collection range consists of mainly tropical and subtropical areas from southern China, covering almost all types of specimens of all kinds of ecological environments in Southwest China, and it has strong characteristics of Southwest China. At the same time, according to the needs of animal research groups from KIZ, the collection also contains many species collected from Xinjiang, Inner Mongolia, Heilongjiang and other regions, covering 1,200 counties from 33 provinces of China, as well as some specimens from more than 10 countries such as Laos, Myanmar, Vietnam, Thailand, Kenya, Japan and the United States. The KNHMZ had provided specimen reviews and consultation services to thousands of scholars at home and abroad each year relying on its talented team and excellent collection, in order to support and serve scientific research and science popularization.

Special Collections

Yunnan contains the highest number of species in China and is known as the "Animal Kingdom". Yunnan is rich in animal species and is home to 2,242 species of vertebrates, accounting for 51.4% of vertebrates across the country. Yunnan has also accommodated 351 endemic animal species, including 270 endemic fish species and 236 nationally key protected terrestrial wild animals, accounting for 55.6% of all in China. Divided by protection level, there are 58 species of terrestrial wildlife under national first-level priority protection and 178 species of terrestrial wildlife under national second-level priority protection, respectively. The collection resources of KNHMZ are based on the characteristics of Yunnan animals. For example, Yunnan has recorded 118 species of 30 genera of Timaliinae, accounting for 94% of recorded genera in China and 82% of recorded species. The KNHMZ's painting collection contains 117 species 29 genera of Timaliinae subfamily.

滇金丝猴标本
Specimen of Yunnan golden monkey
(*Rhinopithecus bieti*)

高黎贡白眉长臂猿（天行长臂猿）（雄）标本
Specimen of Skywalker hoolock gibbon
(*Hoolock tianxing*) (♂)

家在西南地区进行生物多样性考察的历史，记载了科学家在艰苦岁月中的科研成果，也逐渐揭开了云南生物多样性的面纱。

中国科学院成都生物研究所两栖爬行动物标本馆

基本信息

隶属于：中国科学院成都生物研究所
地址：中国四川省成都市武侯区人民南路四段 9 号
建馆年份：1965 年
收藏类群：两栖类、爬行类
标本藏量：11.7 万号
馆长 / 联系人：李家堂，研究员
电子邮箱：lijt@cib.ac.cn
联系电话：+86 28 82890788

标本馆简介

中国科学院成都生物研究所两栖爬行动物标本馆包括两栖爬行动物标本室和两栖爬行动物科普馆，总面积约 3700 平方米，为中国科学院成都生物研究所的支撑系统。该标本馆是中国唯一的两栖爬行动物标本馆。

标本馆自建立以来，就以为国内外广大的科学工作者、国家有关部门、政府相关机构、社会大众等提供科学的、专业的标本查询、标本鉴定、物种分类、动物进化、动物地理、生物多样性保护等各项需求服务为己任，为国内外同行、国家有关部门、社会大众等提供了各项优质的、专业的科学服务，起到了国家不可或缺的两栖爬行动物研究平台的作用，获得了国内外同行和国家有关部委的肯定及高度赞誉。

标本馆还开展了大量的科学研究，其研究重点是系统分类与系统演化、动物地理、物种形成与分化等重要的科学热点问题，这些科学研究为进一步提高标本馆的服务功能起到了关键性的支撑作用。自建馆以来，以标本馆为依托发表科研论文 5000 余篇，出版专著近百部，这些成果建立了中国两栖爬行动物的分类体系，促进了中国两栖爬行动物的科学研究。其中，“中国两栖动物系统学研究”获 2014 年度国家自然科学奖二等奖。

中国科学院成都生物研究所
Chengdu Institute of Biology, CAS

两栖爬行动物标本馆标本库内部
Inside of Specimen Storeroom of the Herpetological Museum

KNHMZ preserved tens of thousands of vertebrate specimens since the 1950s, which has important scientific value and historical significance. These specimens came from several large-scale and significant field trips, which witnessed the history of the first batch of KIZ's scientists conducting biodiversity expeditions in the southwestern region. The veil of biodiversity in Yunnan has been opened.

Herpetological Museum of Chengdu Institute of Biology, CAS

Basic Information

Affiliated to: Chengdu Institute of Biology, CAS
Address: No.9 Section 4, Renmin Nan Road, Chengdu, Sichuan Province, P. R. China
Established in: 1965
Species groups of collection: Amphibians, reptiles
Collection quantity: 117,000
Curator/Contacts: Dr. LI Jiatang, Professor
E-mail: lijt@cib.ac.cn
Tel: +86 28 82890788

Introduction

The Herpetological Museum of the Chengdu Institute of Biology, CAS, is composed of the Herpetological Collection and Herpetological Exhibition, with a total area of 3,700 m^2, belongs to the support system of Chengdu Institute of Biology, CAS. The Herpetological Museum is the unique museum for herpetology in China.

Since its establishment, the museum had offered capable views and services in specimen index, specimen identification, taxonomy, evolutionary biology, biogeography, and conservation, etc. These services provide assistance to native and international researchers, relevant governmental departments, and the public. The museum has played an indispensable role as a national amphibious and reptile research platform, honored domestically and internationally by researchers and the relevant.

The museum has also established a considerable amount of research, we focused on systematics, phylogenetics, biogeography, speciation, and other essential scientific hot spots. These researches had critically improved and supported the service abilities of the museum. Since its establishment, more than 5,000 research papers and near 100 monographs were published base on the herpetological collections, which established the framework and style of taxonomic research and the general view of Chinese herpetology. Among them, "The systematic study on amphibians of China" won the second prize of National Natural Science Prize in 2014.

History

The Herpetological Museum was established in 1965, but the earliest history can be traced back to the Herpetological Research Group established by academician LIU Ch'eng-Chao in West China Union University in 1938. It was officially called the Herpetological Museum in 1998. In 2008, the museum completed the transformation and passed the acceptance, and the Herpetological Exhibition also officially opened. The previous curators (directors of Herpetological Research Group before 1998) of the museum were Prof. HU Shuqin, academician ZHAO Ermi, Prof. FEI Liang, and Prof. WANG Yuezhao. The current curator is Prof. LI Jiatang.

两栖爬行动物科普馆
Herpetological Exhibition of the Herpetological Museum

Overview of Collections

Currently, the collection of the museum holds 854 species, of which 435 in amphibians and 419 in reptiles, with more than 110,000 specimens, accounting for more than

标本馆奠基人刘承钊院士（1900-1976）
The founder of the museum, academician LIU Ch'eng-Chao

历史沿革

两栖爬行动物标本馆最初为1965年成立的两栖爬行动物研究室的标本室（组），1998年正式称为两栖爬行动物标本馆，其最早的标本收藏历史可追溯到1938年刘承钊院士在华西协和大学建立的两栖爬行动物研究组。标本馆于2008年完成改造并通过验收，两栖爬行动物科普馆也正式开馆。标本馆历任馆长（1998年之前为两栖爬行动物研究室主任）为胡淑琴研究员、赵尔宓院士、费梁研究员和王跃招研究员，现任馆长为李家堂研究员。

收藏概况

目前标本馆馆藏标本11万余号，包括总物种数854种，其中两栖类435种、爬行类419种，占中国已知两栖爬行类物种数量的85%以上。标本采集地覆盖中国全部省（自治区、直辖市）和特别行政区。此外，馆藏标本中还有来自国外的标本53种400余号，包括美国、俄罗斯、菲律宾、老挝、马来西亚、缅甸、越南、泰国、南非、巴西、土库曼斯坦、乌兹别克斯坦、吉尔吉斯斯坦、伊朗等国。

特色收藏

模式标本库

模式标本是分类学研究的重要依据和关键材料，标本馆历来重视模式标本保藏，专门建立了模式标本库，用于保藏模式标本（尤其是正模、配模和副模）以及部分珍稀标本。目前，模式标本库保藏有两栖爬行类正模标本、配模标本和副模标本共计199种539号，其中两栖类164种404号，爬行类35种135号。这些模式标本绝大多数为标本馆采集，少数为外单位赠送。此外，标本馆从标本库中逐步整理出地模标本，并且每年都对地模标本进行有计划的补充采集。

已灭绝物种标本

滇螈（*Hypselotriton wolterstorffi*）曾分布于云南省昆明市滇池、阳宗海等水域，由于其生存环境发生巨大改变，现已被IUCN评估为“灭绝”，也是中国唯一被评估为“灭绝”的现生两栖爬行动物。标本馆保藏有滇螈标本78号，大多采于20世纪50年代。除本馆外，国内仅极个别标本馆还有滇螈馆藏标本。因该物种已灭绝，所以其馆藏标本非常珍贵，并在野生动物保护、生态环境保护等方面具有重要警示意义。

馆藏两栖爬行动物标本
Amphibian and reptile specimens in the museum

模式标本库
Type specimen storeroom

85% of the known reptile and amphibian species in China. The collection was collected from all provinces, autonomous regions, municipalities, and special administrative regions of China. In addition, there are about 400 specimens representing 53 species collected from outside China, including the United States, Russia, the Philippines, Laos, Malaysia, Myanmar, Vietnam, Thailand, Republic of South Africa, Brazil, Turkmenistan, Uzbekistan, Kyrgyzstan, Iran, etc.

Special Collections

Type specimens collection

Type specimens are essential evidence and key materials for taxonomic research. The museum has always appreciated the preservation of the type specimens, also particularly established a type specimens' collection for the preservation of type specimens (especially the holotypes, allotypes, and paratypes), as well as rare specimens. Currently, the type specimens' collection had preserved 539 specimens of 199 species, containing 404 individuals of 164 species in amphibians and 135 individuals of 35 species in reptiles. Most of these specimens were collected by the museum, while the rest were donated by other organizations. In addition, the museum gradually organized topotypes from the collections,

滇螈标本
Specimens of *Hypselotriton wolterstorffi*

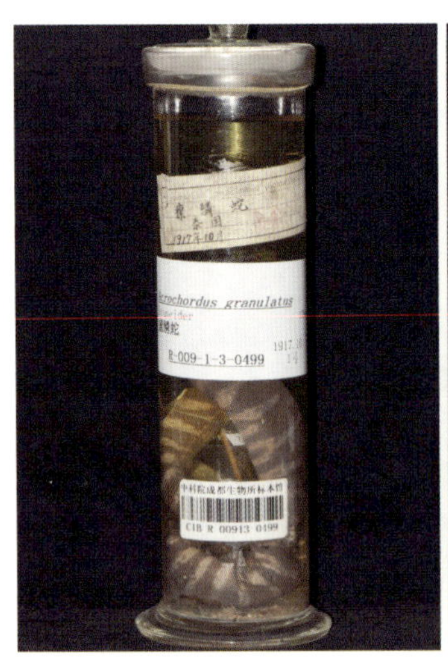
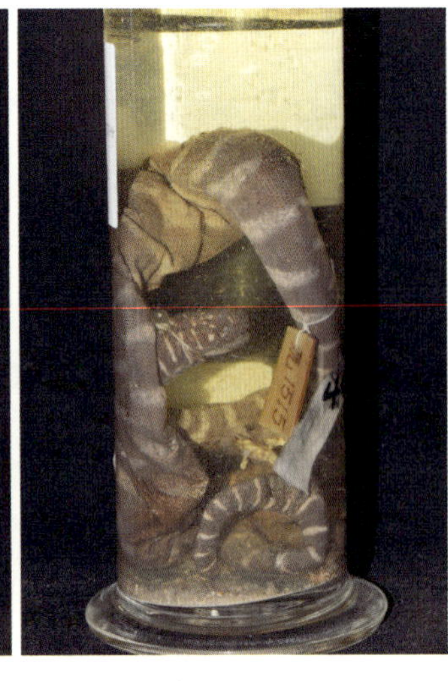

采自 1917 年的瘰鳞蛇标本
The specimen of *Acrochordus granulatus* collected in 1917

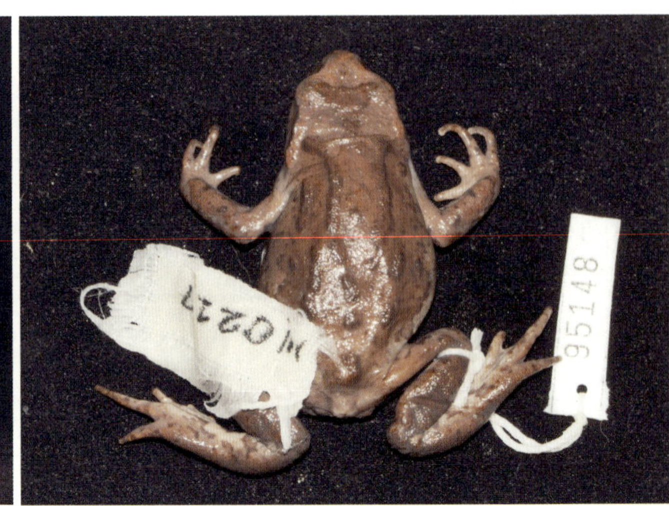

1939 年采集金顶齿突蟾标本
The specimen of *Scutiger chintingensis* collected in 1939

最早的馆藏标本

馆藏标本中，年代最早的是外单位赠送的 1 号瘰鳞蛇（*Acrochordus granulatus*）标本，于 1917 年 10 月采自泰国。该标本至今已超过 100 年，目前状态良好。

本馆最早采集的标本

标本馆的采集历史可追溯到 20 世纪 30 年代末刘承钊院士组织的野外考察。馆藏标本中，由标本馆采集的年代最早的标本是 1 号雌性金顶齿突蟾标本，于 1939 年采自四川峨眉山。其次是 1940 年 8 月 1 日采自峨眉山金顶的峨山掌突蟾。这两号标本至今已保藏 80 年左右，目前状态良好。

标本馆建立后的第一批新种模式标本

标本馆正式成立于 1965 年。次年，胡淑琴和赵尔宓描述并命名新种四川龙蜥 *Japalura szechwanensis*

1940 年采集峨山掌突蟾标本及其原始标签信息
The specimen of *Leptobrachella oshanensis* collected in 1940, and the information is on the tag

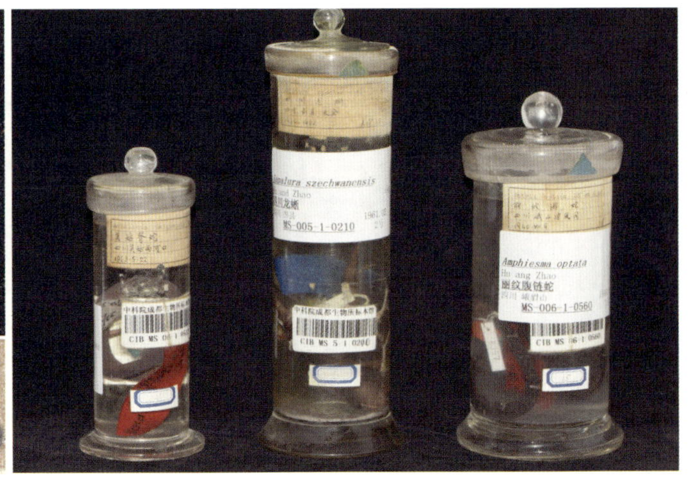

标本馆建立后的第一批新种模式标本
The type specimens of the first three new species after the Herpetological Museum established

as well as systematically replenish the collection of topotypes annually.

Specimens of extinct species

Hypselotriton wolterstorffi distributed in Dianchi Lake and Yangzonghai Lake of Kunming, Yunnan, because of the great changes in its habitat, it was evaluated as "extinct" by IUCN, and it is also the unique officially declared extinct species of living amphibian and reptile in China. There are 78 specimens of *H. wolterstorffi* in the museum, most of which were collected in the 1950s. Except for the museum, there are only a few collections of this species in other institutions. Because the species is extinct, the specimens are very precious, and have important warning significance in wildlife conservation, and ecological environment conservation.

标本馆最早采集的一批温泉蛇标本

The earliest specimens of *Thermophis baileyi* collected by the museum

The oldest specimen

The specimen of *Acrochordus granulatus* is the oldest specimen in the museum, which was collected in Oct. 1917 from Thailand and presented by other institution. This specimen is collected more than 100 years ago, and now the condition is fine.

The earliest specimens collected by the museum

The collection history of the museum can be traced back to the field investigations organized by academician LIU Ch'eng-Chao in the late 1930s. In the collections of the museum, the earliest specimen is a female specimen of *Scutiger chintingensis*, collected from Mt. Emei in 1939. The second is a specimen of *Leptobrachella oshanensis*. These two specimens were collected 80 years ago, and now the condition of them is fine.

The type specimens of the first three new species after the museum established

The museum was established in 1965. In the next year, Prof. HU Shuqin and Prof. ZHAO Ermi described three new species of reptiles. They were *Japalura szechwanensis* (=*Diploderma szechwanensis*), *Amphiesma optatum* (= *Hebius optatus*) and *Achalinus meiguensis*. And they were the first batch of new species described after museum established, with great historical significance. The type specimens are deposited in the Type Specimens Storeroom, the condition of them is fine.

横纹树蛙和墨脱竹叶青蛇的模式标本

The type specimens of *Rhacophorus translineatus* and *Trimeresurus medoensis*

野外记录本
Field notebooks

(= *Diploderma szechwanensis*)、丽纹腹链蛇（原丽纹游蛇）*Amphiesma optatum* (= *Hebius optatus*)、美姑脊蛇 *Achalinus meiguensis*，是标本馆成立后的第一批新种，具有重大历史纪念意义，其模式标本保存于模式标本库，目前保存状态良好。

第一次青藏高原综合科学考察采集的标本

1973 年中国科学院成都生物研究所参与西藏两栖爬行动物考察，仅 1973 年就采集到两栖爬行动物标本近 2900 号，包括两栖类标本约 2500 号，爬行类标本约 400 号。基于本次考察成果，于 1977 年发表两栖类新种 5 个，新亚种 1 个，中国新记录 4 个，西藏新记录 8 个；爬行类新种 1 个，新亚种 1 个，中国新记录 3 个，西藏新记录 5 个。结合后续考察和研究，由胡淑琴研究员主编，于 1987 年出版专著《西藏两栖爬行动物》。本次考察及其研究成果是“第一次青藏高原综合科学考察”的重要组成部分，具有重大历史意义。这批标本（包括模式标本）均妥善保藏于标本馆，现状良好。

历年野外考察和标本采集记录本

野外考察和标本采集过程中的记录，是核查标本信息、了解各物种生物学资料的重要参考资料。标本馆工作人员在长期以来的大量野外工作中，积累了丰厚的标本野外采集记录，这些记录中详细记载了采集标本的产地信息、物种特征、生物学特性、采集路线与日记等信息，是标本馆馆藏标本的重要数据补充，是今后标本馆数字化建设的重要基础资料，也是标本馆野外长期艰苦奋斗的记录和见证。标本馆妥善收藏了自 20 世纪 50 年代起的野外原始记录本，可供来馆学者查阅，让馆藏标本发挥出更大的价值。

中国科学院上海昆虫博物馆

基本信息

隶属于：中国科学院分子植物科学卓越创新中心
地址：中国上海市徐汇区枫林路 300 号
网址：http://www.shem.com.cn
建馆年份：2004 年
收藏类群：昆虫
标本藏量：127 万号
馆长 / 联系人：殷海生，教授级高工
电子邮箱：haishengyin@vip.sina.com
联系电话：+86 21 54924201

标本馆简介

中国科学院上海昆虫博物馆隶属中国科学院分子植物科学卓越创新中心。其前身是法国神父韩伯禄（P. M. Heude）1868 年筹建的“徐家汇博物院”，于 1930 年在吕班路（今重庆南路）重新建立并改名为

Specimens collected in the First Comprehensive Scientific Investigation to Qinghai-Tibet Plateau

In 1973, Chengdu Institute of Biology, CAS participated in the survey of amphibians and reptiles in Tibet. Near 2,900 specimens were collected, including about 2,500 specimens of amphibians, and 400 specimens of reptiles. Based on the results of this survey, in 1977, for amphibians, 5 new species, 1 new subspecies, 4 new records of China, 8 new records of Tibet were published; for reptiles, 1 new species, 1 new subspecies, 3 new records of China, 5 new records of Tibet were published. Combined with continued surveys and researches, edited by Prof. HU Shuqin, the monograph *Amphibia-Reptilia of Tibet* was published in 1987. The survey and its research results are an important part of the "First Comprehensive Scientific Investigation to Qinghai-Tibet Plateau", which is of great historical significance. All the specimens, including type specimens are well deposited in the museum.

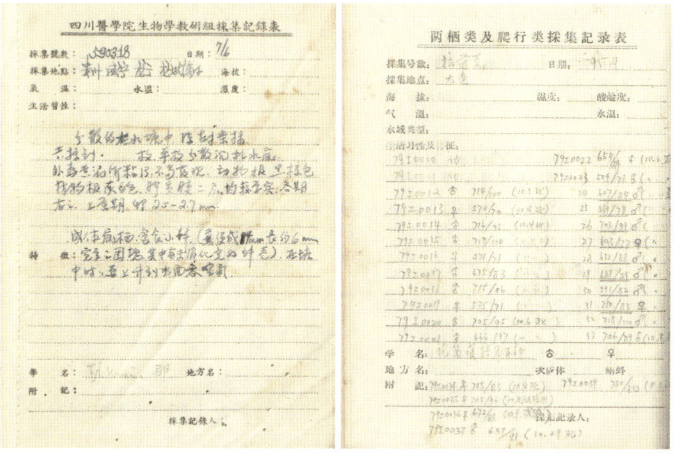

1959 年刘承钊院士等在贵州威宁关于贵州疣螈的野外记录（左）、1979 年赵尔宓院士在辽宁大连关于蛇岛蝮的野外记录（右）
The field notes by academician LIU Ch'eng-Chao about *Tylototriton kweichowensis* from Weining, Guizhou in 1959 (left), and by academician ZHAO Ermi about *Gloydius shedaoensis* from Dalian, Liaoning in 1979 (right)

Field note collections

The record of fieldwork is an important reference for checking specimen information and exploring biological data of species. In the long-term fieldwork, the staff of museum have accumulated abundant fieldwork records of specimens. These records are detailed in localities, species characteristics, biological data, collecting route, diary and other information. They are the important data supplement of specimens deposited in the museum, the important basic data for future digital construction of the museum, as well as the record and witness of long-term hard fieldwork of us. The museum has properly collected the original notebooks since the 1950s, which can be consulted by visitors, to let the specimens play a greater value.

Shanghai Entomological Museum, CAS

Basic Information

Affiliated to: CAS Center for Excellence in Molecular Plant Sciences
Address: No.300 Fenglin Road, Xuhui, Shanghai, P. R. China
Website: http://www.shem.com.cn
Established in: 2004
Species groups of collection: Insects
Collection quantity: 1,270,000
Curator/Contacts: Dr. YIN Haisheng, Professor level senior engineer
E-mail: haishengyin@vip.sina.com
Tel: +86 21 54924201

Introduction

Shanghai Entomological Museum is affiliated to CAS Center for Excellence in Molecular Plant Sciences. In 1868, the museum was prepared to construct by a French priest P. M. Heude and named Insect department of Musée de Zi-Ka-Wei, accomplished in Xujiahui. Then in 1930 it settled in south Chongqing Road (Lyuban Road before) and was named Musée de Heude. In 1953, it belonged to CAS. In 2002, Shanghai Entomological Museum was organized to be founded,

上海昆虫博物馆
Shanghai Entomological Museum

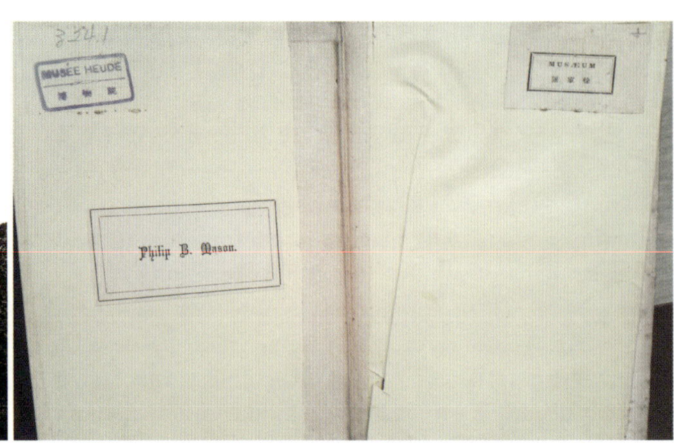

徐家汇博物院和震旦博物院书章
Book seal of Musée de Zi-Ka-Wei and Musée de Heude

“震旦博物院”。1953年归属中国科学院，2002年组建上海昆虫博物馆，由上海市政府和中国科学院、中国科学院上海生命科学研究院植物生理生态研究所共同出资建成，保藏着一大批濒危珍稀昆虫标本及国际和国内的危险性检疫害虫标本，是中国大型的专业昆虫馆，也是华东地区唯一的昆虫标本收藏专业机构。一批著名的分类学家曾在博物馆工作，包括尹文英研究员（中国科学院院士，世界六足纲动物权威专家）、范滋德研究员（国际双翅目协会名誉会员）与一批系统研究直翅目、双翅目和异翅目等昆虫的著名研究人员。

收藏概况

上海昆虫博物馆历史悠长，标本收藏年代久远，现有昆虫标本127万号，定名标本5000余种，模式标本1000余种，收藏类群几乎涵盖昆虫的所有类群，收藏地理范围包括全国以及东南亚地区和南美、非洲等区域，是中国科学院生物标本收藏的重要组成部分，在国内外具有广泛影响和较高知名度。

馆藏标本的特色类群为直翅目、双翅目有瓣蝇类、土壤六足动物，在种类上均覆盖国内已知种类的一半以上，某些类群，如原尾纲、双尾纲、双翅目花蝇科，馆藏物种数达到已知中国种类的70%-90%甚至更多。

上海昆虫博物馆标本收藏部
Collection department of Shanghai Entomological Museum

特色收藏

作为国内唯一的系统从事原尾虫研究的博物馆，自1960年起系统开展了原尾虫的采集、形态、系统分类、分子演化、生态、胚后发育和生物地理等研究，建立了原尾纲三目十科的分类体系。目前馆藏采自中国各地以及日本、俄罗斯和墨西哥的原尾虫9科44属236种，占世界已知原尾虫种类的近30%。

1965年，尹文英院士在上海发现了中国特有的红色原尾虫——红华蚖，建立了华蚖科。红华蚖身披一身棕红色的铠甲，其形态特征同当时全世界已知的原尾虫有很大的区别，该物种的发现被认为是自1907年发现原尾虫以来，原尾虫研究历史上最为激动人心的事件之一。该科的系统分类地位多年来一直是原尾虫系统发生研究的热点。

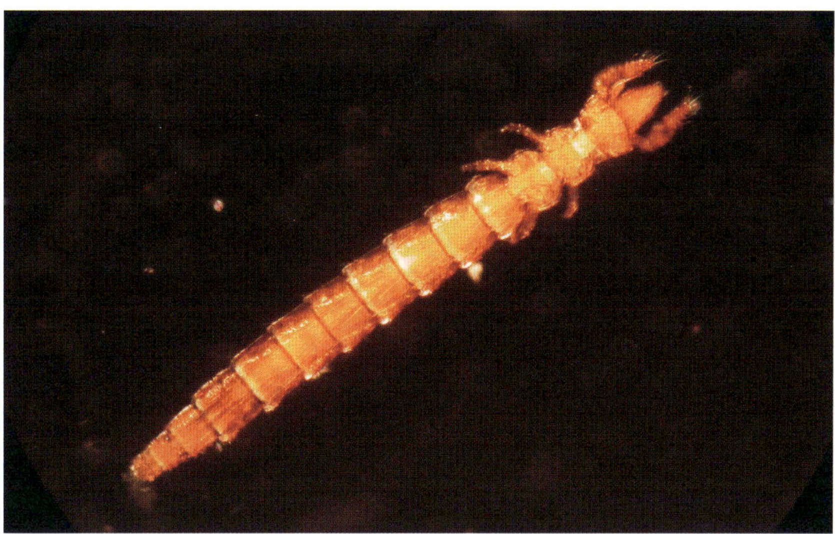

原尾虫特色标本：红华蚖
Characteristic specimen of proturan: *Sinentomon erythranum*

Shanghai government, CAS and SIBS invested special fund to set up the new museum. Shanghai Entomological Museum nationally collects more than 1.2 million insect specimens. Among the collections, there are a great number of animals on the brink of extinction and quarantinable insect specimens. The museum is a large-scale professional entomological museum in China and also the only professional institution on insect specimens' collection in East China. A group of well-known taxonomists work in the museum, including Prof. Yin Wenying (academician of CAS, the world authority on Hexapoda), Prof. FAN Zide (honorary members of the International Congresses of Dipterology) and a group of well-known professors and researchers who have been systematically studying Orthoptera, Diptera, Isoptera and other insect groups.

Overview of Collections

Shanghai Entomological Museum has a long history of specimen collection. There are more than 1.27 million insect specimens, more than 5,000 named specimens, and type specimens for more than 1,000 species. The collection covers almost all the insect groups and from the whole China, Southeast Asia, South America, Africa and other regions. As an important part of the collection of biological specimens of the Chinese Academy of Sciences, it has a wide influence and high popularity at home and abroad.

The characteristic groups of specimens in the collection are Orthoptera, Calyptratae (Diptera), and soil hexapods, covering more than half of the known species in China. Some groups, such as Protura, Anthomyiidae (Diptera), have more than 70%-90% of the known species in China.

Special Collections

As the only museum in China that has been engaged in Protura, the museum has been systematically studying Protura since 1960, including field collection, morphology, systematics, molecular evolution, ecology, post-embryonic development and biogeography, and established the taxonomy system of Protura with three orders and ten families. So far, there are 236 Proturan species (9 families, 44 genera) from China, Japan, Russia and Mexico in the museum, accounting for nearly 30% of the world's known proturan species.

In 1965, a red proturan was first reported in Shanghai by academician YIN Wenying, who named it *Sinentomon erythranum* and established the family Sinontomidae. *S. erythranum* wears a red-brown armor, and its morphological characters are very different from other known Proturan species in the world. This is one of the most exciting events in the history of Protura study since the discovery of proturans in 1907. Due to the unique morphological characters and highly divergent molecular sequence of *S. erythranum*, the systematic position of this family has been a hotspot for many years.

2）植物类馆

中国科学院植物研究所标本馆

基本信息

隶属于：中国科学院植物研究所
地址：中国北京市香山南辛村 20 号
网址：http://pe.ibcas.ac.cn
建馆年份：1928 年
收藏类群：蕨类、苔藓和种子植物
标本藏量：282.9 万号
馆长：孔宏智，研究员
联系人：金效华
电子邮箱：xiaohuajin@ibcas.ac.cn
联系电话：+86 10 62836582

中国科学院植物研究所标本馆
Herbarium of Institute of Botany (PE)

标本馆简介

中国科学院植物研究所标本馆的前身是 1928 年成立的北平静生生物调查所植物部标本室和 1929 年成立的北平研究院植物研究所标本室。建馆之初，馆藏 33.5 万，经过几代植物学家和技术人员的艰辛努力，馆藏植物标本 280 余万号，已成为亚洲最大的植物标本馆，是国家战略生物资源保藏中心之一，是植物多样性的重要遗传资源库和现生植物的信息库，在国内外植物分类学研究领域，特别是东亚植物分类学研究中占有举足轻重的地位。依托标本馆资源，“中国高等植物图鉴及中国高等植物科属检索表”、“中国蕨类植物科属系统排列和历史来源”、“《中国植物志》的编研”等三项科研成果先后获国家自然科学奖一等奖，“中国兰科植物研究”等四项成果先后获国家自然科学奖二等奖。

2019 年 6 月，依托中国科学院植物研究所标本馆建设的“国家植物标本资源库”正式获批。该库联合全国近 20 家单位的标本馆和原“国家标本资源共享平台”（National Specimen Information Infrastructure, NSII）的植物、教学、保护区和极地四子平台共同组建。通过共享中国植物标本资源，完善平台设施，增加收集和保藏国家植物标本资源库，将构建一个集标本馆数据库门户网站于一体的资源共享平台，并面向公众开放，提高植物标本资源在科学研究、生物多样性保护、资源开发、科普等方面的服务水平。

收藏概况

标本采集和收藏历史

标本馆早期收藏包括了静生生物调查所、北平研究院植物研究所、中央研究院自然历史博物馆、自然

2) Herbarium

Herbarium of Institute of Botany (PE), CAS

Basic Information

Affiliated to: Institute of Botany, CAS
Address: No.20 Nanxincun, Xiangshan, Beijing, P. R. China
Website: http://pe.ibcas.ac.cn
Established in: 1928
Species groups of collection: Ferns, bryophytes and seed plants
Collection quantity: 2,829,000
Curator: Dr. KONG Hongzhi, Professor
Contacts: Dr. JIN Xiaohua
E-mail: xiaohuajin@ibcas.ac.cn
Tel: +86 10 62836582

Introduction

Herbarium of Institute of Botany (PE), CAS can be traced back to the herbarium of the Department of Botany of the Fan Memorial Institute of Biology, Peiping in 1928, and the herbarium of the Institute of Botany of the National Academy of Peiping (Beijing) in 1929. In the beginning of its establishment, there were only 335,000 specimens. With the growth of the collections and collection management practices evolved, PE currently contains more than 2.8 million specimens, and it become the largest herbarium in Asia. It is one of the conservation centers of biological resources of national strategy and an important genetic bank for plant diversity and a giant information bank of extant plants, and also plays an important role in the research field of plant taxonomy at home and abroad, especially in East Asia. Relying on the resources of PE, three excellent achievements successively won the first prize of National Natural Science Award, including "Atlas of Chinese Higher Plants and Key to Families and Genera of Chinese Higher Plants", "Systematic Arrangement and Historical Origin of Families and Genera of Chinese Pteridophytes" and "Compilation and Research of *Flora Reipublicae Popularis Sinicae* ", and four achievements, including "Research on Orchidaceae of China", have successively won the second prize.

In June 2019, the National Plant Specimen Resource Center, led by Institute of Botany, CAS, was officially approved. National Plant Specimen Resource Center of China is jointly established by nearly 20 herbaria in China and four sub-platforms of plants, teaching reserves and polar regions of the former "National Specimen Information Infrastructure (NSII)". Through the open of the plant specimen resource of China, improving the facilities of the platform, and increasing the collections and preservation, the National Plant Specimen Resource Center of China will build an open resource platform integrated with herbaria, data information bank and portal sites being available to the public, and to improve the service of plant specimen resources in scientific research, biodiversity conservation, resources development and scientific popularization.

Overview of Collections

Specimen collection and collection history

PE herbarium has a huge number of old collections from institutions such as the Fan Memorial Institute of Biology, the Institute of Botany of the National Academy of Peiping, the Natural History Museum, Academia Sinica and Natural Resources Comprehensive Investigation Committee and individual donation. Since the foundation of P. R. China in 1949, specimen collection has increased rapidly. Many comprehensive scientific investigation projects have been carried out, such as Xizang Complex Investigation, Chinese Herbal Medicine Investigation, South-to-North Water Transfer Project Investigation, which greatly enriched the collection of Herbarium. In particular, Project *Flora Reipublicae Popularis Sinicae* has further spurred the increase of specimens.

Collection of specimens and coverage of taxa

PE has preserved more than 2.8 million specimens, accounting for 28% of the total resources of the National Plant

资源综合考察委员会等机构或个人捐赠的大量馆藏标本。中华人民共和国成立后，标本采集进入了蓬勃发展时期，全国范围开展了规模空前的各种植物资源普查和综合科学考察项目，如西藏综合考察、中草药调查、南水北调考察等，采集了大量的植物标本，极大丰富了标本馆馆藏量。《中国植物志》编写时期，采集了大量标本并保存在标本馆，使标本数量也大幅增加。

馆藏标本量和标本类群覆盖范围

馆藏标本 280 多万号，约占国家植物标本资源库总量的 28%，其中包括 18 万号蕨类植物标本、30 万号苔藓标本和 240 余万号种子植物标本，还妥善保存着 20 000 余份模式标本。馆藏标本类群涵盖了《中国植物志》所记载的 31 142 种植物中约 95% 的蕨类植物和 90% 的种子植物物种，还涵盖了中国 80% 的苔藓植物物种。此外，还收藏了 8.3 万份种子标本，涉及 263 科、3600 余属、2 万余种；收藏了 4.2 万份野生植物 DNA 材料，涉及 323 科、3114 属、6098 种，代表中国被子植物 92% 的属；还收藏了 7 万号植物化石标本。

中国科学院植物研究所标本馆的馆藏标本数量

Collections in Herbarium (PE), Institute of Botany, CAS

类别 Specimens category	总量 Total	科数 Family	属数 Genus	种数 Species
腊叶标本 Dehydrated specimens	2,800,000	367	8,544	86,499
种子标本 Seed specimens	83,000	263	3,600	20,000
模式标本 Type specimens	20,000	207	1,386	8,920
野生植物 DNA 材料 DNA materials of wild plants	42,000	323	3,114	6,098
国外标本 Foreign specimens	270,000	47	3,868	40,070
古植物化石 Fossil specimens	70,000	—	—	—

采集地覆盖范围

植物所标本馆所收藏的标本已覆盖中国各行政区。馆藏标本中数量最多、物种最丰富的省份依次是云南（25.8 万份，1.8 万种）、四川（24 万份，1.3 万种）和西藏自治区（13 万份，1 万种）；收藏最多的类群是菊科和蔷薇科标本，均有 13 万多份，其次是禾本科和豆科标本，均有 9 万余份，毛茛科和广义百合科均为 5 万多份。此外，标本馆通过采集、国际交换等多种途径，收集到来自世界 156 个国家和地区的 27 万份国外标本。近年来随着中国科学院海外科教基地的建立，标本馆积极参与国际合作和联合采集，东南亚（印度尼西亚、缅甸）、南亚（印度、尼泊尔、巴基斯坦）、南美、非洲、大洋洲等生物多样性热点地区标本的数量有大幅增加。

特色收藏

腊叶标本

标本馆馆藏丰富，类型多样，保存有 1949 年以前采集的许多重要标本。其中比较有代表性的是王启无先生 20 世纪 30-40 年代采集的近 5 万份植物标本，涉及类群有苔藓、蕨类和种子植物，模式标本就有 600 多份；中华人民共和国成立后，钟补求是最早在西藏进行科学考察的植物学家。

中华人民共和国成立后，植物研究所作为植物多样性调查的主力，参与了许多项目，标本也主要收藏于 PE，如 1954-1957 年黄河中游水土保持考察标本（1 万余号）、1959 年全国野生经济植物资源普查标本、1973-1976 年青藏综合考察队标本（3500 余号）、1981-1983 年横断山地区综合科学考察标本（4 万号）、1987-2000 年重要地区野生生物资源考察标本（42 000 余号）、2011 年至今泛喜马拉雅地区植物标本（5 万号）。

Specimen Resource Center, including 180,000 fern specimens, 300,000 bryophytes specimens, 2.4 million seed plants. Specifically, there are more than 20,000 types. The species housed at PE cover 80% of mosses, 95% of ferns and 90% of seed plants of the 31,142 species of plants recorded in the *Flora Reipublicae Popularis Sinicae* in China. In addition, there are 83,000 seed specimens in PE, including 263 families, 3,600 genera and 20,000 species; 42,000 DNA materials of wild plants, involving 323 families, 3,114 genera and 6,098 species, representing 92% of the genera of angiosperms in China, 70,000 fossil specimens of plants were also collected.

腊叶标本
Dehydrated specimen

The coverage of biodiversity and geography

The specimens of PE have been collected from 34 provincial administrative regions in China. Approximately 258 thousand sheets (18 thousand species), 240 thousand sheets (13 thousand species), 130 thousand sheets (10 thousand species) have been collected from Yunnan, Sichuan, Tibet, respectively which are the provinces with the largest number of specimens and the most abundant species. Two families, Compositae and Rosaceae, have more than 130 thousand sheets, respectively, followed by Gramineae and Leguminosae, with more than 90 thousand sheets, Ranunculaceae and Liliaceae s. l. with more than 50 thousand sheets. In addition, there are more than 270,000 foreign specimens from 156 countries and regions in the world through the field collection and international exchange. In recent years, with the establishment of the overseas institutions by CAS, PE has actively participated in international cooperation and joint collection. The number of specimens in Southeast Asia (Indonesia, Myanmar), South Asia (India, Nepal, Pakistan), South America, Africa, Oceania and other biodiversity hotspots have increased significantly.

Special Collections

Dehydrated specimens

PE has a rich collection, including many specimens collected before 1949. WANG Chi-wu was one of the most representative botanists in the stage. He collected nearly 50,000 plant specimens in the 1930s and 1940s, including bryophytes, ferns and seed plants. Of these, there are more than 600 types in Wang's collections. Tsoong Pu-chiu was the first botanist who carried out scientific investigations deep in Tibet.

After the foundation of the People's Republic of China, the Institute of Botany participated in many large scale scientific investigation projects, including Investigation on conservation of the Middle Yellow River (1954-1957), 10,000 specimens, National Investigation on wild economic plants (1959), Comprehensive Investigation of Qinghai-Tibet (1973-1976, 35,000 specimens), Comprehensive Investigation of Hengduan Mountains Region (1981-1983, 40,000 specimens), Wildlife Resources Investigation of Important Areas (1987-2000, 42,000 specimens) and Flora of Pan-Himalaya (2011- , 50,000 specimens).

种子标本
Seed specimens

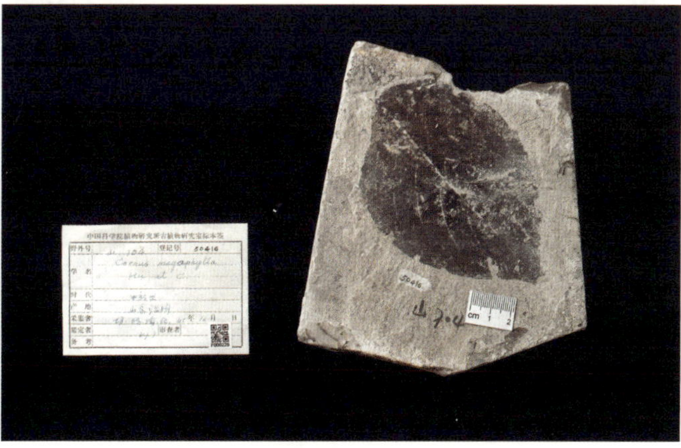
化石标本——山茱萸属 *Cornus megaphylla* 化石
Fossil specimen: *Cornus megaphylla*

种子标本

PE 馆藏有 8 万余号种子标本，覆盖 252 科 3600 属 22 000 余种，位居世界前列，采集范围涉及亚洲、欧洲、非洲等“一带一路”沿线的许多国家。这些标本主要是通过与国内外植物园等相关单位进行交换所得。

化石标本

PE 收藏了约 7 万号植物化石标本。

中国科学院昆明植物研究所标本馆

基本信息

隶属于：中国科学院昆明植物研究所
地址：中国云南省昆明市盘龙区蓝黑路 132 号
网址：http://www.kun.ac.cn
建馆年份：1938 年
收藏类群：种子植物、蕨类植物、苔藓植物、地衣以及大型真菌等
标本藏量：155 万份
馆长 / 联系人：邓涛，副研究员
电子邮箱：dengtao@mail.kib.ac.cn
联系电话：+86 871 65223110

标本馆简介

中国科学院昆明植物研究所标本馆（KUN），由中国现代植物学奠基人之一的胡先骕先生于 1938 年创建，并在蔡希陶、汪发缵、郑万钧、俞德浚、吴征镒等多位植物学家的辛勤耕耘下，历经 80 多年的持续积累，逐步形成了以中国西南地区为标本收藏中心，采集范围覆盖全国、辐射东南亚及“一带一路”沿线等地区，具有国际知名度的区域性标本馆，在中国植物分类、生物地理及生物多样性保护等研究领域中具有重要的学术地位。

标本馆外景
Exterior view of KUN

收藏概况

馆藏最早的标本是蔡希陶先生于 1932 年采自云南的标本（尽管后来国外标本馆赠予的 19 世纪由外国人采自云南的标本时间更早），以及陈谋、吴中伦、蒋英早期在滇不完全的采集。收藏了蔡希陶、王启无、俞德浚、冯国楣、刘瑛、李鸣岗、汪发缵、毛品一、邱炳云、吴征镒、简焯坡、刘德仪等著名植物学家或著名标本采集人在云南所采集的云南植物标本。

在中华人民共和国成立后开展中苏合作的云南植物资源考察、青藏高原综合考察、南水北调综合考察、中草药植物考察、独龙江植物考察及高黎贡

Seed specimens

There are more than 80,000 seed specimens in PE, including 22,000 species in 3,600 genera from 252 families. The seed specimens were collected mainly by exchange with domestic and foreign botanical gardens for decades, including many countries along the B&R in Asia, Europe, and Africa.

Fossil specimens

There are approximately 70,000 fossil specimens of plants in PE.

Herbarium of Kunming Institute of Botany (KUN), CAS

Basic Information

Affiliated to: Kunming Institute of Botany, CAS
Address: Lanhei Road # 132, Panlong District, Kunming, Yunnan Province, P. R. China
Website: http://www.kun.ac.cn
Established in: 1938
Species groups of collection: Seed plants, ferns, bryophytes, lichens and macrofungi, etc.
Collection quantity: 1,550,000
Curator/Contacts: Dr. Den Tao, Associate Professor
E-mail: dengtao@mail.kib.ac.cn
Tel: +86 871 65223110

Introduction

The herbarium KUN was founded in 1938 by Dr. HU Hsen-Hsu, one of the founders of modern botany in China. With the hard work of many botanists such as TSAI Hse-Tao, WANG Fazuan, CHENG Wan-Chun, YU Dejun, and WU Zhengyi, over more than 80 years of continuous accumulation, KUN gradually becomes a specimen collection center in southwest China, collections cover the whole China and some Southeast Asian countries as well as countries along the B&R region. KUN gained its international recognition and leading academic reputation as a regional herbarium through the contribution to plant taxonomy, biogeography and biodiversity conservation.

Overview of Collections

The earliest specimens were collected in 1932 by Dr. Tsai Hse-Tao from Yunnan (although there are some even earlier collections made by foreigners but later donated to KUN by foreign herbariums), along with other collections by CHEN Mou, WU Zhonglun, and JIANG Ying, which were rather incomplete. Early important collections mainly include those by distinguished collectors such as Tsai Hse-Tao, WANG Chi-Wu, YU Dejun, Feng Huo-Mei, LIU Ying, LI Minggang, WANG Fazuan, MAO Pinyi, QIU Bingyun, WU Zhengyi, JIAN Zhuopo, LIU Deyi, etc.

After the founding of the People's Republic of China, through large-scale scientific expeditions such as the Sino-Soviet Cooperation in Yunnan Plant Resource Survey, Qinghai-Tibet Plateau Comprehensive Survey, South-to-North

中华人民共和国成立以后开展的植物资源调查
Plant resources investigation after the founding of the People's Republic of China

山生物多样性考察、武陵山植物考察、墨脱越冬考察等多次大型科学考察活动之后，本馆与国内各植物学研究机构展开广泛交换，使标本藏量大幅度增加。

昆明植物研究所标本馆是中国第二大植物标本馆，目前共收藏除藻类外的植物标本 150 多万份，其中有 120 万份种子植物标本，5 万余份蕨类植物标本，9.9 万号苔藓标本，10 万号真菌标本以及 7 万号地衣标本，木材标本 0.6 万份，浸泡标本 0.3 万份；在这些馆藏标本中，有根据发表时标注和文献统计出来的各类模式标本约 1.3 万份。这些植物标本采自全国各地，以西南及横断山地区的最为丰富；此外，也有部分通过采集和交换获得的东南亚、日本、欧洲、北美以及其他国家和地区的标本。

特色收藏

馆藏标本门类齐全，包括地衣、大型真菌、苔藓植物、蕨类植物和种子植物，是国内标本馆中覆盖生物大类最为丰富的标本馆。

KUN 隐花植物标本库至今已收集保藏了以青藏高原地区为主的地衣标本 7 万余号，是国内少数保藏地衣的标本馆之一。1981 年，昆明植物所新建标本馆大楼落成，全面整理馆藏标本，王立松开始着手整理馆藏地衣标本。当时馆藏地衣标本仅有 371 份，包括钟仁暄（1926 年）在福建采集的标本 11 份，蔡希陶（1932-1934 年）在云南采集的标本 8 份，刘慎鄂、王汉臣（1941-1947 年）在滇西北采集的 252 份，沈祖安（1959 年）在云南采集的标本 47 份，陈邦杰（1955 年）在福建采集的标本 10 份，以及陆定安（1957 年）在安徽采集的标本 43 份。这些是昆明植物研究所最早馆藏地衣标本的宝贵家底。但这些标本当时尚无分类鉴定研究，仅以“地衣”或“Lichen”标注。1981-1983 年，横断山综合考察项目启动，也第一次正式启动了地衣的采集工作，臧穆、黎兴江和王立松参与了地衣的考察，考察范围包括滇西、川西及藏东南等 60 余县，范围之广、时间之长前所未有，采集地衣标本 6304 号，使地衣标本馆藏上了新台阶。2006 年以来，王立松研究员率项目组在多项国家自然科学基金的支持下，重点对青藏高原的横断山地区进行了系统调查，采集了大量的地衣标本，并依托第二次青藏高原综合科学考察研究项目，填补了青海、甘肃南部，以及藏北无人区和高山草甸地衣采集的空白，采集范围覆盖了青藏高原 60% 以上地区，共采集该地区的地衣标本超过 6 万号，使 KUN 成为国内外青藏高原地衣馆藏标本量最多的标本馆。

保存最早的标本：野豌豆属 *Vicia* sp.，1818 年采自法国
The earliest specimen preserved: *Vicia* sp., collected from France in 1818

中国人最早采集的标本：刺参属 *Morina* sp.，邓西蒙 1906 年采自昭通
The earliest specimen collected by Chinese: *Morina* sp., Deng Ximeng collected from Zhaotong in 1906

苔藓植物标本的规模收集始于 1965 年，主要由黎兴江研究员和臧穆研究员负责实施。通过组织野外采集和与国内外研究机构开展标本交换等形式，KUN 收集了以中国西南地区为产地中心的 11 万号标本，其中模式标本 155 号。这些馆藏的苔藓植物标本为 KUN《西藏苔藓志》（1985 年）、《横断山苔藓植物志》（2000 年）、《云南植物志（第 17-19 卷）》（2000-2005 年）、《中国苔藓志（第 1-4 卷）》（1994-2006 年）、*Moss Flora of China*（第 2、4 卷）（2001 年；2007 年）的顺利出版奠定了坚实的基础。

Water Transfer Project Comprehensive Survey, Chinese Herbal Medicine Plant Survey, Dulongjiang Plant Survey, Gaoligong Mountain Biodiversity Survey, Wuling Mountain Plant Survey, and Motuo Wintering Expeditions, and through extensive specimen exchanges activities with various domestic botanical research institutions, the number of collections in KUN has increased significantly.

KUN is the second largest herbarium in China. So far, the KUN has stored more than 1.5 million plant specimens other than algae, including 1.2 million seed plant specimens, more than 50,000 fern specimens, 99,000 bryophyte specimens, 100,000 fungal specimens and 70,000 lichen specimens, 6,000 wood specimens, and 30,000 soaking specimens; among these specimens, there are about 13,000 specimens of various types based on the labeling and literature statistics at the time of publication. These plant specimens are collected from all parts of the country, with the richest from the Southwest China and Hengduan Mountains Region. In addition, some specimens from Southeast Asia, Japan, Europe, North America, Central Asia and other countries and regions were obtained mainly by field trips and exchanges.

Special Collections

The collection has a wide range of specimen categories, including lichens, macrofungi, bryophytes, ferns and seed plants. KUN is the most comprehensive herbarium in China.

Cryptogamic Herbarium of KUN is one of the few herbariums storing lichen specimens in China, focusing on Qinghai-Tibet Plateau. Over 70,000 specimens have been collected till now. With the new herbarium founded in 1981, Prof. WANG Lisong started his work arranging lichen specimens. Only 371 specimens were stored in KUN at that time, including 11 specimens collected by ZHONG Renxuan (1926) from Fujian, 8 collected by TSAI Hse-Tao (1932-1934) from Yunnan, 252 collected by LIU Shen'e and WANG Hanchen (1941-1947) from North-western Yunnan, 47 collected by SHEN Zu'an (1959) from Yunnan, 10 collected by CHEN Pan-Chieh (1955) from Fujian and 43 collected by LU Ding'an from Anhui. These specimens are the first and precious collection of lichens in KUN, but all of them are only labelled with "Lichen" or "Diyi" in Chinese. From 1981 to 1983, with the realization of the scientific expedition of Hengduan Mountains program, the first formal field survey on lichen started. ZANG Mu, LI Xingjiang, and WANG Lisong joined the program. The field survey covered more than 60 counties in West Yunnan, West Sichuan and Southwest Xizang, which was the longest field work covering the widest range ever, during this trip, 6304 numbers of lichen specimens were collected, largely increased the collection of lichen specimens in KUN. Since 2006, with the support of the National Natural Science Foundation of China, the lichen research group led by WANG Lisong collected huge numbers of lichen specimens, focusing on Hengduan Mts. area, and with the start of the Second Tibetan Plateau Scientific Expedition and Research Program, the rang of field survey has been largely increased and covered more areas of Qinghai-Tibet Plateau, including areas without any lichen surveys before, such as Southern part of Qinghai and Gansu, Northern part of Xizang and alpine meadow areas, covering more than 60% of Qinghai-Tibet Plateau, and collected more than 60,000 specimens of lichen from this region, making KUN is largest lichen herbarium from Qinghai-Tibet Plateau worldwide.

The bryophyte collection started growing in 1965, led by Prof. LI Xingjiang and Prof. ZANG Mu. The total number of bryophyte specimens had exceeded 110,000 centered in Southwest China (with 155 type specimens), mainly through field-collecting events and specimen exchange collaborations by KUN. The herbarium collection of those bryophytes laid a solid foundation for the successful publication of the following bryological monographs: *Bryoflora of Tibet* (1985), *Bryoflora of the Hengduan Mountain* (2000), *Flora Yunnanica,* vols. 17-19 (2000-2005), *Flora Bryophytorum Sinicorum*, vols. 1-4 (1994-2006), and *Moss Flora of China*, vols. 2 & 4 (2001; 2007).

最早的采自中国的模式标本：藿香叶绿绒蒿 *Meconopsis betonicifolia*，1866 年采自云南

The earliest type specimen from China: *Meconopsis betonicifolia*, collected from Yunnan in 1866

一份标本，两人采集：A. Heney, 1963 和 F. B. Forbes, 1972 采集

One specimen, collected by two collectors: A. Heney, 1963 and F. B. Forbes, 1972

有历史纪念意义的标本

标本馆还保藏有历史纪念意义的标本，如蔡希陶的烟草标本、第一份橡胶标本、吴征镒的金铁锁标本（纪念吴征镒的老师吴韫珍先生）、纪念臧穆先生的新属模式标本、纪念陈介先生的新种模式标本、纪念陈书坤先生的新种模式标本，等等。

中国科学院武汉植物园标本馆

基本信息

隶属于：中国科学院武汉植物园
地址：中国湖北省武汉市洪山区鲁磨路特 1 号
网址：www.wbg.cas.cn
建馆年份：1956 年
收藏类群：植物
标本藏量：30 万号
馆长 / 联系人：胡光万，研究员
电子邮箱：guangwanhu@wbgcas.cn
联系电话：+86 27 87511510

标本馆简介

中国科学院武汉植物园标本馆地处华中，是华中地区规模最大的植物标本馆，以服务该地区的经济和社会发展为使命。作为武汉植物园的科技支撑部门，标本馆一直以来为植物园的植物分类和植物资源研究提供了重要的技术支撑，并在植物园的引种驯化过程中一直行使着物种鉴定和凭证标本存放等支撑功能。标本馆现有总面积 1800 平方米，建有主标本室、副标本室和非洲植物标本室三个主要标本保存场馆。

武汉植物园标本馆是集植物分类学研究、植物标本收集保存和科普教育为一体的科研及科技支撑机构。自建馆以来，重点研究的植物类群主要有山毛榉科、豆科、葫芦科、猕猴桃科、睡莲科、景天科、凤仙花科、百合科、天南星科、樟科、禾本科、兰科等。重点研究的方向包括华中地区维管植物的区系地理学研究，中

武汉植物园标本馆外景
Exterior view of Herbarium of Wuhan Botanical Garden

Specimens with historical significance

The Herbarium also keeps specimens of historical significance, such as Mr. Tsai Hse-Tao's tobacco specimen, the first rubber specimen, Mr. WU Zhengyi's *Psammosilene tunicoides* specimen (in memory of Mr. WU Zhengyi's supervisor, Mr. WU Yunzhen), new genu type specimen in commemoration of Mr. ZANG Mu, new species type specimen in commemoration of Mr. CHEN Jie, new species type specimen in commemoration of Mr. CHEN Shukun, etc.

Herbarium of Wuhan Botanical Garden(HIB), CAS

Basic Information

Affiliated to: Wuhan Botanical Garden, CAS
Address: Special No.1, Lumo Road, Hongshan District, Wuhan, Hubei Province, P. R. China
Website: www.wbg.cas.cn
Established in: 1956
Species groups of collection: Botany
Collection quantity: 300,000
Curator/Contacts: Dr. HU Guangwan, Professor
E-mail: guangwanhu@wbgcas.cn
Tel: +86 27 87511510

Introduction

The Herbarium of Wuhan Botanical Garden of CAS is located in Central China. It is the largest herbarium in Central China, with the mission of serving the economic and social development of the region. As the scientific and technological support department of Wuhan Botanical Garden, the herbarium has always provided important technical support for the plant classification and plant resources research of the botanical garden, and has been performing the supporting functions of species identification and voucher specimen storage in the introduction and domestication process of the botanical garden. With a total area of 1800 m^2, the herbarium has three main specimen preservation venues: the main herbarium, the sub herbarium and the Herbarium of African Plants.

The Herbarium of Wuhan Botanical Garden is a scientific research and technological support institution integrating plant taxonomy research, plant specimen collection and preservation and popular science education. Since the establishment of the herbarium, the plant groups mainly studied include Fagaceae, Leguminosae, Cucurbitaceae, Actinidiaceae, Nymphaeaceae, Crassulaceae, Balsaminaceae, Liliaceae, Araceae, Lauraceae, Gramineae, Orchidaceae,

武汉植物园标本馆内景
Interior view of Herbarium of Wuhan Botanical Garden

非洲植物标本室
Herbarium of African Plants

国特有植物三大分布中心之一——川东 - 鄂西中国特有属分布中心的保育遗传学和系统进化植物学研究等。

标本馆始建于 1956 年，刚开始仅为一间标本室。1972 年初，湖北省政府拨出专项经费，支持武汉植物园建设独立的标本馆，标本馆于 1972 年 12 月正式建成并投入使用。2001 年 4 月，在财政部专项经费的支持下，标本馆得到扩建，扩建工程于同年 10 月竣工。2013 年，以武汉植物园为依托单位，中国科学院成立了“中国科学院中 - 非联合研究中心”，武汉植物园承担了在非洲组织开展植物资源合作研究和联合考察的任务，并于 2014 年 5 月建成非洲植物标本室，为武汉植物园承担的《肯尼亚植物志》编研项目的开展以及中 - 非联合研究中心开展的有关非洲植物资源的合作研究工作打下了良好的基础。

标本馆还与国内外主要植物标本馆建立了标本交换和标本借阅关系。历年来接待了美国哈佛大学、俄亥俄州立大学、密苏里植物园、英国皇家植物园邱园、日本东京大学等机构开展植物分类学合作研究工作；并与墨西哥国家标本馆、澳大利亚维多利亚国家植物标本馆同行专家建立了长期标本交换合作。

多年来，标本馆在科研和技术人员的不懈努力下，以武汉植物园的科研发展需求为导向，重点围绕植物资源的可持续利用、植物分类学、生态学、生态安全等科学领域开展植物标本收集、保存与研究工作，为学科发展和国家经济建设做出了重要贡献。

收藏概况

标本馆目前收藏保存了维管植物标本 30 万余份，其中蕨类植物标本 9000 余份、裸子植物标本 3000 余份、被子植物标本 25 万份、非洲植物标本 4 万份，包括各类模式标本 652 份。

标本馆自 1956 年开始进行植物标本采集和收藏。60 多年来，一代代植物分类学科技工作者翻山越岭，足迹遍布华中地区，克服重重艰难险阻采集植物标本。近年来，随着各类科研项目的开展，对湖南、四川、云南、西藏、江西以及非洲等国内外的植被和植物资源进行了深入广泛的调查与研究，采集了大量的植物标本，取得了丰硕的研究成果。

自设立非洲植物标本室后，标本馆先后承担了中 - 非联合研究中心设立的“非洲植物资源调查”和“《肯尼亚植物志》编研”等项目，与非洲国家的许多大学和研究机构建立了合作伙伴关系，已在肯尼亚、坦桑尼亚、埃塞俄比亚、乌干达、纳米比亚和马达加斯加等非洲国家开展了植物多样性野外联合科学考察 20 余

粉背绣球模式标本
Type specimen of *Hydrangea glaucophylla*

洪坪杏的模式标本
Type specimen of *Armeniaca hongpingensis*

正式发表前采集的水杉标本
Specimen of *Metasequoia glyptostroboides* collected before official publication

水杉原产地模式标本
Topotype specimen of *Metasequoia glyptostroboides*

etc. The key research directions include floristic geography of vascular plants in Central China, conservation genetics and phylogenetic botany of one of the three distribution centers of endemic plants in China: East Sichuan to West Hubei distribution center of endemic genera in China.

The herbarium was built in 1956. In the beginning, it was only one small collection room. At the beginning of 1972, Hubei provincial government allocated special funds to support the construction of an independent herbarium in Wuhan Botanical Garden, which was officially completed and put into use in December 1972. In April 2001, with the support of the special funds of the Ministry of Finance, the herbarium was expanded, and the expansion project was completed in October of the same year. In 2013, with Wuhan Botanical Garden as the supporting unit, CAS established the "Sino-Africa Joint Research Center, Chinese Academy of Sciences" (SAJOREC). Wuhan Botanical Garden undertook the task of organizing cooperative research and joint investigation on plant resources in Africa. In May 2014, the Herbarium of African Plants was established, and it has laid a good foundation to the compilation and research project of *Flora of Kenya* and cooperative research on African plant resources carried out by SAJOREC.

The herbarium has also established the relationship of specimen exchange and specimen borrowing with major herbarium at home and abroad. Over the years, it has received Harvard University, Ohio State University, Missouri Botanical Garden, Kew garden of Royal Botanical Garden, Tokyo University of Japan and other universities to carry out cooperative research on plant taxonomy, and established long-term specimen exchange cooperation with experts from Mexico National Herbarium and Victoria National Herbarium of Australia.

Over the years, with the unremitting efforts of scientific research and technical personnel, and guided by the scientific research and development needs of Wuhan Botanical Garden, the herbarium has focused on the sustainable utilization of plant resources, plant taxonomy, ecology, ecological security and other scientific fields to carry out the collection, preservation and research of plant specimens, making an important contribution to the discipline development and national economic construction.

Overview of Collections

At present, the herbarium has collected and preserved more than 300,000 vascular plant specimens, including more than 9,000 fern specimens, more than 3,000 gymnosperms specimens, 250,000 angiosperms specimens, 40,000 African plant specimens, including 652 type specimens.

In 1956, the herbarium began to collect and preserve plant specimens. Over the past 60 years, generations of plant taxonomic scientists and technologists have traveled all over central China, overcoming many difficulties and obstacles to collect plant specimens. In recent years, with the development of various scientific research projects, the vegetation and plant resources in Hunan, Sichuan, Yunnan, Tibet, Jiangxi, Africa and other countries have been investigated and studied extensively, a large number of plant specimens have been collected, and fruitful research results have been achieved.

Since the establishment of the Herbarium of African Plants, the herbarium has successively undertaken projects such as "African plant resources survey" and "compilation and research of *Flora of Kenya*" set up by the SAJOREC, established partnerships with many universities and research institutions in African countries, and has been carried out more than 20 joint scientific investigations on plant diversity in Kenya, Tanzania, Ethiopia, Uganda, Namibia and Madagascar, etc., which has accumulated important background data for the research on the distribution pattern and formation mechanism of plant diversity in East Africa, as well as the writing of *Flora of Kenya*. In the process of cooperative research on plant resources in Africa, 11 new African flora have been officially named and published.

中美联合考察队采集的神农峨眉蕨模式标本

Type specimen of *Lunathyrium shennongense* collected by Sino-American Botanical Expedition

中美联合考察队采集的神农架冬青的模式标本

Type specimen of *Ilex shennongjiaensis* collected by Sino-American Botanical Expedition

次，为对东非植物多样性分布格局及形成机制开展研究，以及撰写《肯尼亚植物志》积累了重要的本底资料。在非洲开展植物资源合作研究过程中，正式命名和发表非洲植物新类群 11 个，其正模标本都保存在本标本馆中，副模标本由本标本馆和肯尼亚国家博物馆的东非植物标本馆共同保存。

特色收藏

武汉植物园标本馆长期以来，以采集、收藏和研究华中地区的植物标本为特色，其中收藏的三峡地区和神农架地区植物标本最为丰富和全面，对于研究这些地区的植物区系以及植被的现状和变化历史尤为重要，其重要性不可替代。从 2013 年开始，本馆又增加了非洲植物标本的特色收藏，已成为国内收藏非洲植物标本最多的标本馆，在国内同类植物标本馆中独树一帜，这些特色收藏将使本馆在国际上的影响和地位不断提升。

中国科学院西双版纳热带植物园标本馆

基本信息

隶属于：中国科学院西双版纳热带植物园
地址：中国云南省西双版纳傣族自治州勐腊县勐仑镇
网址：http://hitbc.xtbg.ac.cn
建馆年份：1959 年
收藏类群：维管植物
标本藏量：22.9 万份
馆长 / 联系人：李剑武，高级工程师
电子邮箱：ljw@xtbg.org.cn
联系电话：+86 691 8715480

标本馆简介

中国科学院西双版纳热带植物园标本馆为中国科学院西双版纳热带植物园（以下简称：版纳植物园）的重要支撑机构之一，为本园的学科发展提供技术支撑服务，同时也为地方林业、环保、园林、海关、公安和科普等机构提供技术服务。

西双版纳热带植物园标本馆外景
Exterior view of Herbarium of Xishuangbanna Tropical Botanical Garden

西双版纳热带植物园标本馆收藏以滇南热带地区为中心，以云南亚热带和东南亚热带地区为辐射范围的维管植物标本，同时也收藏地区性森林群落定位研究凭证标本、园林保育凭证标本及种子标本，属于地方性热带植物标本馆，具有鲜明的热带植物收藏特色。目前共收藏有维管植物标本 23.22 万份，约 19 000 种。

标本馆从成立至今，参加了很多地方性和全国性的志书编写。如 20 世纪 70-80 年代的《西双版纳地区植物名录》、《西双版纳高等植物名录》（第一版）、《中国植物志》、《云南植物志》、《西双版纳傣药志》、《中国树木志》等。90 年

The holotypes are all preserved in this herbarium, and the paratypes are jointly preserved by this herbarium and the East African Herbarium of the Kenya National Museum.

Special Collections

For a long time, the Herbarium of Wuhan Botanical Garden has been characterized by collecting, preserving and studying the plant specimens in Central China. Among them, the Three Gorges area and Shennongjia area have the most abundant and comprehensive collection of plant specimens, which is particularly important for the study of the flora and the current situation and change history of vegetation in these areas, and its importance cannot be replaced. Since 2013, the museum has added a special collection of African plant specimens, which has become the largest collection of African plant specimens in China. It is unique among similar plants in China, and these special collections will make the influence and status of the herbarium in the world continue to improve.

Herbarium of Xishuangbanna Tropical Botanical Garden (HITBC), CAS

Basic Information

Affiliated to: Xishuangbanna Tropical Botanical Garden, CAS
Address: Menglun town, Mengla County, Xishuangbanna Dai Autonomous Prefecture Yunnan Province, P. R. China
Website: http://hitbc.xtbg.ac.cn
Established in: 1959
Species groups of collection: Vascular plants
Collection quantity: 229,000
Curator/Contacts: LI Jianwu, Senior Engineer
E-mail: ljw@xtbg.org.cn
Tel: +86 691 8715480

Introduction

The Herbarium of Xishuangbanna Tropical Botanical Garden, CAS (HITBC), is one of the key components of the technical support system for scientific research in Xishuangbanna Tropical Botanical Garden (XTBG). It aims to meet the needs of the botanical research in XTBG, as well as to provide technical services to public organizations, such as forestry, environment protection and management, horticulture, the customs office, security, and popularization of science.

The collections of HITBC mainly include vascular plants from the tropical areas of southern Yunnan, as well as plants from the subtropical areas of Yunnan and tropical areas of Southeast Asia. HITBC also collects voucher specimens for studies on the positioning of the regional forest communities, voucher specimens for gardening and plant conservation, as well as seed specimens. HITBC is a local tropical herbarium with distinctive characteristics of tropical plant collection. Up to the present, HITBC has collected 232,200 vascular plant specimens, which include about 19,000 species.

Since the establishment of HITBC, the staff members have participated in compiling many plant checklists and Floras. For example, during the 1970s and 1980s, they participated in compiling *The Plant Checklist of Xishuangbanna*, *Higher Plants Checklist of Xishuangbanna* (First edition), *Flora Reipublicae Popularis Sinicae*, *Flora Yunnanica*, *Medicinal Plants of the Dai Nationality* and *Tree Flora of China*. After the 1990s, HITBC published several books as the editor, including *Higher Plants Checklist of Xishuangbanna* (Second edition), *List of Seed Plants in the Ailao Mts. of Yunnan Province, China*, *Native Seed Plants in Xishuangbanna of Yunnan*, *Atlas of Native Orchids in China*, and also participated in compiling *Higher Plants of China in Color* and *Rare and Endangered Plants in China*.

西双版纳热带植物园标本馆内景
Inside view of Herbarium of Xishuangbanna Tropical Botanical Garden

The outstanding contributors from HITBC include Prof. TSAI Hse-Tao (the founder of HITBC), Prof. PEI Shengji (taxonomy of Arecaceae), Prof. LI Yanhui (taxonomy of tropical plant, Myristicaceae), Prof. TAO Guoda (taxonomy of tropical plant,

代以后主编《西双版纳高等植物名录》(第二版)、《云南哀牢山种子植物》、《云南西双版纳野生种子植物》、《中国野生兰科植物原色图鉴》;参编了《中国高等植物彩色图鉴》、《中国珍稀濒危植物》。

曾在标本馆工作的蔡希陶教授(创始人)、裴盛基教授(棕榈科分类)、李延辉教授(热带植物分类,肉豆蔻科)、陶国达教授(热带植物分类,龙脑香科)、朱华教授(植物区系地理研究,茜草科粗叶木属)等,为中国植物分类学和版纳植物园标本馆的建设与发展做出了重要贡献。

历史沿革

版纳植物园标本馆于 1959 年由著名植物学家蔡希陶教授创建,在当时称为“植物资源组”。植物资源组致力于西双版纳地区热带植物资源调查和开发利用,随着研究工作的不断开展,标本逐渐增加,到 1966 年,共采集、存储植物标本 14 000 多份。1974 年 6 月,版纳植物园成立了“植物分类研究室”,由裴盛基教授任室主任,直到 1983 年 12 月。1987 年 1 月,版纳植物园归并入中国科学院昆明植物研究所,归属于新成立的“种质资源保护研究室”,由陶国达任室主任直至 1995 年底。1996 年 11 月至 1998 年 12 月,版纳植物园与昆明植物研究所分离,而与中国科学院昆明生态研究所合并,标本馆归属于“植物多样性保护研究室”,由殷寿华任室主任,朱华负责标本馆工作。2000 年,标本馆归属植物地理研究组,由朱华任馆长。2012 年以后挂靠新成立的“综合保护中心”,为全园科学研究、园林园艺和科普教育服务,由殷建涛任常务副馆长直至 2017 年,2018 年起由李剑武任代理副馆长。新的标本馆于 2002 年 4 月竣工并投入使用,建筑面积达到 1200 平方米,设有正、复号标本室,浸泡标本室,标本制作室等。正号标本室 563 平方米,设计容纳标本 50 万份,所有标本柜为可移动密集柜。

收藏概况

标本馆目前为止共收藏有维管植物标本 23.22 万份,大部分采自云南省及周边热带、亚热带地区,通过交换、赠予等方式获得了一批其他地区的标本,如我国广东省、四川省、广西壮族自治区等,以及泰国、老挝、马来西亚等的标本,同时也获得了一些采集时间较早的标本,最老的标本采自 1865 年。

特色收藏

中国的热带雨林直至 1975 年才被世界所承认,且热带雨林代表性植物望天树和版纳青梅的模式标本之一就存放于版纳植物园标本馆。标本馆还收藏了全国包括云南的德宏傣族景颇族自治州、普洱市、红河哈尼族彝族自治州,以及广西壮族自治区、海南省等地发现的龙脑香科植物标本。除龙脑香科植物标本以外,还收集了很多在中国历史上有着重要意义、经济价值的植物标本,如三叶橡胶等。

总之,本馆的收藏是以中国热带和亚热带地区,以及东南亚热带地区的维管植物标本为主,是国内研究热带植物区系的重要平台。

中国科学院成都生物研究所植物标本馆

基本信息

隶属于:中国科学院成都生物研究所
地址:中国四川省成都市武侯区人民南路四段九号
网址:http://cdbi.cib.ac.cn/biology-mis/page/index.jsf
建馆年份:1958 年
收藏类群:苔藓、蕨类、裸子植物、被子植物和部分真菌

Dipterocarpaceae), and Prof. ZHU Hua (study on floristic geography, Rubiaceae: *Lasianthus*). They made not only great achievements in plant taxonomy research, but also important contributions to the construction and development of HITBC.

History

HITBC, initially named the "Plant Resources Group", was founded in 1959 with the establishment of XTBG by Prof. Tsai Hse-Tao, a famous botanist in Chinese history. The staff of the group was devoted to the plant resource investigation, as well as their exploitation and utilization. As a result of the continuous plant investigation, the number of specimens increased gradually. By 1966, more the 14,000 plant specimens had been collected by the group and stored in the herbarium. In June 1974, XTBG established a Department of Taxonomy, and Prof. PEI Shengji was the Director of this until December 1983. In January 1987, XTBG was attached to the Kunming Institute of Botany, CAS. The herbarium was placed under the supervision of the Department of Germplasm Resources and Prof. TAO Guoda was the Director until 1995. From November 1996 to December 1998, XTBG separated from the Kunming Institute of Botany and combined with the Kunming Institute of Ecology, CAS. The herbarium was then a group attached to the Department of Plant Diversity, with Prof. YIN Shouhua as the Director and Dr. ZHU Hua in charge of the herbarium. After 2000, HITBC belonged to the Phytogeography Research Group, with Prof. ZHU Hua as the Director of the herbarium. Since 2012, HITBC has been part of the Center for Integrative Conservation at XTBG and continues to support all research groups, the horticultural department, and popular science. Mr. YIN Jiantao was the executive deputy curator until 2017. Then, from 2018, Mr. LI Jianwu took the position. In April 2002, a new building for HITBC was completed, which is 1,200 m^2 in area, including the proper specimen room, the duplicate specimen room, the pickled specimen room, and the office. The proper specimen room covers an area of 563 m^2, and can store 500 thousand specimens. All the specimen cabinets are on movable racks to save space.

Overview of Collections

So far, HITBC has deposited more than 232,200 vascular specimens in HITBC, most collected from tropical and subtropical Yunnan. The collection also includes some older specimens coming from exchange and donation with other herbaria, such as Guangdong, Sichuan, and Guangxi in China, as well as Thailand, Laos, and Malaysia. At the same time, some specimens collected earlier were obtained, and the earliest specimen in HITBC was collected in 1865.

Special Collections

The rainforests in China were not recognized by the world until 1975. The type specimens of *Parashorea chinensis* and *Vatica xishuangbannaensis*, which are tropical dipterocarps from China, are preserved in HITBC. HITBC also collects dipterocarp specimens from other regions of Yunnan, including Dehong, Pu'er, and Honghe, as well as other provincial districts, like Guangxi and Hainan. HITBC has also collected some precious historical specimens with important economic values, such as *Hevea brasiliensis*.

In summary, HITBC's collections focus on vascular specimens from tropical and subtropical China, as well as Southeast Asia. It becomes the most important platform for tropical floristic research in China.

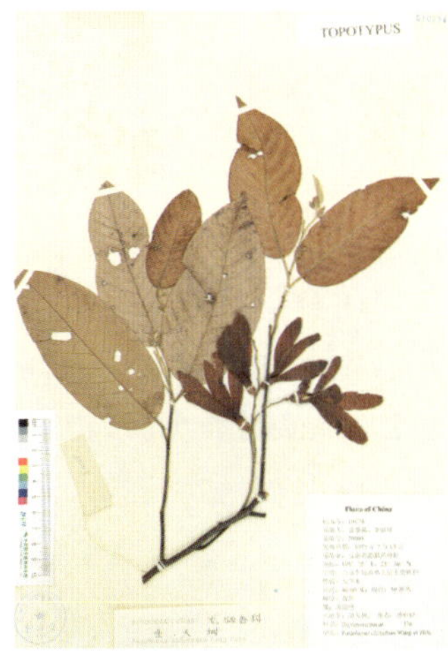

望天树模式标本

Type specimen of *Parashorea chinensis*

Herbarium of Chengdu Institute of Biology (CDBI), CAS

Basic Information

Affiliated to: Chengdu Institute of Biology, CAS
Address: No.9 Section 4, Renmin Nan Road, Wuhou District, Chengdu, Sichuan, P. R. China
Website: http://cdbi.cib.ac.cn/biology-mis/page/index.jsf
Established in: 1958
Species groups of collection: Bryophytes, ferns, gymnosperms, angiosperms and some fungi
Collection quantity: 368,000

标本藏量：36.8 万份
馆长：高信芬，研究员
联系人：徐波
电子邮箱：xubo@cib.ac.cn
联系电话：+86 28 82890757

植物标本馆外景
Exterior view of the Herbarium (CDBI)

标本馆简介

标本馆现有高等植物标本 36.8 万份，9000 余种；收藏的标本主要来源于长江上游、青藏高原、喜马拉雅地区，近十年来增加了大量豆科、蔷薇科、菊科、百合科、伞形科、薯蓣科、蕨类植物标本。

历史沿革

CDBI 的前身是 1958 年成立的中国科学院四川分院农业生物研究所植物标本馆。1962 年研究所更名为中国科学院西南生物研究所。1971 年归省政府管理并更名为四川省生物研究所。1978 年更名为中国科学院成都生物研究所。

收藏概况

植物标本馆的重要收藏来自于历史上一些科考调查项目，如四川经济植物调查（1959-1966 年）、四川植被调查（1971-1984 年）、青藏高原科考调查（1980-1983 年）、贡嘎山植物多样性调查（1981 年、1994 年）、瓦屋山植物多样性调查（1994 年）、攀枝花植物资源调查（2002-2004 年）、藏南植物多样性调查（2002-2003 年）、九顶山植物区系调查（2002-2006 年）、都江堰植物资源调查（2012-2014 年）、画稿溪国家级自然保护区本底调查（2012-2014 年）、崇州市鞍子河省级自然保护区植物多样性调查（2015-2016 年）、康定市陆生植物多样性调查（2017 年）、蜂桶寨国家级自然保护区科考调查（2017-2018 年）、藏东南植物多样性调查（2016-2020 年）、大雪山 – 小相岭植物多样性调查（2019-2020 年）等科考调查。

标本类群覆盖苔藓、蕨类、裸子植物、被子植物和部分真菌。区域植物标本采集范围主要为我国四川和西藏两地，并有部分缅甸北部和尼泊尔标本。重点研究类群的采集范围为全国和东南亚地区。重点收藏了长江上游、青藏高原、喜马拉雅山脉地区的植物标本；重要植物类群收藏了豆科、蔷薇科、菊科、百合科、伞形科、薯蓣科、蕨类植物等。

特色收藏

标本馆的区域特色收藏为横断山脉地区植物标本，占本馆收藏总量的 70%，其中标本较多的为四川稻城、乡城、康定、泸定、九龙、宝兴、茂县、攀枝花、都江堰、叙永、崇州，以及西藏墨脱、察隅、波密、芒康等市县。

成都生物研究所植物标本馆的馆藏特色为四川省区域植物标本，是研究四川植物物种多样性不可或缺的标本馆，重点研究的类群也是其重点收藏类群，如蕨类植物，以及开花植物豆科、蔷薇科、伞形科、百合科、菊科、报春花科、薯蓣科等采集覆盖全国范围。这些馆藏植物标本为《中国植物志》、《四川植物志》的编研，以及重要科属的研究提供了丰富的实物材料。

Curator: Dr. GAO Xinfen, Professor
Contacts: XU Bo
E-mail: xubo@cib.ac.cn
Tel: +86 28 82890757

Introduction

At present, there are approximately 368,000 specimens of higher plants and more than 9,000 species in the herbarium. The collected specimens are mainly from the upper reaches of the Yangtze River, the Qinghai-Tibet Plateau, and the Himalayas. In the past decade, a large number of specimens of Leguminosae, Rosaceae, Asteraceae, Liliaceae, Umbelliferae, Dioscoreaceae, and ferns have been added.

植物标本馆内景
Inside view of the Herbarium (CDBI)

History

CDBI was the successor of the herbarium of the Institute of Agricultural Biology, Sichuan branch of CAS, founded in 1958. The institute was renamed the Southwest Institute of Biology of the CAS in 1962. From 1971, the institute was transferred to the provincial government and renamed the Sichuan Institute of Biology. In 1978, it was renamed the Chengdu Institute of Biology, CAS.

Overview of Collections

The herbarium has many important historical collections from the Sichuan Economic Plant Survey (1959-1966), Sichuan Vegetation Survey (1971-1984), The Qinghai-Tibet Plateau Scientific Expedition (1980-1983), Gongga Mountain Plant Diversity Survey (1981, 1994), Wawu Mountain Plant Diversity Survey (1994), Panzhihua Plant Resources Survey (2002-2004), Southern Tibet Plant Diversity Survey (2002-2003), Jiuding Mountain Flora Survey (2002-2006), Dujiangyan Plant Resources Survey (2012-2014), Huagaoxi National Nature Reserve Expedition (2012-2014), Plant Diversity Survey of Anzihe Provincial Nature Reserve, Chongzhou City (2015-2016), Kangding Terrestrial Plant Diversity Survey (2017), Fengtongzhai National Nature Reserve Scientific Expedition (2017-2018), Southeast Tibet Plant Diversity Survey (2016-2020), Daxueshan-Xiaoxiangling Plant Diversity Survey (2019-2020).

The collection encompasses all major groups of plants (bryophytes, ferns, gymnosperms and angiosperms) as well as fungus. Areas of specialization mainly include the flora of Sichuan and Xizang. Some are from northern Myanmar and Nepal. The collection range of important research taxa includes the whole country and Southeast Asian regions. The collected specimens are mainly from the upper reaches of the Yangtze River, the Qinghai-Tibet Plateau, and the Himalayas. The important research taxa include families of Leguminosae, Rosaceae, Asteraceae, Liliaceae, Umbelliferae, Dioscoreaceae, and ferns.

Special Collections

The regional characteristics of the collection are specimens in Hengduan Mountains Region, accounting for 70% of the total collection, of which the majority are from Daocheng, Xiangcheng, Kangding, Luding, Jiulong, Baoxing, Maoxian, Panzhihua, Dujiangyan, Xuyong, Chongzhou in Sichuan province, and Motuo, Chayu, Bomi and Mangkang in Tibet.

A significant part of the herbarium is the specimens with special strengths in the Sichuan province, which is indispensable for the study of plant species diversity in Sichuan. The groups which the herbarium focuses on are also the key collections, such as ferns and flowering plant families of Leguminosae, Rosaceae, Umbelliferae, Liliaceae, Asteraceae, Primulaceae, Dioscoreaceae and so on, covering the whole country. These collections provide abundant material for the compilation of *Flora Reipublicae Popularis Sinicae* and *Flora of Sichuan*, as well as the research of important families and genera.

中国科学院华南植物园标本馆

华南植物园标本馆
Herbarium of South China Botanical Garden, CAS

基本信息

隶属于：中国科学院华南植物园
地址：中国广东省广州市天河区兴科路 723 号
网址：http://herbarium.scbg.cas.cn
建馆年份：1928 年
收藏类群：维管植物和苔藓
标本藏量：111.1 万份
馆长：张奠湘，研究员
联系人：曾飞燕，汤银珠
电子邮箱：zengfeiy@scbg.ac.cn, yinzhu@scbg.ac.cn
联系电话：+86 20 37252692

标本馆简介

华南植物园标本馆（IBSC）的前身是中山大学植物研究室的标本室，由著名植物学家陈焕镛院士于 1928 年创建，曾用代码 SCIB、HC 等。华南植物园鼎湖山树木园标本室于 2007 年并入本馆。本馆主要收藏维管植物和苔藓标本，特别是热带、亚热带地区的全球植物标本。

历史沿革

中山大学植物研究室（1928）
中山大学植物研究所（1929）
中山大学农林植物研究所（1930）
中山大学农林植物研究所香港办事处（1938）
广东植物研究所（1942）
中山大学农林植物研究所（1945）
中国科学院华南植物研究所（1954）
广东省农林水科技服务站经济作物队（1968）
广东省植物研究所（1972）
中国科学院华南植物研究所（1978）
中国科学院华南植物园（2003）

收藏概况

华南植物园创建伊始就十分重视植物标本的采集。研究所的《第一次五年报告》（1934 年）明确指出，研究所的设立以调查广东植物种类为首要任务，并对采集任务做出全面规划。至 1934 年初，馆藏标本已达 15 万份，可供交换的标本 20 余万份。此后，农林植物研究所的采集工作不仅遍及广东、海南各地，还逐渐扩展到邻近各省。1936 年，在中国植物采集史上书写了最为悲壮的一页：邓世纬等 7 人于 8 月 23 日赴贵州开展植物标本采集。夏秋之交，盛行时疫，10 月 13 日助理杨昌汉、技工徐方才不幸染病去世。邓世纬将他们的棺木运回时不幸感染，到了 17 日，他和助理黄孜文相继病亡，其他三人后来亦病危。噩耗传至广

Herbarium of South China Botanical Garden (IBSC), CAS

Basic Information

Affiliated to: South China Botanical Garden, CAS
Address: No.723 Xingke Rd, Tianhe District, Guangzhou, Guangdong, P. R. China
Website: http://herbarium.scbg.cas.cn
Established in: 1928
Species groups of collection: Vascular plants and bryophytes
Collection quantity: 1,111,000
Curator: Dr. ZHANG Dianxiang, Professor
Contacts: ZENG Feiyan, TANG Yinzhu
E-mail: zengfeiy@scbg.ac.cn, yinzhu@scbg.ac.cn
Tel: +86 20 37252692

Introduction

The herbarium IBSC was founded in 1928 by academician Prof. Chun Woon-young, a famous botanist. It was formerly known as the herbarium of the Botanical Research Office of Sun Yat-Sen University, and has used SCIB, HC and other codes. The herbarium of Dinghushan Arboretum of South China Botanical Garden (SCBG) was incorporated into IBSC in 2007. Collections are mainly of vascular and bryophyte specimens, especially specimens from tropical and subtropical regions globally.

History

Botanical Research Office, Sun Yat-sen University (1928)
Botanical Institute, Sun Yat-sen University (1929)
Institute of Agriculture and Forestry, Sun Yat-sen University (1930)
Hong Kong Office, Institute of Agriculture and Forestry, Sun Yat-sen University (1938)
Botanical Institute of Kwangtung (1942)
Institute of Agriculture and Forestry, Sun Yat-sen University (1945)
South China Institute of Botany, CAS (1954)
Team of Economic Crops, Guangdong Science and Technology Service Station of Agriculture, Forestry and Water (1968)
Botanical Institute of Guangdong Province (1972)
South China Institute of Botany, CAS (1978)
South China Botanical Garden, CAS (2003).

Overview of Collections

Since the establishment of the Institute of Agriculture and Forestry of Sun Yat-sen University, great efforts have been made to the collection of plant specimens. The first Five-Year Report (1934) clearly pointed out that investigating the plant species in Guangdong Province was the primary task, and a comprehensive plan for the collection task was also made. By the beginning of 1934, the number of collections had reached 150,000, and more than 200,000 specimens were available for exchange. Since then, the collection work had not only reached all regions in Guangdong and Hainan, but also gradually expanded to the neighboring provinces. In 1936, the most solemn and stirring page was written in the history of plant collection in China: DENG Shiwei and six other people went to Guizhou Province on August 23 to collect plant specimens. At the turn of summer and autumn, epidemics prevailed. On October 13, assistant YANG Changhan and skilled worker XU Fangcai died of the disease. When DENG Shiwei brought their coffins back, he was unfortunately infected. On October 17, DENG and his assistant HUANG Ziwen died one after another, and the other three people were also critically ill. When the bad news reached Guangzhou, there was no different voice of mourning! Later, Prof. CHUN Woon-Young found a new

世纬苣苔模式标本
Type specimen of *Tengia scopulorum*

州，众人无不同声哀悼！后来，陈焕镛教授从邓世纬采集的标本中发现了苦苣苔科一新种，便以世纬苣苔（*Tengia scopulorum*）命名以纪念之。

中华人民共和国成立前，先辈们进行了大量的采集工作，为标本馆奠定了雄厚的基础。中华人民共和国成立后，标本馆除继续派遣采集人员赴各地采集外，还与研究课题和国家重大项目相结合，组织多次综合性考察和植物资源普查，调查的范围也逐步扩展到越南、缅甸等周边国家，乃至大洋彼岸的拉丁美洲、非洲。随着采集工作的深度和广度都进一步提高，标本的数量也迅速增加。截至目前，馆藏标本超过 115 万份，采自国内 33 个省，以及其他 107 个国家与地区，涵盖物种 48 700 多种，其中种子植物 87 万份、苔藓植物 4 万份、蕨类植物 4 万份、液浸标本 6400 份，复份标本近 20 万份。馆内收藏的东南亚国家植物标本约占馆藏量的 8%，特别是有一批采自菲律宾的标本保存完好，而其本国标本已被毁于第二次世界大战。

特色收藏

模式标本

模式标本是新种发表的依据标本，对于植物鉴别有着非常重要的作用。本馆收藏的模式标本有超过 7000 份，涉及已经发表的 4300 多个分类群，包括活化石银杉（*Cathaya argyrophylla*）、中药材巴戟天（*Morinda officinalis*）、名贵红木原材料植物降香黄檀（*Dalbergia odorifera*）、为纪念著名学者任鸿隽而以其名字命名的任豆（*Zenia insignis*）等。

银杉模式标本
Type specimen of *Cathaya argyrophylla*

液浸标本

为了满足科研的需要，一些植物的花、果、茎等不宜制作成干标本，改用适当的药液浸泡，并储存在密封的玻璃瓶内，这样就能长久保存植物体或器官的原来形态。馆内收藏的液浸标本共 6400 份。

木材标本

近年来，红木产业的发展十分迅猛，而红木的主要原材料是豆科植物黄檀、紫檀等，这是我园的传统重要研究类群。经过多年的调查与收集，目前已收集 30 余种 50 多份红木的木材标本。另外，2018 年 9 月 16 日的超强台风“山竹”过后，标本馆收集了 92 份风灾倒木的木材标本，这些木材标本都是活生生的科普材料。

中国科学院庐山植物园标本馆

基本信息

隶属于：中国科学院庐山植物园
地址：中国江西省九江市濂溪区威家镇中国科学院庐山植物园鄱阳湖分园
建馆年份：1934 年
收藏类群：被子植物、裸子植物和蕨类植物
标本藏量：19 万份

species of Gesneriaceae from the specimens collected by DENG Shiwei, and named it as *Tengia scopulorum* in memory of them.

Before the founding of P. R. China, the predecessors had achieved tremendous contribution in collecting specimens, which laid a solid foundation for the herbarium IBSC. After that, in addition to collecting in various places, they also combined with the major national funding and some research projects to carry out many investigations and general survey of plant resources. The herbarium also paid much attention to the exploration of the neighboring countries such as Vietnam, Myanmar, etc., and other tropical countries in Latin America and Africa. With the further improvement of the depth and breadth of collection, the collection has been increasing rapidly. Up to now, the herbarium has accumulated a collection of over 1.15 million specimens from 33 provinces in China and 107 countries and regions, covering more than 48,700 species, including 870,000 seed plants, 40,000 bryophytes, 40,000 pteridophytes, 6,400 liquid immersed specimens, and have nearly 200,000 duplicates available for exchange. The specimens collected from Southeast Asian countries account for about 8% of the total, and among them, those collected from the Philippines were well preserved and especially valued, as the other duplicates of the same collections preserved in the Philippines were all destroyed in the Second World War.

Special Collections

Type specimens

The type specimen is the basis for the publication of new species, which plays an important role in plant identification. There are more than 7,000 type specimens in the herbarium, involving more than 4,300 published taxa, including living fossil *Cathaya argyrophylla*, traditional Chinese medicine *Morinda officinalis*, the raw material plant of "Hongmu" *Dalbergia odorifera*, *Zenia insignis* named after the famous scholar REN Hongjun and so on.

Liquid immersed specimens

In order to meet the needs of scientific research, some plants' flowers, fruits, stems, etc. should not be made into dry specimens. Instead, they should be preserved in appropriate liquid media like alcohol, and stored in sealed glass bottles, so that the original forms of plants or organs can be preserved for a long time. A total of 6,400 liquid immersed specimens were collected in the herbarium IBSC.

红木标本
Specimens of "Hongmu"

Wood specimens

In recent years, the development of "Hongmu" furniture industry is very rapid, and their raw materials are mainly Leguminous plants such as *Dalbergia*, *Pterocarpus* and so on, which are important traditional research groups in SCBG. After years of investigation and collection, more than 30 species and over 50 samples of "Hongmu" have been collected. Moreover, after the super typhoon "Shanzhu" on September 16, 2018, the herbarium collected 92 pieces of wood samples from the wind disaster. These wood specimens are all popular science materials.

Herbarium of Lushan Botanical Garden (LBG), CAS

Basic Information

Affiliated to: Lushan Botanical Garden, CAS
Address: Poyang lake branch, Lushan botanical garden, Chinese academy of sciences, Weijia town, Lianxi district, Jiujiang city, Jiangxi province, P. R. China
Established in: 1934
Species groups of collection: Angiosperms, gymnosperms and ferns
Collection quantity: 190,000

馆长：彭焱松，副研究员
联系人：胡菀
电子邮箱：hwan603@163.com
联系电话：+86 792 8282223

标本馆简介

中国科学院庐山植物园标本馆成立于1934年，由我国著名植物学家胡先骕、秦仁昌和陈封怀等老一辈植物学家为创建庐山植物园而成立，首任标本室主任由秦仁昌教授兼任。早期曾用代号LUS，现用代号LBG。

庐山植物园标本馆是江西省目前最大的植物标本馆，保存有被子植物、裸子植物和蕨类植物标本19万份，其中包括模式标本500多份。这些标本采自全国，尤其以江西省植物标本收集最为全面。现保存有20世纪20-50年代我国早期植物学家秦仁昌、王启无、方文培、陈少卿、蔡希陶、冯国楣、郑万钧、胡启明、熊耀国、熊杰、赖书绅、聂敏祥等在我国西南、华中及华东、华南等地采集的大量标本。标本馆还保存有少量国外标本，如日本、越南、韩国、美国、丹麦、瑞典、新加坡、缅甸、朝鲜、俄罗斯及部分东南亚国家的标本。另有少量种子标本和木材标本。已经完成标本数字化14万份。

依托标本馆出版的专著有《江西经济植物志》、《江西树木志》、《中国植物志》（大风子科和旌节花科）、《江西植物志》、《江西植被》、《江西杜鹃花》、《庐山树木》、《江西中草药》、《庐山中草药》、《庐山植物名录》等。

中国科学院庐山植物园标本馆
Herbarium of Lushan Botanical Garden

标本馆内部
Inside the herbarium

历史沿革

1934-1949年：静生生物调查所庐山森林植物园标本室。主任：秦仁昌、陈封怀
1949-1953年：中国科学院植物分类研究所庐山工作站标本室。主任：陈封怀
1953-1970年：中国科学院庐山植物园标本室。主任：陈封怀、熊耀国、赖书绅
1970-1996年：江西省庐山植物园标本室。主任：赖书绅、詹选怀
1996-2019年：江西省、中国科学院庐山植物园标本馆。馆长：詹选怀、彭焱松
2019年至今：中国科学院庐山植物园标本馆。馆长：彭焱松

Curator: Dr. PENG Yansong, Associate Professor
Contacts: HU Wan
E-mail: hwan603@163.com
Tel: +86 792 8282223

Introduction

In 1934, a team of renowned botanists, namely Prof. HU Hsen-Hsu, Prof. CHING Ren-Chang and Prof. CHEN Feng-Hwai established the herbarium at Lushan Botanical Garden, and Prof. CHING Ren-Chang was the first director of the herbarium. Previously it was known as LUS, but now it is known as Lushan Botanical Garden (LBG).

Currently, LBG is the largest herbarium in Jiangxi Province and preserves 190,000 specimens, including angiosperms, gymnosperms, and ferns. More than 500 type specimens were collected too. All these specimens were collected all over China, especially from Jiangxi Province. At present, a large number of specimens of Southwest, Central, East and South China from 1920s to 1950s are preserved, which are collected by early Chinese botanists, such as CHING Ren-Chang, WANG Chi-Wu, FANG Wen-Pei, CHEN Shao-Qing, TSAI Hse-Tao, FENG Huo-Mei, CHENG Wan-Chun, HU Chi-Ming, HSIUNG Yao-Kuo, XIONG Jie, LAI Shushen, NIE Minxiang etc. Besides this, LBG has also collected and preserved some specimens from foreign countries such as Japan, Vietnam, the Republic of Korea, the United States, Denmark, Sweden, Singapore, Myanmar, the Democratic People's Republic of Korea, Russia and some other Southeast Asian countries. A small number of seed and wood specimens have also been preserved at LBG. 140,000 specimens have been digitized.

The monographs published by the herbarium include *Economic Flora of Jiangxi*, *Tree Records of Jiangxi*, *Flora Reipublicae Popularis Sinicae* (Flacourtiaceae and Stachyuraceae), *Flora of Jiangxi*, *Vegetation of Jiangxi*, *Rhododendron of Jiangxi*, *Tree of Lushan*, *Herbal Medicine of Jiangxi*, *Herbal Medicine of Lushan*, *Plant List of Lushan*, etc.

History

1934-1949: Herbarium of Lushan Forest Botanical Garden, the Fan Memorial Institute of Biology. Curator: CHING Ren-Chang and CHEN Feng-Hwai

1949-1953: Herbarium of Lushan Workstation, Institute of Plant taxonomy, CAS. Curator: CHEN Feng-Hwai

1953-1970: Herbarium of Lushan Botanical Garden, CAS. Curator: CHEN Feng-Hwai, XIONG Yaoguo and LAI Shushen

1970-1996: Herbarium of Lushan Botanical Garden, Jiangxi Province. Curator: LAI Shushen and ZHAN Xuanhuai

1996-2019: Herbarium of Lushan Botanical Garden of Jiangxi province and CAS. Curator: ZHAN Xuanhuai and PENG Yansong

2019 till date: Herbarium of Lushan Botanical Garden, CAS. Curator: PENG Yansong

Overview of Collections

LBG has been collecting plant specimens from Mt. Lushan area since its establishment. From 1934 to 1936, more than 30 expeditions were conducted by researchers and collected over 2,000 specimens and more than 20 new species and new records. The collection activities outside Lushan mainly includes TSAI Hse-Tao's collection in Yunnan in 1934, collections at Mt. Jiuhuashan and Mt. Huangshan in Anhui Province from 1935 to 1937, collections at Mt. Taibaishan, Mt. Zhongnanshan, Mt. Nanwutaishan, Chengdu, Tianquan, Mt. Omei in 1936 and Yunnan by Yu Te-Town in 1936-1937. More than 30,000 specimens were collected during the period.

After the outbreak of the Anti-Japanese War, all the staff of Lushan Forest Botanical Garden removed to Lijiang, Yunnan in December 1938. In order to facilitate the work, Prof. CHING Ren-Chang set up a workstation of Lushan Forest Botanical Garden and opened up a specimen room in his office to preserve the specimens which collected by himself and donated by other institutions. In April 1939, the Lushan Botanical Garden was occupied by the Japanese army. Considering that the Lushan Botanical Garden belonged to the Fan Memorial Institute of Biology, the Japanese army transported some of the 120 boxes of articles (including 30,000 specimens) stored in the Guling School of American to Peiping and put them in the same place. In 1939, FENG Huo-Mei led a team to collect specimens in Zhongdian and Lijiang. A total of 6391 number specimens were collected, 4 samples per number, saved in PE, KUN, IBSC and LBG, respectively. By 1942, a total of 20,000 specimens had been collected in LBG.

From 1946 to 1949, a total of 14,000 specimens were collected during this period from three different areas, namely Mt. Lushan, Mt. Mufushan and southern Yunnan. From 1950 to 1953, Lushan Botanical Garden was transferred to CAS and affiliated to the Institute of Plant Taxonomy of CAS. During this period, a total of 10,000 samples were collected

收藏概况

庐山植物园标本室从成立开始就着手收集庐山地区植物标本。1934-1936 年，在庐山调查 30 多次，采集标本 2000 多号，发现 20 多个新种和新记录。在庐山以外的采集主要有 1934 年蔡希陶在云南的采集，1935-1937 年在安徽九华山、黄山地区的采集，1936 年在太白山、终南山、南五台山、成都、天全、峨眉山的采集，1936-1937 年俞德浚在云南的采集，共采集到标本超过 3 万份。

抗战爆发后，1938 年 12 月，庐山森林植物园全体员工撤到云南，在丽江设立庐山森林植物园丽江工作站。为方便工作，秦仁昌在仅有的办公室内开辟出一间作为标本室，保存自己采集和其他机构捐赠的标本。1939 年 4 月庐山植物园被日军占领，考虑到庐山植物园隶属于静生生物调查所，日军将保存于庐山美国牯岭学校的 120 箱物品（包含标本室的 3 万份标本）部分运到北平，和静生生物调查所的标本放于一处。1939 年在云南丽江，由冯国楣带队分别在中甸和丽江进行标本采集，共采集到标本 6391 号，每号 4 份，分别保存于 PE、KUN、IBSC 和 LBG。到 1942 年，庐山森林植物园标本室共采集标本 2 万号。

1946-1949 年，植物标本采集分三块区域进行：庐山、幕阜山脉地区和云南南部，其间共采集到标本 1.4 万号。1950-1953 年，庐山植物园收归到中国科学院，隶属于中国科学院植物分类研究所。其间标本采集范围主要集中于庐山及周边区域，共采集 1 万号。1953-1970 年，庐山植物园隶属于中国科学院，多次参与中国科学院组织的全国植物资源调查，其间采集到标本达 4 万份，并完成众多专著的编写。1970-1996 年，庐山植物园改隶到江西省，之后主要参与江西地区的植物资源调查，采集的标本有 3 万多份。1996 年后，庐山植物园改由中国科学院和江西省共同领导，其间标本采集主要集中于长江以南及江西周边地区，采集的标本有 5 万份。2019 年起，庐山植物园标本馆重新回归中国科学院。

特色收藏

庐山植物园标本馆以收藏江西地区植物标本为主，标本数量超过 10 万份，采集地包含了江西地区所有的山区、县和大部分乡镇。本馆最古老的标本收藏是由静生生物调查所早期交换来的灰柳（*Salix cinerea*）标本，采于 1792 年。同时，本馆也保存了胡先骕、秦仁昌、陈封怀、钟观光、蔡希陶、吴征镒、王文采、郑万钧、方文培、张宏达、胡启明等著名植物学家所采集或鉴定过的标本。

3）微生物类馆

中国科学院微生物研究所菌物标本馆

基本信息

隶属于：中国科学院微生物研究所

地址：中国北京市朝阳区北辰西路 1 号院 3 号

网址：http://biaobenguan.im.ac.cn

建馆年份：1953 年

收藏类群：壶菌、接合菌、子囊菌、担子菌、半知菌、地衣、卵菌和黏菌等

标本藏量：53.6 万号

馆长 / 联系人：蔡磊，研究员

电子邮箱：cail@im.ac.cn

联系电话：+86 10 64806123

植物科学绘画
Botanical scientific illustration

馆藏最早标本，采于 1792 年
The earliest specimen collected in 1792

was mainly concentrated in Mt. Lushan and its surrounding area. From 1953 to 1970, LBG was subordinated to CAS and participated in the different national surveys that were organized by CAS and more than 40,000 specimens were collected and some monographs were written during this period. From 1970 to 1996, Lushan Botanical Garden was transferred to Jiangxi Province. After that, LBG mainly participated in the investigation of plant resources of Jiangxi, with more than 30,000 specimens were collected during this period. After 1996, Lushan Botanical Garden was jointly led by CAS and Jiangxi Province. During this period, specimen collection was mainly concentrated in the south of the Yangtze River and the surrounding areas of Jiangxi, with 50,000 specimens collected so far. From 2019, LBG has returned to CAS.

Special Collections

The LBG mainly focused on the plant specimens' collections from Jiangxi Province. So far more than 100,000 specimens have been collected from all mountain areas, counties and most towns of Jiangxi Province. *Salix cinerea* specimen was the first specimen to the LBG, which was collected in 1792 from the early exchange of the Fan Memorial Institute of Biology. At the same time specimens collected or identified by famous botanists such as HU Hsen-Hsu, CHING Ren-Chang, CHEN Feng-Hwai, TSOONG Kuan-Kuang, TSAI Hse-Tao, WU Zheng-yi (Wu C.Y.), WANG Wencai (Wang W. C.), CHENG Wan-Chun, FANG Wenpei, ZHANG Hongda and HU Chi-Ming are also preserved in LBG.

3) Fungarium

Fungarium of Institute of Microbiology(HMAS), CAS

Basic Information

Affiliated to: Institute of Microbiology, CAS
Address: No.1 West Beichen Road, Chaoyang District, Beijing, P. R. China
Website: http://biaobenguan.im.ac.cn
Established in: 1953
Species groups of collection: Chytridiomycetes, zygomycetes, ascomycetes, basidiomycetes, deuteromycetes, lichens, oomycetes and myxomycetes, etc.
Collection quantity: 536,000
Curator/Contacts: Dr. CAI Lei, Professor
E-mail: cail@im.ac.cn
Tel: +86 10 64806123

Introduction

The Fungarium of Institute of Microbiology, CAS was founded in 1953. It is the only fungarium with a comprehensive collection of taxa in China and is the largest fungarium in Asia. It has established a collaboration and exchange program

标本馆简介

中国科学院微生物研究所菌物标本馆创建于1953年，是我国唯一一个收藏类群全面的菌物标本馆，标本收藏量为亚洲之最。自建馆以来，已经与十几个国家22个国外标本馆建立了标本交换、借阅等业务联系。标本馆建筑面积1204平方米，保藏条件优良，管理制度完备，在我国菌物研究和科普领域发挥着重要的作用。

中国科学院微生物研究所菌物标本馆
Fungarium of Institute of Microbiology, CAS

历史沿革

中央研究院植物研究所标本馆真菌部、北平研究院植物研究所标本馆真菌部、清华大学农学院植物病理系标本室—中国科学院真菌植物病理研究室标本室（1953年）—中国科学院应用真菌研究所标本室（1956年）—中国科学院微生物研究所真菌标本室（1958年）—中国科学院菌物标本馆（2000年）—中国科学院微生物研究所菌物标本馆（2016年至今）。

中央研究院院士合影(1948)，戴芳澜（第四排右2）和邓叔群院士（后排左1）
The academicians of Academia Sinica (1948), Prof. TAI Fang-Lan (No. 2 from right in line 4), Prof. TENG Shu-Chün (No. 1 from left in the back row)

收藏概况

目前已收藏菌物标本53万余号，其中模式标本3000余号，馆藏标本来自全国34个省及世界111个国家和地区。菌物标本馆保藏范围包括壶菌、接合菌、子囊菌、担子菌、半知菌、地衣、卵菌和黏菌等，所藏标本隶属于2000余属约1.56万种，其中保藏最早的标本采自1786年。

特色收藏

在第二次世界大战时期，为使宝贵的科研材料免遭浩劫，我国菌物学先驱邓叔群先生将2000余号菌物标本寄往美国康奈尔大学保存。这些标本为20世纪初采集的珍贵标本。2009年11月7日，“康奈尔大学中国菌物标本赠回仪式”在中国科学院微生物研究所举行，为推进两国和两机构间的合作与交流创造了新的契机。此次赠回的标本共计2278份，为我国菌物系统学及其发展史的研究提供了重要的

with 22 foreign herbariums or fungariums of more than a dozen countries. The Fungarium is of excellent condition for the preservation of fungal collections, with 1204 m^2 floor area, and operated under a sophisticate management system. It plays an important role in mycological researches and popularization.

部分康奈尔大学赠回标本

Some specimens returned from Cornell University

History

The Fungarium was formerly the combination of several institutions including the Department of Mycology, Herbarium of Institute of Botany, Academia Sinica; Department of Mycology, Herbarium of Institute of Botany, National Academy of Peiping; and Herbarium, Department of Phytopathology, Tsinghua University (1953), which later became the Herbarium, Institute of Applied Mycology, CAS (1956), and then Herbarium, Institute of Microbiology, CAS (1958), Herbarium Mycologicum, Academiae Sinicae (HMAS, 2000), and now Fungarium of Institute of Microbiology, CAS (HMAS, 2016-).

Overview of Collections

The Fungarium currently preserves more than 530,000 specimens including over 3,000 type specimens. The specimens were collected from 34 provinces of China and 111 foreign countries and regions. The collections preserved in HMAS contain chytridiomycetes, zygomycetes, ascomycetes, basidiomycetes, deuteromycetes, lichens, oomycetes and slime molds, representing more than 2,000 genera and about 15,600 species. The earliest specimen was collected in 1786.

Special Collections

During World War II, Prof. Teng Shu-Chün, the pioneer of Mycology in China, sent more than 2,000 Chinese fungal collections to Cornell University, USA to protect these precious materials from destruction. These specimens are collected in the early 20th century and precious. On November 7, 2009, a ceremony was held in the Institute of Microbiology, CAS, in which Cornell University returned the prized Chinese specimens to the home HMAS. The 2,278 precious collections are key certificate materials to mycological research in China. This starts new scientific cooperation between mycologists in

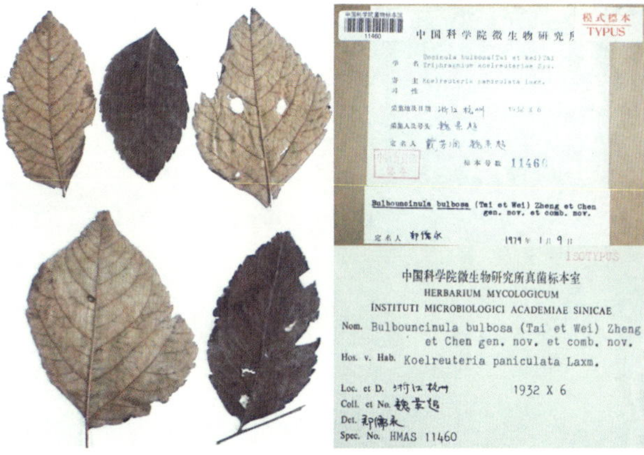

戴芳澜院士 1932 年发表的病原真菌球钩丝壳（*Uncinula bulbosa*）的正模标本

Holotypes of *Uncinula bulbosa* published by academician TAI Fang-Lan in 1932

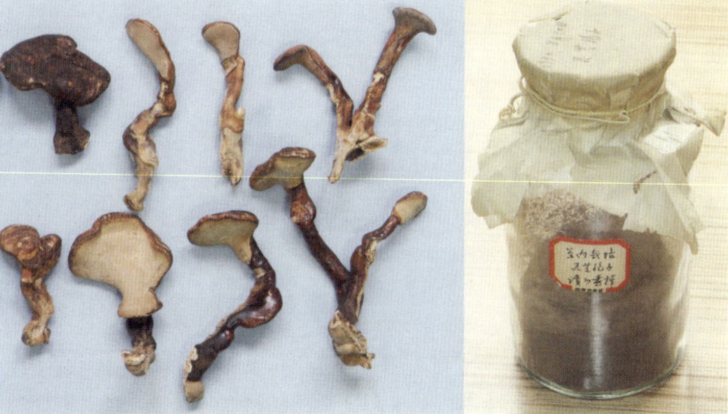

重要馆藏标本。左：1959 年，邓庄博士（邓叔群之女）收集的在中国科学院微生物研究所首次人工栽培的灵芝标本；右：1970 年，中国科学院微生物研究所获得第一瓶人工栽培的灵芝孢子粉

The key collections in the Herbarium. Left, the specimen of the first artificially cultivated *Ganoderma lucidum*, collected by TENG Zhuang (the daughter of Prof. TENG Shu-Chün) from the Institute of Microbiology, CAS in 1959; right, the first bottle of basidiospore powder from artificially cultivated *G. lucidum* in the Institute of Microbiology, CAS, in 1970

凭证材料。此外，标本馆拥有美国农业部等国外科研机构赠送的标本，包括产自世界各地的标本 1 万余份。

菌物标本馆拥有大批具有重要学术和应用价值的珍贵藏品，涵盖了重要的植物病原真菌和经济真菌物种。这些重要材料包括戴芳澜院士和邓叔群院士 20 世纪 30 年代发表新种时所依据的正模标本，以及人类首次人工栽培的灵芝和世界第一瓶人工培养的灵芝孢子粉。菌物标本馆为菌物物种鉴定、菌物学研究及相关产业发展提供了重要的支撑。

4）水生生物类馆

中国科学院海洋生物标本馆

基本信息

隶属于：中国科学院海洋研究所
地址：中国山东省青岛市南海路 7 号
网址：http://www.qdio.ac.cn/mbm/
建馆年份：1950 年
收藏类群：海藻、海洋微生物、原生动物、无脊椎动物、脊椎动物等
标本藏量：85 万号
馆长 / 联系人：王少青，高级工程师
电子邮箱：mbm@qdio.ac.cn
联系电话：+86 532 82898911

标本馆简介

中国科学院海洋生物标本馆始建于 1950 年，是我国规模最大的海洋生物标本馆，海洋生物标本收藏量居亚洲之首。标本馆与中国科学院海洋研究所海洋生物分类与系统演化实验室并行，汇聚成了一支国际上少有的建制完整、门类齐整、研究力量雄厚的海洋生物分类与多样性研究和标本保藏团队，是我国海洋生物多样性研究的中心和策源地。

中国科学院海洋生物标本馆
Marine Biological Museum, CAS

历史沿革

1950 年中国科学院水生生物研究所青岛海洋生物研究室成立，原北平研究院动物研究所部分从事海洋生物研究的科研人员来到青岛，同时将原静生生物调查所和北平研究院动物研究所的 2000 余号海洋生物标本带来青岛，这是标本馆最早的标本收藏。

1959 年中国科学院海洋研究所成立，分别组建了海洋植物、海洋无脊椎动物和海洋脊椎动物等研究室，同时建立各自的标本室，分别收藏各自类群的标本。

1996 年海洋研究所以海洋植物研究室、海洋无脊椎动物研究室和海洋脊椎动

the two countries. In addition, the Fungarium has a collection of specimens donated by the U.S. Department of Agriculture and other foreign scientific research institutions, including more than 10,000 specimens from all over the world.

Vast multitudes of precious collections refer to key plant pathogenic and economic fungi, with important academic and applied values, which are housed in the Fungarium. The crucial materials include the holotypes of some new species proposed by Prof. TAI Fang-Lan and Prof. TENG Shu-Chün in 1930s, the first specimen and the first basidiospore powder of artificially cultivated *Ganoderma lucidum* ("Lingzhi", traditional Chinese medicine). The Herbarium provides indispensable support for fungal identification, mycological research and fungal industry.

4) Hydrobiont Collection / Museum

Marine Biological Museum, CAS

Basic Information

Affiliated to: Institute of Oceanology, CAS
Address: No.7 Nanhai Road, Qingdao, Shandong, P. R. China
Website: http://www.qdio.ac.cn/mbm/
Established in: 1950
Species groups of collection: Seaweed, marine microorganism, protozoa, invertebrate, vertebrate, etc.
Collection quantity: 850,000
Curator/Contacts: WANG Shaoqing, senior engineer
E-mail: mbm@qdio.ac.cn
Tel: +86 532 82898911

Introduction

The Marine Biological Museum of CAS, founded in 1950, is the largest museum of marine organisms in China, with the largest marine collection in Asia. The museum, together with the Department of Marine Organism Taxonomy and Phylogeny, Institute of Oceanology, CAS, has formed a rare international team of marine biological classification and diversity research and specimen preservation with the complete system, complete categories and strong research force. It is the center and source of marine biodiversity research in China.

History

In 1950, Qingdao Marine Biology Research Laboratory of Institute of Hydrobiology of CAS was established. Some researchers from the former Fan Memorial Institute of Biology and Institute of Zoology of the National Academy of Peiping came to Qingdao. At the same time, more than 2,000 marine biological specimens were brought to Qingdao, which is the earliest specimen collection for the museum.

In 1959, the Institute of Oceanology of CAS was established, and the research laboratories of marine plants, marine invertebrates and marine vertebrates were established, respectively. At the same time, their own specimen rooms were established to collect specimens of their own groups.

In 1996, based on the researchers of classification discipline in the laboratory of marine plants, marine invertebrates and marine vertebrates, the Division of Classification and Phylogeny of Marine Organisms was established, and all the collections merge to establish the Marine Biological Museum, CAS.

In 2002, with the support of key construction projects of knowledge innovation project of CAS, the new building of the museum was officially opened and renamed as the Marine Biological Museum, CAS. Prof. LI Xinzheng was the first director of the museum.

In 2007, on the basis of the Marine Biological Museum, CAS, the Department of Marine Organism Taxonomy and Phylogeny was established by integrating the researchers engaged in the research of marine biological taxonomy. It has become a team of marine taxonomy and biodiversity research and specimen preservation with the most complete and powerful research force in China, involving almost all the important categories of marine biology in China. Academician LIU Ruiyu is the honorary curator, researcher XU Kuidong is the director of the laboratory, and senior engineer WANG Shaoqing is the deputy director, deputy curator and curator of the museum.

Overview of Collections

The total construction area of the museum is 4,920 m^2, which is divided into four areas: storage area, research area,

物研究室的分类学科研人员为基础，组建了海洋生物分类与系统演化研究室，同时将各标本室合并，成立中国科学院海洋研究所海洋生物标本馆。

2002 年在中国科学院知识创新工程重点建设项目的支持下，新馆正式启用，更名为中国科学院海洋生物标本馆，李新正研究员担任首任馆长。

2007 年在中国科学院海洋生物标本馆基础上，将从事海洋生物分类学研究的科研人员整合，成立中国科学院海洋研究所海洋生物分类与系统演化实验室。汇聚成了一支我国海洋生物门类最为齐整、研究力量最为雄厚的海洋生物分类学和生物多样性研究及标本保藏团队，涉及了我国几乎所有重要海洋生物门类的分类研究。刘瑞玉院士任名誉馆长，徐奎栋研究员任实验室主任，王少青高级工程师任实验室副主任、标本馆副馆长、馆长至今。

收藏概况

海洋生物标本馆建筑总面积为 4920 平方米，分为储藏区、研究区、技术区、展示区四个区域。分别设有模式标本库、珍贵（历史）标本库、海藻腊叶标本库、海藻液浸标本库、浮游生物标本库、无脊椎动物标本库、软体干壳标本库、鱼类标本库、极地生物标本库、DNA 凭证标本库、深海生物标本库、超低温标本库等。按海洋生物分类系统，分别存放着液浸、腊叶、干壳、干制（剥制）、冷冻标本等。此外，标本馆还设有分类学实验室、标本处理室、图书资料室、数据库机房和学术报告厅等辅助设施。

截至 2019 年底，海洋生物标本馆收藏了各类海洋生物标本 85 万号，最早的标本采集于 1889 年。标本的采集范围包括整个中国海域、西太平洋以及南、北极的 57 个国家和地区。包含了潮间带、浅海、深海、极地、珊瑚礁、红树林、河口等各种海洋环境类型，底栖、浮游、游泳等各种海洋生态类型的海洋生物标本。收藏主体是我国 2000 年前进行过的几乎所有大型海洋生物调查采获的全部或大部分标本，以及 1950 年建馆以来通过其他调查项目采集及国际合作、交换等补充的世界各地的标本，还有 2014 年以来开展深海大洋调查所采获的深海生物标本等。

标本馆馆藏标本门类齐全，现行的生物六界分类系统中，除真菌界外，其余五界的标本均有收藏。收藏的主要类群有海藻、海洋微生物、原生动物（纤毛虫、有孔虫、放射虫等）、无脊椎动物（浮游动物、多孔动物、刺胞动物、环节动物、螠虫动物、星虫动物、软体动物、节肢动物、苔藓动物、腕足动物、棘皮动物、半索动物等）、脊椎动物（鱼类、哺乳类、海鸟、爬行类等）等。这些珍贵的标本显示了我国丰富的海洋生物资源蕴藏和高生物多样性特点。它们是全球生态变化的实物见证，是当前几个热点研究领域的重要材料。由于当今海洋生态环境发生了很大变化，许多标本即使是在同一地点、同一季节也不一定能再采集到，更加显示出海洋生物标本馆所收藏标本的珍贵。

海洋生物标本馆已陆续对 504 440 号馆藏标本进行了数字化整合，占馆藏标本总量的 62.94%。分别建立了中国沿海海洋生物信息数据库、模式标本数据库、海藻标本数据库、全国海洋普查标本、无脊椎动物标本数据库、海洋鱼类标本数据库、极地标本数据库、海洋生物 DNA 条形码数据库等专业数据库。使用“中国科学院海洋生物标本管理信息系统”，采用条形码及分类码技术对馆藏标本实现了数字化管理。通过“中国科学院海洋生物标本馆标本资源共享平台”，向国内外用户提供网络共享服务，是目前国内海洋生物学领域规模最大、最为权威的数据库服务系统。

在标本信息的数字化过程中，标本馆与国际接轨，进行了联合国教育、科学及文化组织 - 政府间海洋学委员会 (UNESCO-IOC) 领导下的“海洋生物地理信息系统 (OBIS)”建设，是 OBIS 中国节点所在地。标本馆还开发完成了“海洋生物标本馆标本资源 APP 平台”移动客户端，科研人员可以使用手机等移动设备，通过扫描条形码和二维码，快速访问“海洋生物标本馆标本资源共享平台”，获得与电脑版相同的采集、鉴定、图像、文献以及 DNA 等全部标本信息。

海洋生物标本馆标本库
The collection of the Marine Biological Museum

海洋生物标本馆超低温标本库
The ultra-low temperature specimen room of the Marine Biological Museum

technology area and exhibition area. There are different specimen rooms to preserve different types of specimens such as type specimens, precious (historical) specimens, dehydrated algal specimens, algal liquid immersed specimens, plankton specimens, invertebrate specimens, mollusk specimens, fish specimens, polar organism specimens, DNA voucher specimens, abyssal organism specimens, ultra-low temperature specimens, etc. According to the classification system of marine life, there are liquid immersed, dehydrated, dried shells, dried skin (peeled), frozen specimens, etc. In addition, the collection also has taxonomic laboratory, specimen processing room, library, database room, academic report hall and other auxiliary facilities.

By the end of 2019, the museum has collected 850,000 specimens of various marine organisms, the earliest of which was collected in 1889. The collection range of specimens includes 57 countries and regions, including the whole China Sea area, the Western Pacific Ocean, the South and the Arctic. It includes various types of marine environment such as intertidal zone, shallow sea, deep sea, polar region, coral reef, mangrove, estuary, benthos, plankton, swimming and other types of marine ecological specimens. The collection subjects are all or most of the specimens collected in almost all the large-scale marine biological surveys conducted in China before 2000, specimens collected through other survey projects and supplemented by international cooperation and exchange since the establishment of the museum in 1950, and specimens of deep-sea organisms collected in the deep-sea ocean survey conducted since 2014.

The collection of the museum is complete. In the current six Kingdom Classification System of biology, the specimens of five kingdoms are collected except for the fungi. The main collection groups are seaweed, marine microorganisms, protozoa (ciliates, foraminifera, radiolaria, etc.), invertebrates (floating, porous, prickly cell, link, *Urechis*, sipunculan, mollusk, arthropod, moss, brachiopod, echinoderm, hemichordata, etc.), vertebrates (fish, mammals, seabirds, reptiles, etc.). These precious specimens show the rich marine biological resources and high biodiversity in China. They are physical witnesses of global ecological change and important materials in several hot research fields. Due to the great changes in the marine ecological environment, many specimens can not be collected even in the same place and season, which shows the value of the specimens collected by the museum.

在西沙晋卿岛潜水采集生物标本
Collecting biological specimens in diving on Xisha Jinqing Island

The collection has digitized 504,440 specimens, accounting for 62.94% of the total collection. The database of China's coastal marine life information, type specimen database, seaweed specimen database, national marine census specimen, invertebrate specimen

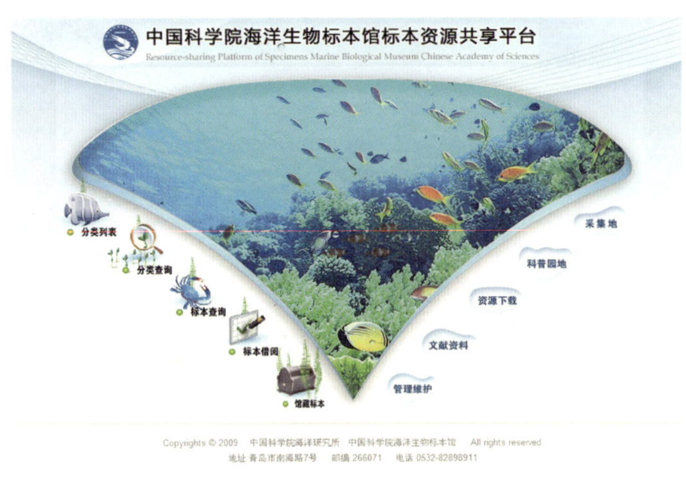

海洋生物标本馆标本资源共享平台
Specimen Resource Sharing Platform of Marine Biological Museum

特色收藏

深海生物标本库建设

依托中国科学院先导A类专项“热带西太平洋海洋物质能量交换及其影响”和国家科技基础调查专项“西太平洋典型海山生态系统科学调查”等，先后在西太平洋马里亚纳、雅浦、卡罗琳、麦哲伦海山区和马努斯海盆、南海冷泉区、冲绳海槽热液区等采集了5922号、超过800种大型深海生物标本和8000余株微生物样品，最深的标本采集于马里亚纳海沟水深7443米处，建立了目前国内最大规模的深海生物标本库。深海生物标本库的建立和运行，为从事深海特殊环境生物多样性的研究提供了支撑。依靠库藏的深海标本，先后发表了1个新科、4个新属、50多个新种和46个深海细菌新种，实现了对深海环境和资源的新认知。成为国际上有影响的深海生物资源库。已有美国、澳大利亚等10多个国家的专家到访过深海生物标本库。

采集于1898年的藻类标本

1950年中国科学院海洋生物研究室在青岛建立，从国立北平研究院和静生生物调查所带来了一批珍贵的历史标本，从中发现了这号采集于1898年4月10日，盖有“静生生物调查所植物标本室”收藏章的标本，这是标本馆收藏最早的海藻标本。

大伞藻

大伞藻数量稀少，在我国台湾恒春半岛和海南有分布。通常群生于浅水区或中度浪袭礁岩的隐蔽面，尤其是生长在鹅卵石上的大伞藻，就像一个人头上长满了白发，又称“生发石”。

深海生物标本库
Deep-sea Biological Specimen Bank

database, marine fish specimen database, polar specimen database, marine biological DNA barcode database and other professional databases have been established separately. By using the "Information System of Marine Biological Specimen of the Chinese Academy of Sciences", the digital management of collected specimens is realized by using bar code and classification code technology. It is the largest and most authoritative database service system in the field of marine biology in China by providing online sharing services to users at home and abroad through the "Specimen Resource Sharing Platform of Marine Biological Museum".

In the process of digitalization of specimen information, the museum is in line with the international standards and has carried out the construction of the "Ocean Biogeographic Information System (OBIS)" under the UNESCO-IOC, which is the location of the OBIS China node. The museum has also developed a mobile client of the "Specimen Resource Platform of Marine Biological Museum" app. Researchers can use mobile phones and other mobile devices to quickly access the "Specimen Resource Sharing Platform of Marine Biological Museum" by scanning bar codes and QR codes, and obtain all specimen information, such as collection, identification, images, documents and DNA, which are the same as the computer version.

Special Collections

Construction of Deep-sea Biological Specimen Bank

Based on the Class A loading project of CAS "Tropical Western Pacific Ocean Material and Energy Exchange and its Impact" and the national science and technology foundation survey project "Scientific Investigation on the Ecosystem of Typical Seamounts in the Western Pacific" led by CAS, studies were carried out in Mariana, Yapu, Caroline, Magellan sea mountains and Manus Sea basin, cold spring area of the South China Sea, Okinawa trough hydrothermal area, etc. The number of 5,922 and more than 800 kinds of large-scale deep-sea biological specimens and more than 8,000 strains of microbial samples were collected. The deepest specimens were collected at a depth of 7,443 m in the Mariana Trench, and the largest deep-sea biological specimen bank in China was established. The establishment and operation of the deep-sea biological specimen bank provide support for the research on the biodiversity of the deep-sea special environment. Based on the deep-sea specimens, one new family, four new genera, more than 50 new species and 46 new species of deep-sea bacteria have been published successively, realizing the new cognition of deep sea environment and resources. It has become an influential deep sea biological resource pool in the world. Experts from more than 10 countries, such as the United States and Australia, have visited the deep sea specimen bank.

Algae specimen collected in 1898

In 1950, the Laboratory of Marine Biology of CAS was established in Qingdao. A number of precious historical specimens were brought from the National Academy of Peiping and the Fan Memorial Institute of Biology. The specimen collected on April 10, 1898, with the seal of the "Herbarium of Fan Memorial Institute of Biology", was found. It is the earliest algae specimen in the museum.

采集于 1898 年的藻类标本

Algae specimen collected in 1898

Acetabularia major

The amount of *Acetabularia major* is rare, and it is distributed in Hengchun Peninsula of Taiwan and Hainan. It is usually found on the hidden surface of shallow water or moderate wave striking reef rocks, especially on the cobblestone, just like a person's head is covered with white hair, also known as "Hair Stone".

This specimen was purchased by a stone enthusiast in Anshan, Liaoning Province, in 1987. He has never known its name. After winning the national rare stone Grand Prix, it attracted widespread attention. CCTV science and education channel reporter and the stone holder took the specimen with them to look for experts all over the country and failed to identify it. In 2005, he came to Qingdao and was identified by experts of the museum as a marine organism *Acetabularia major*, which cracked the secret of this strange stone. CCTV recorded the whole process of searching for identification, filmed and broadcasted a documentary named "White Haired Devil", which caused a sensation. As this is a precious specimen integrating ornamental and scientific features, it was collected by the museum at the end of 2006 after many times of communication and consultation.

大伞藻
Acetabularia major

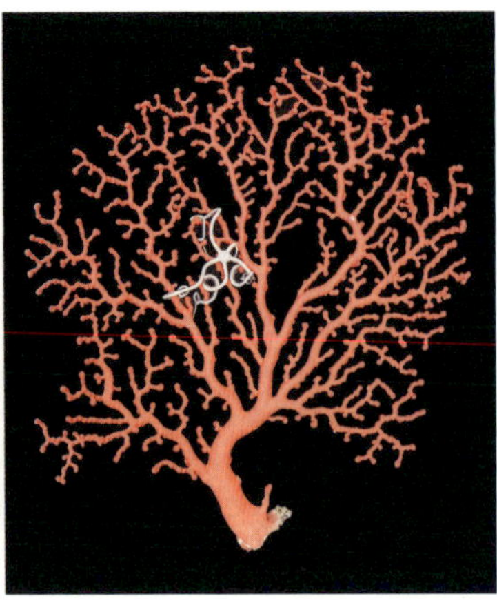

红拟柳珊瑚标本
The specimen of *Paragorgia rubra*

生活在海底的活体红拟柳珊瑚
Living *Paragorgia rubra* on the sea floor

该标本是 1987 年辽宁鞍山一位奇石爱好者在市场购得的，一直不知其名。在全国奇石大奖赛获奖后引起广泛关注，中央电视台科教频道记者和这块石头的持有人带着这块标本在全国各地寻找专家鉴定未果。2005 年辗转来到青岛，经标本馆专家鉴定为海洋生物大伞藻，破解了这块奇石的秘密。中央电视台将寻求鉴定的整个过程记录下来，拍摄播出了以“白发魔头”为名的纪录片，引起轰动。由于这是一块集观赏性和科学性于一体的珍贵标本，经多次沟通协商，2006 年底其被标本馆收藏。

红拟柳珊瑚

群体为单平面结构，红色，高 53 厘米，宽 44 厘米。生活状态下珊瑚虫伸展，白色，分布于迎流面，以最大限度捕食水流中浮游生物。是发现于热带西太平洋的拟柳珊瑚属的第四个新物种。该科珊瑚与名贵的红珊瑚科（Coralliidae）亲缘关系最近，区别在于拟柳珊瑚科的轴由较为松散的骨片组合而成，而红珊瑚的轴是一个致密的整体。该标本于 2014 年 12 月 24 日由“科学”号采集于雅浦海山水深 373.00 米处。

网状小尖柳珊瑚

标本高 1.14 米，宽 1.45 米。群体呈单平面结构，主分枝连接成网状，生活时黄褐色至灰白色，珊瑚虫明显，不收缩；珊瑚骨骼黑色，由珊硬蛋白和非骨片钙质物质构成，分枝核心中空。珊瑚表面嵌有大量钙质白色骨片。通过同位素测年显示该珊瑚年龄约 1000 岁，是深海中个体较大、寿命较长的一类柳珊瑚，也是目前国内保

网状小尖柳珊瑚标本
The specimen of *Muricella reticulate*

生活在海底的活体网状小尖柳珊瑚
Living *Muricella reticulate* on the sea floor

Paragorgia rubra

This specimen has a single plane structure, red, height 53 cm, width 44 cm. Under the living condition, the corals are white and distributed on the upstream surface to catch plankton in the water as much as possible. It is the fourth new species of *Paragorgia* found in the tropical western Pacific Ocean. The corals of this family are closely related to the precious Coralliidae family. The difference is that the axis of Paragorgiidae is composed of relatively loose bone fragments, while the axis of red coral is a compact whole. This specimen was collected by the "Science" research ship on December 24, 2014 at the depth of 373.00 m in the Yapu sea mountain.

Muricella reticulate

The specimen is 1.14 m high and 1.45 m wide. The main branches of the colony are connected into a network with a single plane structure. The life of the colony is yellowish-brown to grayish-white, and the corals are obvious and non shrinking. The coral skeleton is black, which is composed of gorgonin and non-bone calcium material, and the branch core is hollow. The coral surface is embedded with a large number of calcareous white bone pieces. The isotopic dating shows that the coral is about 1000 years old. It is a kind of gorgonians with large size and long life in the deep sea. It is also the oldest invertebrate specimen in China. The specimen was collected by the "Science" research ship on March 22, 2016, at a depth of 159.80 m in Mariana sea mountain.

Entemnotrochus rumphii

Entemnotrochus rumphii appeared in the Cambrian of Paleozoic, about 570 million years ago, and it is known as a "living fossil in the sea". Its main feature is that there is a crack in the mouth of the shell, and the crack reaches half of the area around the bottom of the shell. The mouth cover is larger than the mouth of the shell, the shell is yellow with flame color, and the diameter is about 10-23 cm. Because of the scarcity of *E. rumphii*, the beautiful shell and pattern, and the reputation of "King of Shellfish", it is the goal of shellfish lovers to pursue and collect, and also the most expensive one in the shellfish market at present. This specimen was collected in the South China Sea in September 2013.

龙宫翁戎螺
Entemnotrochus rumphii

Grimpoteuthis sp.

Grimpoteuthis is a kind of very special octopus, which is widely distributed in the continental shelf of temperate, tropical and cold sea areas. There are some deep-sea species in the Arctic Sea, the deepest of which can reach several thousand meters. They live in the deep sea from 100 m to more than 7,000 m. There are more than ten kinds of records in the world. The typical feature is that the carcass has a pair of fins similar to the elephant's ears. When swimming, it is similar to the Dumbo in Disney cartoon, so it gets its name. It is commonly known as "Deep-sea Dumbo" or "Dumbo Octopus". This specimen is the only one in China. On August 21, 2017, it was collected by the "Science" research ship in the Caroline seamount in the Western Pacific Ocean, at a depth of 1240 m.

存的最老的无脊椎动物标本。该标本于2016年3月22日由“科学”号采集于马里亚纳海山水深159.80米处。

龙宫翁戎螺

龙宫翁戒螺出现于古生代寒武纪，距今约五亿七千万年前，有着“海中活化石”之称。其主要特征为壳口有一道裂缝，罅裂达壳底周围之一半。口盖比壳口大，壳呈黄色带火焰色彩，直径10-23厘米。由于龙宫翁戎螺极为稀少，又有美丽的壳体和花纹，有“贝类之王”的美誉，是贝类爱好者争相收藏的目标，也是目前贝类市场中最贵的一种螺。该标本于2013年9月采集于南海。

烟灰蛸

烟灰蛸是一类非常特殊的章鱼，它们广泛分布于温带、热带和寒带海域的大陆架，北冰洋还有一些深海性品种，最深可达几千米。生活在100米到7000多米的深海区域。全世界目前记录有十几种。其典型特点是胴部具有一对类似大象耳朵的鳍，游泳时酷似迪士尼动画片中的小飞象（Dumbo）而得名，俗称“深海小飞象”或“小飞象章鱼”(Dumbo Octopus)。该标本是目前国内仅有的一号烟灰蛸标本，2017年8月21日由“科学”号采集于西太平洋卡罗琳海山水深1240米处。

盲鼬鳚

目前世界上仅有的2号标本之一（另一号标本保存于丹麦国家博物馆），该标本于1959年7月13日通过拖网采集于南海水深1100米处，由我国海洋鱼类学家成庆泰先生鉴定。由于在水体200米以下就没有可见光了，盲鼬鳚的眼睛已退化成2个白点。

中国科学院水生生物博物馆

基本信息

隶属于：中国科学院水生生物研究所
地址：中国湖北省武汉市武昌区东湖南路7号
网址：http://mhbs.ihb.ac.cn
建馆年份：2005年
收藏类群：藻类、原生动物、水生无脊椎动物、底栖动物、鱼类和水生植物等
标本藏量：40万号
馆长 / 联系人：张先锋，研究员
电子邮箱：zhangx@ihb.ac.cn
联系电话：+86 27 68780068

标本馆简介

水生生物博物馆收藏有40万号标本，包括我国淡水鱼类标本1000余种，30余万号；鱼类模式标本310种；产自国外34个国家和地区的鱼类标本600余种；藻类标本2万多号；以及部分水生无脊椎动物标本。本博物馆的鱼类收藏中，最具特色的是鲤形目和鲤科的标本最为完整，反映了东亚鱼类区系的特点，具有广泛的国际影响力。由于独具特色的收藏和高水平的研究，本博物馆已成为亚洲淡水鱼类多样性研究的中心。

水生生物博物馆展厅面积1000平方米，展示有白鲟、白暨豚等灭绝物种；中华鲟、扬子鳄、长江江豚、胭脂鱼等国家重点保护野生动物；被誉为活化石的矛尾鱼标本；此外还设置了7个主题展示，包括馆藏特

烟灰蛸标本
The specimen of *Grimpoteuthis* sp.

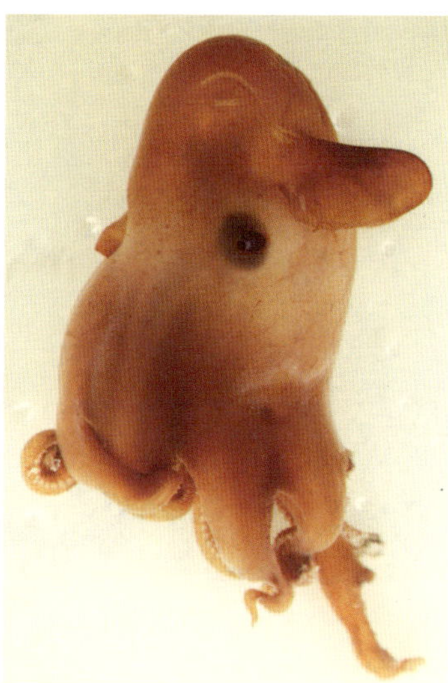

烟灰蛸
Grimpoteuthis sp.

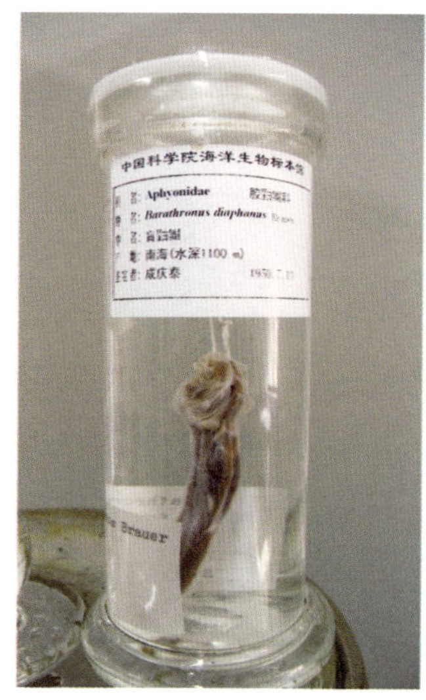

盲鼬鳚标本
The specimen of *Barathronus diaphanous*

Barathronus diaphanous

At present, there are 2 *Barathronus diaphanous* specimens in the world. This is one of them. The other one is preserved in the National Museum of Denmark. The specimen was collected in the South China Sea on July 13, 1959 at a depth of 1,100m through trawling, and was identified by Mr. CHENG Qingtai, a marine ichthyologist in China. Since there is no visible light 200 m below the water, the eyes of the *Barathronus diaphanous* have degenerated into two white spots.

The Museum of Hydrobiological Sciences (MHBS), CAS

Basic Information

Affiliated to: The Institute of Hydrobiology (IHB), CAS
Address: No.7 of East Lake South Road, Wuchang District, Wuhan, Hubei Province, P. R. China
Website: http://mhbs.ihb.ac.cn
Established in: 2005
Species groups of collection: Algae, protozoa, aquatic invertebrates, benthos, fish and aquatic plants, etc.
Collection quantity: 400,000
Curator/Contacts: Dr. ZHANG Xianfeng, Professor
E-mail: zhangx@ihb.ac.cn
Tel: +86 27 68780068

Introduction

At present, MHBS collects 400,000 specimens, which include over 300,000 specimens belonging to more than 1,000 species of freshwater fishes; 310 type specimens of fishes; over 600 species of fishes from 34 foreign countries/regions; over 20,000 algae specimens; and some aquatic invertebrate specimens. Among the fish collections, the most distinctive are the complete collections of Cypriniformes and Cyprinidae, which reflect the characteristics of the fish fauna in East Asia and have extensive influence worldwide. Due to the unique collections and high-level studies, MHBS has become the center for Asian freshwater fish diversity research.

MHBS houses an exhibition hall with area of 1,000 m^2 and displays some extinct species including the Chinese paddlefish (*Psephurus gladius*) and the Chinese river dolphin (*Lipotes vexillifer*); some national protected animals including the Chinese Sturgeon (*Acipenser sinensis*), the Chinese alligator (*Alligator sinensis*), the Yangtze finless porpoise (*Neophocaena asiaeorientalis*), the Chinese sucker (*Myxocyprinus asiaticus*), etc.; the so-called "living fossil" animal coelacanths (*Latimeria chalumnae*); some theme exhibitions including 7 special specimens exhibition, animal evolution

水生生物博物馆外景
Exterior of the MHBS Building

水生生物博物馆门厅
Entrance hall of the MHBS

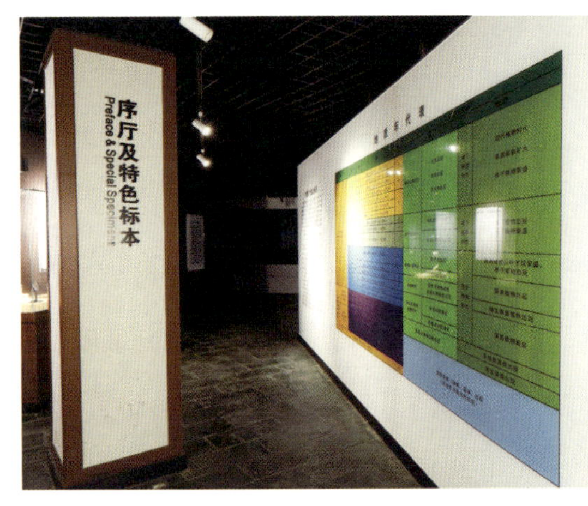

序厅及特色标本
Preface and special specimens

矛尾鱼与动物演化
Coelacanth and animal evolution

殊标本展示、动物进化展示、长江中下游湿地展示、西藏高原湖泊鱼类展示、洞穴鱼类展示、长江鱼类多样性展示、长江珍稀特有鱼类展示、模式标本展示、四大家鱼展示、常见水生经济鱼类展示、最新研究成果展示、照片墙展示等，以及长江十年禁渔、水生植物、渔文化等特色展示。

博物馆通过多媒体方式，展示了藻类、水生植物、水生无脊椎动物、鲸类等水生生物类群。博物馆还展示了水生生物方面的最新研究成果，如鱼类养殖新品种“中科三号（异育银鲫）”、“全雄黄颡鱼”和转基因黄河鲤（冠鲤）等。

水生生物博物馆也肩负有科普教育的重任，设有科普课堂，不定期讲座。博物馆还开通了展厅现场 WiFi，推出了微信公众号和展品二维码。现场通过扫码，可进一步了解展品信息。此外，博物馆还开发了手机 APP，并开通了网站。

水生生物博物馆全年对外开放，采取预约参观方式。可通过电话（027-68780985）、微信公众号（中国科学院水生生物博物馆订阅号）、APP（水生生物博物馆）及网站等几种方式预约参观。

水生生物博物馆发展的过程中，几代科学家先后编著出版了多项水生生物学专著，包括：《湖泊调查基本知识》，《中国动物志·硬骨鱼纲·鲤形目》中卷、下卷，《中国淡水藻志》8 卷册，《中国动物志·淡水枝角类》，《中国动物志 · 环节动物门 · 蛭纲》，《中国经济动物志 - 淡水鱼类》，《中国鲤科鱼类志（上、下）》，《中国鞘藻目专志》，《中国淡水轮虫志》，《长江鱼类》，《中国鱼类系统检索》，《中国淡水鱼类原色图集（1-3）》，《西

exhibition, exhibition of the wetland in the middle and lower reaches of the Yangtze River, Tibet Plateau fishes exhibition, cave fishes exhibition, exhibition of the diversity of Yangtze River fishes, exhibition of rare and endemic fishes of the Yangtze River, type specimens exhibition, four major Chinese carps exhibition, common aquatic economic organism exhibition, exhibition of recent researches findings, exhibition of old photos of IHB, CAS, etc.; and special exhibitions of 10-year fishing ban in Yangtze, aquatic plants, and fishing culture.

Other groups of aquatic organisms, such as algae, aquatic plants, aquatic invertebrates, and cetaceans, and some of the latest research findings, such as studies of culture species "Zhongke III (allogyogenetics silver crucian carp)", "all-male yellow catfish", and transgenic Yellow River carp, are displayed by using multi-media approaches.

大型浸制标本
Large liquid immersion specimens

MHBS holds important responsibilities for science education and communication. In particular, MHBS sets up some science education classes and nonscheduled or on demand lectures and museum online. In order to link the museum and the publics, MHBS sets WiFi and WeChat Subscription for publics to scan two-dimensional code to gain more information or stories behind the exhibited specimens. Besides, MHBS develops smartphone APPs and builds official websites to present more information about MHBS online.

MHBS is open to the public all year round. It can be visited by appointment. You can make an appointment by phone (027-68780985), WeChat public number (Subscription: The Museum of Hydrobiological Sciences, CAS), APP (The Museum of Hydrobiological Sciences) and official website, etc.

In the development process of MHBS, several generations of scientists have published a number of monographs on hydrobiology, including *Basic Knowledge of Lake Survey* (named in Chinese only), *Fauna Sinica Osteichthyes Pisces: Cypriniformes II*, *Fauna Sinica Osteichthyes Cypriniformes III*, eight volumes of *Flora Algarum Sinicarum Aquae Dulcis*, *Fauna Sinica Crustacea Freshwater Cladocera*, *Fauna Sinica Annelida Hirudinea*, *Chinese Economic Fauna: Freshwater Fishes* (named in Chinese only), *Ichthyofauna Sinica Cyprinidae I* (named in Chinese only), *Ichthyofauna Sinica Cyprinidae II* (named in Chinese only), *Monographia Oedogonisles Sinica*, *Fauna Rotiferarum Sinicarum Aquae Dulcis* (named in Chinese only), *Yangtze River Fishes* (named in Chinese only), *Systematic Synopsis of Chinese Fishes*, *Original Color Illustrated Book of Chinese Freshwater Fishes* (named in Chinese only), *Tibet aquatic invertebrates* (named in Chinese only), etc.

History

The predecessor of MHBS was the Natural Historical Museum, Academia Sinica established in Nanjing in 1930, which was renamed the Institute of Zoology and Botany, Academia Sinica in 1934. During World War II, the institute was moved to Nanyue County, Hunan Province, then to Yangshuo County, Guangxi Province, and finally to Beibei District, Chongqing City. After the war, the institute was moved to Shanghai, and then reorganized as the Institute of Hydrobiology (IHB) of CAS in 1950. In 1954 the IHB, CAS was moved to Wuhan and later, the freshwater fish specimen collection room was founded. In 2005, MHBS was constituted through combining the collection room with other specimen collections.

Overview of Collections

After the establishment of the People's Republic of China, as one of the most important units, IHB, CAS has organized a series of field investigations nationwide and collected a large number of specimens and survey data, which contain various groups of aquatic organisms, including algae protozoa, aquatic invertebrates, zoobenthos, fish, and aquatic plants. The details are as follows:

Both the investigation of the Yangtze River Basin including Taihu Lake, Liangzi Lake, Minjiang River, Three Gorges, Danjiangkou, Qingjiang River, etc. and the investigation of the Heilongjiang River Basin, Yongding River in Haihe River Basin, Huaihe River Basin, Yellow River Basin, Pearl River Basin, etc. were carried out in the 1950s; the

藏水生无脊椎动物》等。

历史沿革

水生生物博物馆的前身是1930年成立于南京的中央研究院自然历史博物馆。1934年自然历史博物馆更名为中央研究院动植物研究所。第二次世界大战期间，动植物研究所先后迁至湖南南岳、广西阳朔，最后转移到重庆北碚。抗战胜利以后，该研究所迁到上海。1950年，该研究所在上海改组为中国科学院水生生物研究所。1954年，迁至武汉，原标本室发展为淡水鱼类标本馆。2005年，以淡水鱼类标本馆为主体，整合其他标本收藏构成了水生生物博物馆。

收藏概况

中华人民共和国成立后，中国科学院水生生物研究所（以下简称“水生所”）作为最主要单位之一，组织了一系列的野外调查，考察采集区域涵盖全国，采集了大量标本，积累了大量的水生生物调查数据，水生生物类群涵盖藻类、原生动物、水生无脊椎动物、底栖动物、鱼类和水生植物等水生生物类群，具体如下。

20世纪50年代进行考察的长江流域，如太湖、梁子湖、岷江、三峡、丹江口、清江等，黑龙江流域，海河流域永定河水系，淮河流域，黄河流域，珠江流域等的水生生物调查；60年代进行的西藏水系科学考察；70年代进行的横断山脉水系、湘江流域调查；80年代进行的西藏高原、湘西、三江源、新疆水系、杭州西湖调查；90年代进行的湖北各大湖泊、两湖（鄱阳湖和洞庭湖）流域、黄淮流域、云南滇池、南极调查。2000年至今，依托于水生生物博物馆的不同研究团队对西藏雅鲁藏布江流域及高原湖泊、林芝尼洋河、金沙江、赤水河、嘉陵江、三峡水库、丹江口水库、南水北调东线湖群、南水北调中线沿线、汉江中下游、湖北各湖泊、巢湖、太湖、傀儡湖、千岛湖、杭州西湖、东钱湖、滇中各湖、青海湖、博斯腾湖、乌梁素海、呼伦湖、东江、赣江、白洋淀，以及上海、海南、青岛、天津、晋城、韶关、广州、深圳等城市水体进行了水生生物调查。中国科学院对水生所在梁子湖的工作很重视，曾配备一艘带帆的机动船，取名“梁子号”。

特色收藏

中国淡水鱼类中，鲤形目有857种，是最大的淡水鱼类类群。鲤形目中，鲤科是最大的科，有579种，其中许多是中国和东亚的特有类群。水生生物博物馆的标本收藏中，以鲤形目鲤科鱼类收藏最为完整，收

“梁子号”考察船
“Liangzi” Research vessel

1957年7月在水生所梁子湖工作站的部分人员及武汉大学来实习的部分大学生
Part of the staffs and interns from Wuhan University, Liangzi Lake workstation, IHB, CAS, July 1957

水生生物博物馆馆藏模式标本
Type specimens in MHBS

部分馆藏鲤科鱼类标本
Parts of the specimens of Cyprinid in MHBS

scientific survey of Tibet's water system was in the 1960s; surveys of the Hengduan Mountain water system and the Xiangjiang River Basin were in the 1970s; Investigations on Tibet Plateau, Xiangxi, Sanjiangyuan, Xinjiang River System, and West Lake in Hangzhou were in 1980s; researches of major lakes in Hubei, Poyang Lake Basin and Dongting Lake Basin, Huanghe and Huaihe River Basin, Dianchi Lake, and Antarctic were in the 1990s; and surveys of the Yarlung Zangbo River Basin and plateau lakes, Linzhi Niyang River, Jinsha River, Chishui River, Jialing River, Three Gorges Reservoir, Danjiangkou Reservoir, eastern and middle routes of South-North Water Transfer Project, the middle and lower reaches of the Han River, lakes in Hubei, Chaohu Lake, Taihu Lake, Puppet Lake, Thousand Islands Lake, West Lake in Hangzhou, Dongqian Lake, lakes in central Yunnan, Qinghai Lake, Bosten Lake, Wuliangsuhai Lake, Hulun Lake, Dongjiang River, Ganjiang River, Baiyangdian Lake, and urban water bodies in Shanghai, Hainan, Qingdao, Tianjin, Jincheng, Shaoguan, Guangzhou, and Shenzhen were from 2000 till now, conducted by several research teams relying on MHBS. The CAS attaches great importance to the work of IHB, CAS in Liangzi Lake. Well-equipped boats called "Liangzi" were offered to help the research work at that time.

馆藏中华鲟标本
The specimen of Chinese Sturgeon in MHBS

馆藏白鲟标本
The specimen of Chinese paddlefish in MHBS

Special Collections

Among the freshwater fishes in China, there are 857 species belonging to cyprinid, which is the largest group of freshwater fishes. Among Cypriniformes, Cyprinidae is the largest family with 579 species, many of which are endemic to China and East Asia. MHBS possesses complete collections of the cyprinid fish of the order cyprinid, which cover more than 90 percent of China's cyprinid fish and reflect the characteristics of the fish fauna in East Asia. Due to the unique collections and high-level studies, MHBS has become the center for Asian freshwater fish diversity research. In recent years, MHBS is gradually expanding from the preservation of freshwater fish specimens to other aquatic specimens and from traditional ichthyology to aquatic biodiversity, resources, and water environment. With the latest research

馆藏白暨豚标本
The specimen of Chinese river dolphin in MHBS

藏了中国 90% 以上的鲤形目鱼类，反映了东亚鱼类区系特点，也成为亚洲淡水鱼类多样性研究的中心。近年来，水生生物博物馆正逐步由淡水鱼类标本保藏向其他水生生物标本保藏拓展，由传统鱼类学向水生生物多样性、资源、水环境学科延伸。设有最新研究成果，如水产养殖种类、转基因鱼类等展示。将把水生生物博物馆打造为水生生物知识库、传播中心。

馆藏特色标本有展示有白鲟、白暨豚等灭绝物种；中华鲟、扬子鳄、长江江豚、胭脂鱼等国家重点保护野生动物；被誉为活化石的矛尾鱼标本；长江鱼类 420 多种，标本馆收藏有近 400 种，约占 90%，是国内收藏最为系统完整的博物馆之一；高原鱼类收藏等。其中最具历史纪念意义的收藏为高原鱼类，从 20 世纪 50 年代开始，在极端艰苦的条件下水生所学者们先后二十余次进入有“世界屋脊”之称的青藏高原进行鱼类生物学的科学考察，发现并命名了大量高原鱼类，阐述其演化过程的三个阶段与青藏高原隆起的三个主要阶段存在着对应关系，丰富了动物地理学研究。

中国科学院南海海洋生物标本馆

基本信息

隶属于：中国科学院南海海洋研究所
地址：广东省广州市海珠区新港西路 164 号
建馆年份：1979 年
收藏类群：浮游生物、软体动物、鱼类等
标本藏量：22.6 万号
馆长：谭烨辉，研究员
联系人：连喜平
电子邮箱：tanyh@scsio.ac.cn, xplian@scsio.ac.cn
联系电话：+86 20 89021159, +86 20 89023198

标本馆简介

中国科学院南海海洋生物标本馆创建于 1979 年，建筑面积 2200 平方米，其中标本库 1500 平方米，科普展厅 700 平方米，是目前国内收藏标本种类和数量繁多并富有热带海洋特色的大型海洋生物标本馆。标

标本馆外景
Exterior view of the MBCSCS

科普展厅
Exhibition hall

information, such as aquaculture species, GMO fish display, MHBS has become a knowledge base of aquatic creatures and a communication center of the scientific trends.

The special collections of MHBS are specimens of extinct species including Chinese paddlefish and Chinese river dolphin; national key protected wild animals including the Chinese Sturgeon, the Chinese alligator, the Yangtze finless porpoise, the Chinese sucker, etc.; the so-called "living fossil" animal coelacanths; nearly 400 out of more than 420 kinds of Yangtze River fishes, which account for 90% fishes in Yangtze River and become one of the most systematic and complete collections in China; and plateau fishes, which are the most significant in historical collections. In the 1950s, under extremely harsh conditions, the scientists of IHB, CAS entered the Qinghai-Tibet Plateau, also called "the roof of the world", for over 20 times and carried out scientific research on the biology of fish in the Qinghai-Tibet Plateau. As a result, a large number of plateau fishes were discovered and named. Then, three stages of their evolution were expounded, which revealed a corresponding relationship between their evolution and the three main stages of the Qinghai-Tibet Plateau uplift and enriched the studies of zoogeography.

Marine Biodiversity Collections of South China Sea (MBCSCS), CAS

Basic Information

Affiliated to: South China Sea Institute of Oceanology, CAS
Address: No.164 West Xingang Road, Haizhu District, Guangzhou, Guangdong, P. R. China
Established in: 1979
Species groups of collection: Plankton, mollusks, fish, etc.
Collection quantity: 226,000
Curator: Dr. TAN Yehui, Professor
Contacts: LIAN Xiping
E-mail: tanyh@scsio.ac.cn, xplian@scsio.ac.cn
Tel: +86 20 89021159, +86 20 89023198

Introduction

Marine Biodiversity Collections of South China Sea (MBCSCS), CAS was founded in 1979, with a construction area of 2,200 m^2, of which the specimen library is 1,500 m^2, and the exhibition hall is 700 m^2. It is a large marine biological collection with a wide variety and quantity of specimens and rich tropical marine characteristics of China. MBCSCS contains more than 220,000 marine biological specimens, including specimens collected since the establishment of the South China Sea Institute of Oceanology, CAS in 1959 and exchanges and presented by international friends and other museums. The collection mainly focus on marine life in the South China Sea and the Indo-Western Pacific waters. The specimens not only provide support for tropical marine biodiversity-related research, but also provide scientific basis for understanding the South China Sea biological resources and declaring sovereignty. There is a group with have 6 researchers now, mainly engaged in taxonomy and phylogenetic research of marine organisms such as marine plankton, mollusks and marine fish. More than 50 books were published such as "Fauna Sinica Crustaceans Amphiptera, Amphipoda", "Fishes from Nansha Islands to South China Coastal waters", "Research on Seagrass in the South China Sea", "Species list of Biodiversity in the Nansha Islands", "Identification Manual of Protective wild water Animals", "Regional Oceanography of China seas-Biological Oceanography", "Research on Red Tide in the South China Sea", "Red Tide in the South China Sea", "Sanya Bay Ecological Environment and Biological Resources", "Research on Marine Taxonomy and Biogeography in the Nansha Islands and its adjacent waters", "Zooplankton of China Seas. Volume 1-2", "Fishes from Nansha Islands to South China Coastal

海洋科学走廊
Marine science gallery

出版专著
Published monographs

本馆收藏了1959年中国科学院南海海洋研究所建所以来采集以及来自馆际交流和国际友人赠送的海洋生物标本逾22万号，门类齐全。立足南海、面向印度-西太平洋是标本馆的收藏定位，馆藏标本不仅为热带海洋生物多样性相关研究提供支撑，同时也为摸清南海生物资源以及宣示主权提供最有力的科学依据。标本馆现有研究人员6人，主要从事海洋浮游生物、软体动物、鱼类等海洋生物的分类学和系统进化研究。标本馆编写和参与编写专著等50余本，如《中国动物志 甲壳动物亚门 端足目 蛾亚目》、《珊瑚礁鱼类——西沙群岛及热带观赏鱼》、《中国南海海草研究》、《南沙群岛海区生物多样性名典》、《水生野生保护动物识别手册》、《中国区域海洋学——生物海洋学》、《中国南海赤潮研究》、《南中国海赤潮》、《三亚湾生态环境与生物资源》、《南沙群岛及其邻近海域海洋生物分类区系和生物地理研究》、*Zooplankton of China Seas. Volume 1-2*、《南沙群岛至华南沿岸的鱼类》（一、二）、《中国海洋生物多样性的保护》、《台湾周围水域和南海北部浮游动物：种类与分布》等。

收藏概况

南海海洋生物标本馆立足南海、面向印度-西太平洋，收藏了自1950年以来的从海藻和海草到海洋大型哺乳动物等门类齐全的热带海洋生物标本22.6万号，主要类群从海洋藻类、海草、海洋无脊椎动物、海洋鱼类、爬行类到大型鲸豚类动物标本，达5700余种。

特色收藏

馆藏丰富的南海珊瑚礁、海草床和红树林等生态系统的生物标本，造礁石珊瑚、珊瑚礁鱼类以及其他礁栖生物和潮间带生物标本，凸显了标本馆在收集南海热带、亚热带海洋生物多样性热点区域标本方面的特色和优势；另外馆内不乏国家一级保护动物中华白海豚，国家二级保护动物海龟、玳瑁和砗磲，广东省

浸制标本库
Liquid immersion specimens room

珊瑚标本
Coral specimens

Waters 1-2", "Protection of marine biodiversity in China", "Zooplankton in the waters around Taiwan and the northern South China Sea: species and distribution", etc.

Overview of Collections

MBCSCS mainly focuses on marine organisms in the South China Sea and the Indo-Western Pacific water. The collection of tropical and subtropical marine features with a wide range from seaweed and seagrass to large marine mammals since 1950, more than 226,000 specimens in the collection. The main groups include marine algae, seagrass, marine invertebrates, marine fish, reptiles, and large cetacean specimens, over 5,700 species.

Special Collections

The collection is rich in biological specimens from South China Sea coral reefs, seagrass beds, mangroves and other ecosystems in the tropical and subtropical regions, especially reef building stone corals, reef fish and other reef organisms and intertidal organisms, which highlights the characteristics and advantages of the collecting specimens from the hot spots of tropical and subtropical marine biodiversity in the South China Sea; in addition, there are preserved specimens of national protected animals of China, such as national first class protected animal Chinese White Dolphin, national second

中华白海豚
Sousa chinensis

海龟
Chelonia mydas

玳瑁
Eretmochelys imbricata

大砗磲
Tridacna gigas

级保护动物中国鲎，以及海洋爬行类有毒生物海蛇等珍稀物种标本。馆藏国内外友人赠送的珍稀柳珊瑚和菲律宾海产贝类标本，大多数为标本馆首次记录，具有较高的科研和观赏价值。标本馆对南海和印度洋海域海洋生物标本的连续采集积累，对我国海洋长远发展具有战略性意义。

热带生物特色标本

馆藏丰富的热带海洋生物标本，如干制石珊瑚标本 8594 号、浸制柳珊瑚和软珊瑚标本 1650 号；此外还有我国海域分布的种类齐全的砗磲科、海龙科及海蛇科标本，这些标本为热带海洋生物资源和生物多样性研究提供科研支撑。

珍稀濒危物种标本

馆藏国家一级保护动物中华白海豚标本 2 号、江豚标本 3 号；国家二级保护动物海龟标本 7 号、玳瑁标本 7 号、大砗磲标本 6 号；这些标本都具有极大的科研价值和科普价值。

5）古生物类馆

中国科学院古脊椎动物与古人类研究所标本中心

基本信息

隶属于：中国科学院古脊椎动物与古人类研究所
地址：北京市西城区西直门外大街 142 号
建馆年份：1956 年
收藏类群：脊椎动物化石、人类化石、旧石器时代文化遗物、现代脊椎动物骨骼和现代人骨骼
标本藏量：23.5 万件
馆长 / 联系人：李淳，研究员
电子邮箱：lichun@ivpp.ac.cn
联系电话：+86 10 88369438

标本馆简介

中国科学院古脊椎动物与古人类研究所标本中心始建于 1956 年，初名标本室，是在中国古脊椎动物学和研究所的开创者杨钟健先生的亲切关怀与悉心指导下成立的。1983 年更名为标本馆，2018 年更名为标本中心。其前身可追溯至 1922 年农矿部地质调查所在地质矿产陈列馆内增设的古生物化石展室，距今已近百年。标本中心是研究所进行标本管理、收藏、修复、保护、数字化建设和开发利用的机构，是研究所的重要组成部分。中心下设馆藏部、技术服务部和数字可视化研发部，其中馆藏部现有职员 7 人。

丰富的藏品是研究所创新发展的原动力和有力保障。在研究所几代同仁的不懈努力下，中心现有各类藏品 24 万件，主要包括脊椎动物化石、人类化石、旧石器时代文化遗物、现代脊椎动物骨骼和现代人骨骼五大类。特别是在古脊椎动物和古人类标本的收藏方面，标本中心是亚洲该领域收藏门类最齐全、数量最丰富的机构，在国际上也享有盛誉。为促进学科立足国际前沿、探索生物演化、追寻人类起源、传播科学知识做出了积极贡献。

中心编著的主要著作有《山海遗珍》（2019），该书收录标本四百多件，对“从鱼到人”波澜壮阔的生命演化历程和旧石器时代人类社会的技术演变模式予以介绍及展示。

class protected animals sea turtle, hawksbill, giant clam as well as specimens of Guangdong provincial protected animal Chinese horseshoe crab and marine reptile sea snake etc. Most of the rare gorgonians and Philippine marine seashell specimens donated by the other country are the first recorded for the collection and have high scientific research and ornamental value. The collection and accumulation of marine biological specimens from the South China Sea and the Indian Ocean is of strategic significance to the long-term development of China's oceans.

Biological specimens of tropical

The collection is rich in tropical marine biological specimens, such as 8,594 dry stone coral specimens, 1,650 liquid immersed gorgonian and soft coral specimens; in addition, there are a wide range of species of Tridacnidae, Syngnathidae and Hydrophiidae from China's sea area, which provide scientific research support for the research of tropical marine biological resources and biodiversity.

Specimen of rare and endangered species

The collection includes 2 specimens of *Sousa chinensis*, 3 specimens of *Neophocaena asiaeorientalis*, 7 specimens of *Chelonia mydas*, 7 specimens of *Eretmochelys imbricata* and 6 specimens of *Tridacna gigas*, all of which are of great scientific and popular science value.

5) Paleontology Collection/Museum

Collection Center of Institute of Vertebrate Paleontology and Paleoanthropology (IVPP), CAS

Basic Information

Affiliated to: Institute of Vertebrate Paleontology and Paleoanthropology (IVPP), CAS
Address: No.142 Xizhimenwai Street, Xicheng District, Beijing, P. R. China
Established in: 1956
Species groups of collection: Vertebrate fossils, human fossils, Paleolithic cultural relics, modern animal skeletons and modern human skeletons
Collection quantity: 235,000
Curator/Contacts: Dr. LI Chun, Professor
E-mail: lichun@ivpp.ac.cn
Tel: +86 10 88369438

Introduction

Collection Center of IVPP was founded in 1956 as Collection House, which was established under the careful guidance of C. C. Young, founder of Chinese vertebrate paleontology and IVPP. It was renamed Collection Museum in

古脊椎动物与古人类研究所标本中心
Collection Center, IVPP

标本中心馆藏部
Department of storing, Collection Center

标本观察室
Specimen observation room

《山海遗珍》
Treasures from the Mountains and Seas

收藏概况

标本中心目前有证可查最早收藏的标本，是地质调查所顾问、瑞典地质学家安特生于 1918 年 12 月在河南新安采集的晚上新世哺乳动物化石。1929 年农矿部地质调查所新生代研究室（今研究所前身）成立，当年便在北京周口店发现了震惊世界的“北京人”头盖骨化石。第二次世界大战期间，研究所前辈不畏艰辛，坚持不懈采集标本，在云南、四川等地都留下了他们的足迹，为今日标本中心的海量藏品打下了坚实的基础。

中华人民共和国成立后，随着我国古生物事业的蓬勃发展，标本中心收藏的标本在时空跨度和物种门类方面都有了显著的提高。现有的藏品门类涵盖了鱼类、两栖类、爬行类、鸟类、哺乳类动物以及人类，从寒武纪到当今的全新世，范围覆盖全国。特别是我国境内一些国际古生物学研究热点区域的标本，如燕辽生物群、热河生物群、云南曲靖古生代鱼类、云贵三叠纪海生爬行动物、甘肃临夏盆地晚新生代哺乳动物群和青藏高原冰期哺乳动物群等，每年都会吸引大量国际古生物学者前来交流、学习。

特色收藏

众多藏品中，有许多具有重要的科学研究价值和历史纪念意义。相关标本的研究文章多次发表在 *Nature*、*Science* 等国际顶级刊物，为研究脊椎动物的起源和演化、人类的起源与行为变迁、生物 - 环境协

浙江曙鱼
Shuyu zhejiangensis

孟氏中生鳗
Mesomyzon mengae

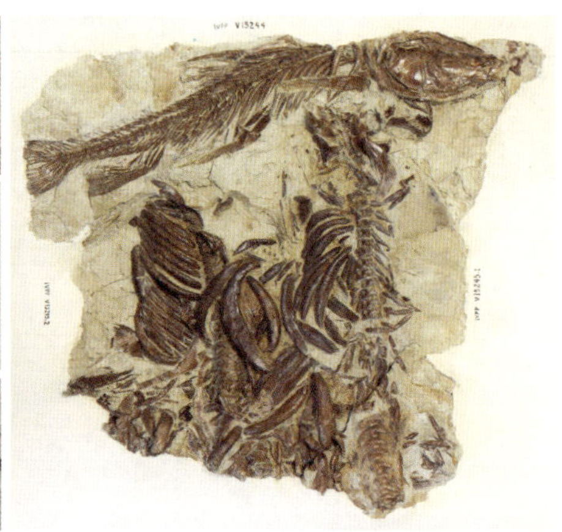
伍氏献文鱼
Hsianwenia wui

1983 and Collection Center in 2018. It can be traced back to the palaeontological exhibition room added in the Geological and Mineral Museum, Geological Survey of China (GSC) in 1922, nearly a hundred years ago. As an important part of IVPP, Collection Center is responsible for specimen management, storing, preparation, protection, digital construction and utilization, and consists of three departments: storing, technical service and R&D in digital visualization, among which there are seven staff members in storing department.

Abundant collections are the driving force and strong guarantee for innovation and development of IVPP. With the unremitting efforts of staff members in IVPP, the center collected about 240,000 specimens currently, consisted with vertebrate (including human) fossils, Paleolithic artifacts, and modern vertebrate (including human) skeletons. The center is Asia's top collection in diversity and quantity of vertebrate and human fossils, and also very famous in the world. It has made positive contributions to promoting the discipline to be at the international forefront, exploring biological evolution, tracing the human origin, and disseminating scientific knowledge.

Treasures from the Mountains and Seas is a representative book compiled by the center in 2019, and more than 400 specimens were collected in this book, which introduces and displays the magnificent evolution "from fish to human" and human technological mode change in the Paleolithic age.

Overview of Collections

At present, the earliest documented collection is a late Pliocene mammal fossil collected by J. G. Anderson, a Swedish geologist and consultant of GSC, in Xin'an, Henan, December 1918. When the Cenozoic Research Laboratory (the predecessor of IVPP), GSC was established in 1929, the skull fossil of Peking Man, which shocked the world, was discovered in Choukoutien (Zhoukoudian), Beijing. During World War II, the predecessors of IVPP feared hardships and persisted in collecting specimens, leaving their footprints in Yunnan, Sichuan and other places, laying a solid foundation for the massive collections of the center.

After the founding of the People's Republic of China, with the vigorous development of paleontology in China, specimens collected by the center have significantly improved in terms of space-time and species. Now, collections came from Cambrian to Holocene all over China, covering fish, amphibian, reptile, bird, mammal and human. In particular, specimens from some international palaeontological hotspots in China, such as Yanliao Biota, Jehol Biota, Paleozoic fish in Qujing, Yunnan, Triassic marine reptiles in Guizhou and Yunnan, late Cenozoic Mammalian Faunas in Linxia Basin, Gansu and Ice Age Mammalian Fauna in Qinghai-Tibet Plateau, attract a large number of international paleontologists to communicate and study every year.

Special Collections

Many collections have important scientific value and historical significance. Related researches have been published in *Nature*, *Science* and other top international journals for many times, which provides strong support for research on origin and evolution of vertebrates, origin and behavior change of human, and coevolution of life and environment.

Shuyu zhejiangensis, Silurian Period, Changxing, Zhejiang. This specimen provides key fossil evidence to solve the mystery of the origin of vertebrate jaws. It is considered a missing link in life evolution as important as *Tiktaalik* and *Archaeopteryx*.

Mesomyzon mengae, Early Cretaceous, Ningcheng, Inner Mongolia. It fills the blank of lamprey fossil record since Carboniferous, adds to the knowledge of lampreys' evolutionary history. *Mesomyzon* not only documents larval stages of fossil lampreys but also indicates a unique three-phased life cycle in modern lampreys emerged no later than the early Cretaceous.

Hsianwenia wui, Pliocene Epoch, Qaidam Basin, Qinghai. It reflects fish's dramatic physiological adaptation to severe environmental pressure and witnesses Qaidam Basin's aridification, and provides a convincing link between environmental changes on the Tibetan Plateau and biological adaptation of its inhabitants.

Guiyu oneiros, Silurian Period, Qujing, Yunnan. *Guiyu* is the oldest and best-preserved fossil osteichthyans, possibly even the oldest near-complete gnathostome (jawed vertebrate). It is also the only completely preserved jawed vertebrate of the Silurian Period, pushing the oldest, almost complete fossil record of bony fish by about 8 million years, filling the morphological gap between osteichthyans and other jawed vertebrates.

梦幻鬼鱼
Guiyu oneiros

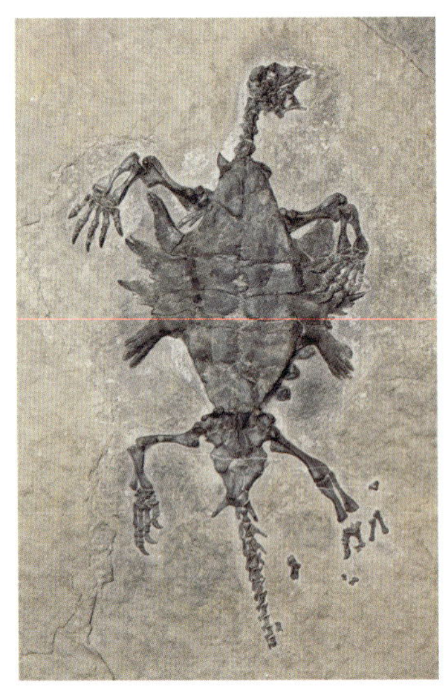

半甲齿龟
Odontochelys semitestacea

同演化提供了有力支持。

浙江曙鱼，志留纪，浙江长兴。为解开脊椎动物颌的起源之谜提供了关键的化石证据，被认为是与提塔利克鱼、始祖鸟等一样重要的生命演化缺失环节。

孟代中生鳗，早白垩世，内蒙古宁城。填补了自石炭纪以来七鳃鳗化石纪录的空白，同时增进了对七鳃鳗演化历史的认识。首次记录了化石七鳃鳗的幼体和变态期幼体特征，并且显示现代七鳃鳗独特的三期生命史早在距今1.25亿年前的早白垩世晚期就已成型。

伍氏献文鱼，上新世，青海柴达木盆地。不仅显示了鱼类对极端环境的生理适应能力，也是柴达木盆地干旱化过程的见证，并提供了环境变化与生物适应之间令人信服的案例。

梦幻鬼鱼，志留纪，云南曲靖。目前全球最古老且保存完整的硬骨鱼化石，甚至有可能是最古老的有颌脊椎动物化石。它也是志留纪唯一完整保存的有颌类，将硬骨鱼化石纪录向前推进了约800万年，填充了硬骨鱼类与其他有颌类之间的形态学鸿沟。

半甲齿龟，晚三叠世，贵州关岭。世界上最原始的龟类之一，也是填补达尔文演化理论空缺的重要物种，其完整的腹甲和缺失的背甲与龟类胚胎发育过程相互呼应，为破解200年来令人费解的龟类起源之谜提供了至关重要的化石证据。

天山哈密翼龙，早白垩世，新疆哈密。大量雌雄天山哈密翼龙个体、蛋和胚胎共同在化石中富集保存，包括全球第一枚三维立体保存的翼龙蛋和第一个三维立体保存的翼龙胚胎。这一发现被认为是“翼龙研究

天山哈密翼龙
Hamipterus tianshanensis

许氏禄丰龙
Lufengosaurus huenei

奇异帝龙
Dilong paradoxus

难逃泥潭龙
Limusaurus inextricabilis

Odontochelys semitestacea, Late Triassic, Guanling, Guizhou. It is one of the most primitive turtles in the world and an important species to fill gaps in Darwinian evolution theory. Its complete plastron and missing carapace echo the development process of turtle embryos. It provides vital fossil evidence for the mysterious origin of turtles lasting for two hundred years.

Hamipterus tianshanensis, Early Cretaceous, Hami, Xinjiang. A large number of male and female *Hamipterus tianshanensis* fossil individuals, eggs, and embryos have been preserved, including the world's first three-dimensionally preserved pterosaur egg and embryo. This discovery is considered to be "one of the most exciting discoveries in the 200 years of pterosaur research".

顾氏小盗龙
Microraptor gui

Lufengosaurus huenei, Early Jurassic, Lufeng, Yunnan. It is the first complete dinosaur fossil skeleton discovered, excavated, prepared, studied, named, and mounted by Chinese researchers, and is known as "China's first dinosaur", which was found in 1938.

Dilong paradoxus, Early Cretaceous, Beipiao, Liaoning. The discovery proves that the early ancestor type of *Tyrannosaurus* was small. The specimen covered with feathers proves that theropod dinosaurs had common ancestors with birds.

Limusaurus inextricabilis, Late Jurassic, Junggar Basin, Xinjiang. This specimen provides the most powerful evidence for the mystery of finger homology between birds and non-bird dinosaurs.

Microraptor gui, Early Cretaceous, Chaoyang, Liaoning. It is the first four-winged dinosaur in the world. It provides the most direct fossil evidence for the "arboreal origin theory" of bird's flight.

胡氏耀龙
Epidexipteryx hui

李氏源掠兽
Origolestes lii

圣贤孔子鸟
Confuciusornis sanctus

胡氏辽尖齿兽
Liaoconodon hui

五尖张和兽
Zhangheotherium quinquecuspidens

200 年来最令人激动的发现之一”。

许氏禄丰龙，早侏罗世，云南禄丰。发现于 1938 年，这是第一具由中国人发现、发掘、修复、研究、命名并装架的完整恐龙化石骨架，被誉为“中国第一龙”。

奇异帝龙，早白垩世，辽宁北票。证明了霸王龙类早期的祖先类型是小型的。同时，化石标本覆盖着羽毛的事实再一次证明兽脚类恐龙与鸟类有共同的祖先。

难逃泥潭龙，晚侏罗世，新疆准噶尔盆地。为揭开鸟类与非鸟恐龙手指同源性之谜等科学问题提供了最有力的证据。

顾氏小盗龙，早白垩世，辽宁朝阳。世界上最早发现的有确凿证据的四翼恐龙。它的发现为鸟类飞行的“树栖起源说”提供了最直接的化石证据。

胡氏耀龙，中侏罗世，内蒙古宁城。带状尾羽在兽脚类恐龙中的首次发现，为揭开鸟类起源、飞行起源、羽毛起源之谜提供了新的证据。

圣贤孔子鸟，早白垩世，辽宁北票。是比始祖鸟进步的鸟类，也是最早长有与现生鸟类类似的无齿角质喙的鸟类之一。

李氏源掠兽，早白垩世，辽宁北票。其多件标本首次展示了一个关键特征，即听骨与麦克尔软骨之间无骨质链接，代表了哺乳动物演化中听觉与咀嚼模块分离的关键节点，弥合了过渡型中耳和典型哺乳动物中耳在演化过程中表型特征的空缺。

胡氏辽尖齿兽，早白垩世，辽宁建昌。此标本首次呈现了较完整且相互关联的外鼓骨、锤骨和砧骨，为早白垩世哺乳动物中耳由附着于下颌的“听小骨”，到完全的哺乳动物听小骨的过渡演化提供了形态学证据。

五尖张和兽，早白垩世，辽宁北票。这是世界上第一件较完整的对齿兽类标本，也是我国第一具中生代哺乳动物骨架。它的颅基部和下颌形态特征表明，其听觉器官的发育在哺乳动物演化历史中仍处在比较原始的阶段。它的发现揭开了辽西原始哺乳动物研究的序幕。

Epidexipteryx hui, Middle Jurassic, Ningcheng, Inner Mongolia. This is the first discovery of elongated ribbon-like tail feathers in theropod dinosaurs, providing new evidence for the bird, flight, and feather origin.

Confuciusornis sanctus, Early Cretaceous, Beipiao, Liaoning. It is one of the first birds with toothless horny beaks like modern birds, and it has primitive and derived characters that means more "advanced" than *Archaeopteryx*.

Origolestes lii, Early Cretaceous, Beipiao, Liaoning. Specimens of this species display auditory bones, including the surangular, have no bone contact with the ossified Meckel's cartilage. This configuration probably represents the initial morphological stage of the definitive mammalian middle ear showing that hearing and chewing apparatuses have evolved in a modular fashion.

Liaoconodon hui, Early Cretaceous, Jianchang, Liaoning. The specimen displays the first unambiguous and correlate ectotympanic, malleus and incus of an early Cretaceous eutriconodont mammal form, which provides morphological evidence for the transition of the middle ear from the "ossicle" attached to the mandible to the complete ossicle in Early Cretaceous mammals.

Zhangheotherium quinquecuspidens, Early Cretaceous, Beipiao, Liaoning. This is the first symmetrodont mammal specimen discovered worldwide and is the first Mesozoic mammal skeleton ever found in China. Moreover, its basal part of the skull and mandible shows auditory organ is still in a primitive stage during mammal evolution. The discovery of the fossil unveils the beginning of primitive mammal research in West Liaoning, China.

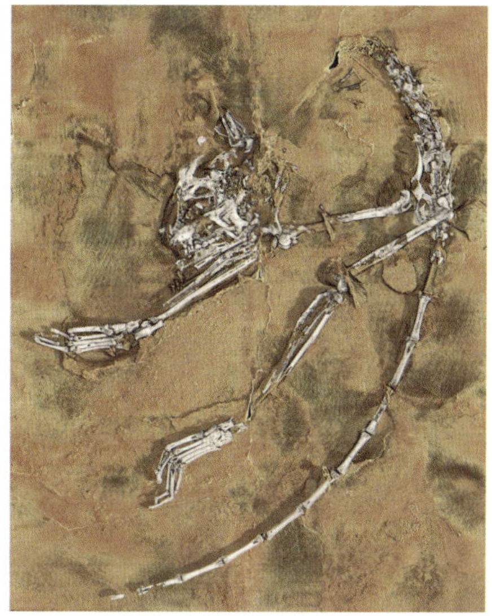

阿喀琉斯基猴
Archicebus achilles

沙拉木伦始巨犀
Juxia sharamurenensis

西藏披毛犀
Coelodonta thibetana

北京直立人
Homo erectus pekinensis

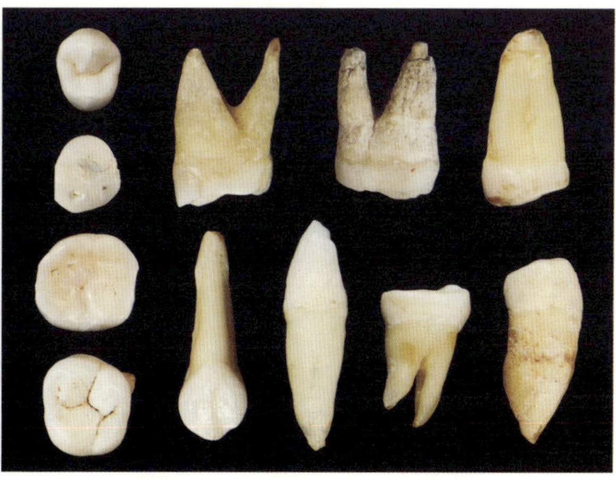

道县古人类牙齿化石
Teeth fossils of Daoxian hominid

阿喀琉斯基猴，早始新世，湖北松滋。这是目前已知最古老且保存最完整的灵长类化石骨架，为确定类人猿与其他灵长类的分异时间和早期演化模式提供了关键证据。

沙拉木伦始巨犀，中始新世，内蒙古四子王旗。小型、原始的巨犀类，其骨骼特征表明它们很可能动作灵活，并有较高的奔跑速度。这是现存最完整、最古老的巨犀骨架。

西藏披毛犀，中上新世，西藏札达。这些化石是已知最早和最原始的披毛犀，它的发现表明披毛犀起源于青藏高原。披毛犀在上新世中期以青藏高原为"训练营"对寒冷气候进行预适应，并在距今 280 万年前随着冰期寒冷气候的扩展，从青藏高原扩散到其他地区。

北京直立人，中更新世(距今 60 万 -20 万年)，北京周口店。这是目前仅存的头盖骨化石，发现于 1966 年。"北京人"是北京地区最早的拓荒者，它的发现奠定了人类演化进程中直立人阶段的存在，是古人类学上最重要的发现之一。

道县古人类牙齿化石，晚更新世（距今 12 万 -8 万年），湖南道县。共计 47 枚牙齿化石，发现于 2011-2013 年。道县古人类被认为是东亚目前已知最早的具有完全现代形态的人类，为东亚现代人起源和演化提供了新的线索。

中国科学院南京地质古生物研究所科学传播中心化石标本馆、南京古生物博物馆

基本信息

隶属于：中国科学院南京地质古生物研究所
地址：中国江苏省南京市北京东路 39 号
建馆年份：1998 年
收藏类群：古生物化石
标本藏量：46.2 万件
馆长：王永栋，研究员
联系人：袁文伟
电子邮箱：wwyuan@nigpas.ac.cn
联系电话：+86 25 83282289

标本馆模式标本楼
Building of type specimen collection

标本馆简介

中国科学院南京地质古生物研究所科学传播中心化石标本馆成立于 1998 年，是在 1928 年成立的中央研究院地质研究所标本室的基础上发展起来的，其前身是中国科学院南京地质古生物研究所标本室。2018 年起隶属于中国科学院南京地质古生物研究所科学传播中心。馆藏标本约 46.2 万件，其中 16.5 万件模式标本。

标本馆历史悠久，藏品丰富，既有我国古生物学的早期开拓者李四光、葛利普、孙云铸、黄汲清、尹赞勋、赵亚曾、斯行健等教授采集研究的标本，更汇集了瑞典、德国、美国、英国、加拿大、澳大利亚、波兰、捷克、苏联、伊朗、日本等国家交流

Archicebus achilles, Early Eocene, Songzi, Hubei. This fossil specimen is the most basal and complete primate skeleton providing key support for the dichotomy of anthropoids and other primate and early evolution mode.

Juxia sharamurenensis, Middle Eocene, Siziwang Banner, Inner Mongolia. This fossil represents a small-sized and primitive giant rhinoceros. They are likely to be agile and have a high running speed. This skeleton is the most complete and oldest paraceratheres in the world.

Coelodonta thibetana, Middle Pliocene, Zanda, Tibet. These fossils represent the oldest primitive woolly rhino in the world and suggest that their original place is the Qinghai-Tibet Plateau. Woolly rhinos dispersal from the Qinghai-Tibet Plateau to other places with the expansion of the Ice Age's cold climate 2.8 million years ago, after they got preadaptation for cold climate from the Qinghai-Tibet Plateau, which played as a training ground during the middle Pliocene.

Homo erectus pekinensis, Middle Pleistocene (600,000-200,000 yBP), Zhoukoudian, Beijing. This is the only skull fossil in the collection, found in1966. Peking Man is the earliest pioneer in Beijing, whose discovery established the existence of *Homo erectus* in the process of human evolution, and is one of the most important discoveries in paleoanthropology.

Teeth fossils of Daoxian hominid, Late Pleistocene (120,000-80,000 yBP), Daoxian, Hunan. A total amount of 47 teeth fossils were found in 2011-2013. Daoxian hominid is considered the earliest fully modern humans in morphology in East Asia, which provides a new clue for the origin and evolution of modern humans in East Asia.

Fossil Collection Department of the Science Communication Center of Nanjing Institute of Geology and Palaeontology, Nanjing Museum of Palaeontology, CAS (NIGPAS)

Basic Information

Affiliated to: Nanjing Institute of Geology and Palaeontology, CAS
Address: No.39, Beijing East Road, Nanjing, Jiangsu Province, P. R. China
Website: http://159.226.74.248/, www.nmp.ac.cn
Established in: 1998
Species groups of collection: Palaeontological fossil
Collection quantity: 462,000
Curator: Dr. WANG Yongdong, Professor
Contacts: YUAN Wenwei
E-mail: wwyuan@nigpas.ac.cn
Tel: +86 25 83282289

Introduction

The Fossil Collection Department of the Science Communication Center of Nanjing Institute of Geology and Palaeontology, Nanjing Museum of Palaeontology, CAS was founded in 1998, which was developed on the basis of collection house of Institute of Geology, Academia Sinica established in 1928. Its predecessor is the herbarium of Nanjing Institute of Geology and Palaeontology, CAS. Since 2018, it has been under the Scientific Communication Center of Nanjing Institute of Geology and Palaeontology, CAS. There are about 462,000 specimens in the collection including 165,000 type specimens.

南京古生物博物馆
Nanjing Museum of Palaeontology

The department has a long history and rich collections. It not only includes the specimens collected by professors LI Siguang, Grabau, SUN Yunzhu, HUANG Jiqing, YIN Zanxun, ZHAO Yazeng, SI Xingjian, etc., the early pioneers of Paleontology in China, but also collections from the exchanges and cooperation researches of dozens of countries, including Sweden, Germany, the United States, the United Kingdom, Canada, Australia, Poland, the Czech Republic, Soviet Union, Iran, Japan, etc. After several generations of hard work, the Collection Department (fossil herbarium) has accumulated more

合作研究的标本。经过几代人的艰苦努力，标本馆已积累了3000余篇（册）研究成果的标本，地域分布广，时代范围宽，化石门类全，是国内外有关地质古生物学科研、生产、教育部门在科研和实际工作中参考对比和引用的重要基本资料。

南京古生物博物馆隶属于中国科学院南京地质古生物研究所，由中国科学院和江苏省人民政府共建，是一个集展览、收藏、研究和教育为一体的现代化博物馆，是自然科学普及和教育的重要基地，具有科学性、知识性、观赏性和趣味性。2005年12月31日正式对外开放。博物馆的前身为南京古生物所“史前生物展”古生物陈列室。此陈列室于2000年被中国古生物学会授牌为首批“中国古生物学会全国科普教育基地”之一。

南京古生物博物馆场馆建筑面积8000平方米，展厅面积4200平方米，分上下两层，馆内展陈以古生物化石为本，尤以古无脊椎动物、古植物和微体古生物为主，是目前世界上最大的古生物专业博物馆之一。展示的精品化石标本980件，馆藏标本1200余件。内容包括一个“生命的进化”主题展和14个专题展。

收藏概况

标本馆收藏的标本都是中国古生物工作者长期以来在野外考察中采集的各类化石标本。本馆还继承和保管了1928年成立的中央研究院地质研究所标本室的标本。本馆主要收藏科研人员正式发表的学术文章中所列的模式标本，总数已经超过16万件，标本采集自全国各地。此外，本馆近年来也开始收藏科研人员尚未正式发表的在研标本。

南京古生物博物馆以古生物化石为本，是目前世界上最大的古生物专业博物馆之一，馆藏标本1200余件，产自全国各地。博物馆的标本收藏前身来自1952年成立的南京古生物所“史前生物展”的古生物陈列室的标本。目前，馆里收藏的化石以古无脊椎动物、古植物和微体古生物为主，藏品丰富，展品精美，其中尤以云南“澄江动物群”、包括“中华龙鸟”在内的辽西“热河生物群”、贵州“关岭生物群”等化石标本最为珍贵，堪称国宝级的化石精品。

这些标本分别在博物馆的各个展厅收藏或者展示，包括生物进化主题展厅的门厅、上山之路、化石奥秘、地球的起源、前寒武纪、古生代、中生代、新生代的地质时代为轴线。古生物化石的专题展厅，包括澄江动物群、恐龙天地、微观世界、生物登陆、无脊椎动物化石大观、二叠纪生物大灭绝、古植物园、海百合化石墙、热河生物群、鸟的起源与演化、恐龙时代的海洋、南京地史演变、南京直立人和“我从哪里来？”。

模式标本库
Type specimen collection

在研标本楼外景
Exterior of the building of specimen under research

than 3,000 volumes of published research results, with wide geographical distributions, wide age ranges and complete fossil categories. It is an important reference, comparison and reference basic data for relevant scientific research, production and education departments at home and abroad in scientific research and practical work for geology and palaeontology.

Nanjing Museum of Palaeontology subordinates to the Nanjing Institute of Geology and Palaeontology, CAS, was jointly sponsored by CAS and the People's Government of Jiangsu Province. It is a modern museum of palaeontology with integrative functions in exhibition, collection, research and education, and opens to the whole public, especially the youngsters. The museum was established and operated from December 31, 2005. The museum was originated from the Exhibition Hall of "Prehistorical Life Exhibition" founded by NIGPAS. It was appointed as one of the National Science and Education Base approved by the Palaeontological Society of China in 2000.

The total construction area of the museum is about 8,000 m^2, including an exhibition area of about 4,200 m^2, which comprises two floors. Nanjing Museum of Palaeontology is one of the largest palaeontology museums in the world and hosts rich and diverse fossil exhibitions covering invertebrate palaeontology, palaeobotany and micropalaeontology. The museum has collections of over 1,200 specimens, including 980 well-preserved fossils. The exhibition consists of a topic hall of "Evolutionary History of Life" and other 14 special exhibits.

Overview of Collections

The specimens housed in the department include all kinds of fossil specimens collected by Chinese paleontologists in the field for a long time. The department also inherits and keeps the specimens of the Geological Research Institute, Academia Sinica, which was established in 1928. The department mainly houses more than 160,000 type specimens listed in academic articles officially published by scientific researchers. These collections came from all localities and horizons over the country. In addition, in recent years, the department also collected the specimens that are under research and have not yet been officially published by researchers.

The Nanjing Museum of Palaeontology is exclusively to the exhibition of fossils of ancient life in deep time. It is also one of the largest specialized museums for the topic of paleontology in the world. The museum owns more than 1,200 items in collections from all over the country. The museum was originated from the exhibition hall named "Prehistorical Life Exhibition" which founded in 1952 by NIGPAS. It majors in invertebrate palaeontology, palaeobotany and micropalaeontology. Among them, the most remarkable exhibit issues are national treasures of fine fossils from "Chengjiang fauna" in Yunnan, *Sinosauropteryx prima* from "Jehol biota" in western Liaoning, and sea reptiles from "Guanling biota" in Guizhou.

The museum exhibition covers two parts, biological evolution section and fossils of ancient life section. The former section includes: The Introduction Hall, The Road to Climbing, Mystery ff Fossils, The Origin of Earth, Pre-Cambrian, Paleozoic, Mesozoic and Cenozoic. The special exhibit sections of ancient life include: Chengjiang Biota, Dinosaur World, Microscopic Fossil World, Life onto Land, Variety of Invertebrate Fossils, Permian Mass Extinction, Palaeobotanic Garden, Crinoid fossil wall, Jehol Biota, Origin and Evolution of Birds, Ocean in the Age of Dinosaurs, Geological Evolution of the Nanjing Area, Nanjing *Homo erectus*, and Where do I come from.

The exhibition contents are rich and colorful, delivering plenty of enjoyable information about the origin, diversity and evolution of ancient life. The museum offers a wide variety of creative, fun, and educational programs that bring the prehistoric past to life to the public, including a large number of rare specimens, beautiful pictures, simulation models and landscape restorations. It fully demonstrates the mystery of fossils and the diversity of paleontology, reemerges the history of biological evolution and its co-evolution with the environment, reveals the major events in the process of biological evolution, and also reveals the major discoveries and research achievements of Chinese paleontology, inspiring and enhancing the public's interest and understanding of natural history.

澄江生物群展厅及标本

Exhibition hall and specimen of Chengjiang biota

展览内容丰富多彩，提供了大量关于古代生命起源、多样性和进化的有趣信息。这些展览以大量珍稀标本、精美图片、仿真模型和景观复原，并提供各种创意、娱乐和教育项目，让公众了解史前历史。充分展示了化石的奥秘和古生物的多样性，再现了生物进化及其与环境协同演化的历史，揭示了生物进化过程中的重大事件，同时也展示了中国古生物学界的重大发现和研究成果，启发、增进了公众对自然历史的兴趣和认识。

特色收藏

本馆收藏了大量中国古无脊椎动物和古植物化石的模式标本。本馆收藏的最早的模式标本有葛利普在20世纪20年代发表的中国南方泥盆纪腕足动物化石和李四光在20世纪20年代研究的中国北方蜓类化石薄片。

此外，本馆于1984年开始就致力于澄江动物群标本的收藏。在云南澄江县帽天山被发现的寒武纪早期澄江动物群是一个举世罕见的化石宝库。澄江动物群中化石丰富且保存精美，生动再现了距今5亿2000万年前海洋生物的真实面貌，充分显示出寒武纪早期生物的多样性，将绝大多数现生动物门类的演化历史追溯到寒武纪开始，为揭示生物“寒武纪大爆发”的奥秘提供了极其珍贵的证据，因而在国际上被誉为“20世纪最惊人的科学发现之一”。本所于1998年在澄江建立了集野外研究、国际学术交流和科普宣传等功能为一体的澄江古生物研究站。除此之外，一些特色收藏的标本还包括：热河生物群的中华龙鸟模式标本，辽宁古果模式标本，这些成果发表在英国的 *Nature* 和美国的 *Science* 杂志上，其中后者还是 *Science* 的封面成果。还包括世界上最早的银杏植物化石——义马银杏，以及燕辽生物群、道虎沟生物群和缅甸琥珀生物群的标本。

南京古生物博物馆依托南京古生物所标本馆（拥有超过46万件化石标本），展览了许多特色展品，如澄江动物群化石、热河生物群化石、山旺生物群化石、关岭生物群化石等。精品化石有中华龙鸟化石、辽宁古果化石、鱼龙、云南虫、海口虫、微网虫、奇虾、纳罗虫、灰姑娘虫、硅化木等。

昆虫化石
Fossils of insect

微网虫
Microdictyon sp.

灰姑娘虫

灰姑娘虫体长2.5厘米，由头和躯干两部分组成。头部像戴着一个半圆形的头盔（头甲）。头盔很大，躯干前端的5个体节都能一并得到保护。头盔的后部形成了一个漂亮的W形弯曲。灰姑娘虫的口接近正方形。在头部前端，即口的两侧前缘的地方长有一对细长的触须。在腹前长着带柄的大眼睛。背甲分26节，前5节甲都藏在头盔里。有许多附肢，每一个附肢包括内外两肢，内肢由13个肢节和1个爪组成。滤食性，游泳型。

奇虾

奇虾是寒武纪海洋的大型掠食者，位于食物链金字塔的顶端。奇虾是一类体型巨大、身体造型奇特

Special Collections

A large number of type specimens of Chinese fossil invertebrates and plant fossils are collected in this museum. The earliest type specimens housed in this collection museum include the Devonian brachiopod fossils of southern China published by Grabau in the 1920s, and the thin sections of the fusulinids of northern China studied by LI Siguang in the 1920s.

The department started to house the specimens from the Chengjiang fauna in 1984. The Early Cambrian Chengjiang fauna from Chengjiang, Yunnan, is a very unusual fossil lagerstaette. It contains extraordinarily well-preserved fossils in great abundance, vividly reflecting the true composition and appearance of marine communities of 520 million years ago. The discovery of the Chengjiang fauna extends the fossil records of most living animal phyla back to the beginning of the Cambrian, providing evidence for deciphering the mystery of the "Cambrian Explosion" of animals. The Chengjiang fauna has been appraised as one of the most exciting discoveries of the 20th century. The Chengjiang Field Station of Palaeontology was established in 1998 to support field research, international academic exchange, and scientific promotion and education. Other highlighted and special specimens in the collection include: the type specimens of *Sinosauropteryx prima* and *Archaefructus liaoningensis* from the Jehol biota. These results were published in journals of *Nature* and *Science*. It also includes the world's first ginkgo plant fossils - *Ginkgo yimaensis*, the Yanliao biota, the Daohugou biota and the Burmese amber biota.

Founded by and affiliated to the Nanjing Institute of Geology and Palaeontology, CAS (NIGPAS), which owns over 460,000 specimens of fossils, Nanjing Museum of Palaeontology possesses large collections of fossils, such as fossils from Chengjiang Fauna, Jehol biota, Shanwang biota and Guanling biota. Famous collections include *Sinosauropteryx prima*, *Archaefructus liaoningensis*, *Ichthyosauria*, *Yunnanozoon*, *Haikouella*, *Microdictyon*, *Anomalocaris*, *Naraoia*, *Cindarella*, and silicified wood.

Cindarella eucalla

Cindarella eucalla's body is 2.5 cm long and consists of head and trunk. The head is like wearing a semicircle helmet. The helmet is large, and all five segments in the front of the torso can be protected at the same time. The back of the helmet forms a beautiful W-bend. *Cindarella eucalla*'s mouth is close to a square. There are a pair of long and thin antennae in the front of the head, i.e., the front of both sides of the mouth. There are big eyes with handles in front of the abdomen. The back armor is divided into 26 sections, and the first 5 sections are hidden in the helmet. There are many appendages, each of which consists of two internal and external limbs. The internal limb consists of 13 segments and one claw. Filter feeding, swimming type.

Anomalocaris

Anomalocaris is a large predator of the Cambrian ocean, at the top of the pyramid of the food chain. And it is a kind of fossil arthropod with a huge size and peculiar body shape. It has large and fine compound eyes, special predation appendages, mouth cones and swimming paddle-like limbs, with a body length of up to 2 meters. It is considered to be the first top predator in the Phanerozoic marine ecosystem, and also one of the most representative star animals in the "Cambrian explosion". In the Cambrian Chengjiang biota of 520 million years ago in Yunnan Province, China, at least four kinds of shrimp fossils have been found, including broom shrimp and Hede shrimp. The discovery of many types of *Anomalocaris* fossils in the Chengjiang biota shows that large predatory *Anomalocaris* have been highly diversified in the early stage of the Cambrian explosion, which proves that complex food chains and ecosystems similar to modern oceans have been formed in the Cambrian explosion.

奇虾

Anomalocaris

Psilophyton

The oldest terrestrial vascular plant is Psilophyton. As the name suggests, it is an early fern with bare planting objects and no leaf cover. Compared with the later higher plants and modern plants, the morphological characteristics and organs of Psilophyton are very simple and primitive. Among them, there is the *Cooksonia*, the earliest terrestrial vascular plant on

的化石节肢动物，具有大而精细的复眼、特化的捕食前附肢、口锥和游泳桨状肢，体长最大可达 2 米，被认为是显生宙海洋生态系统中最早的顶级捕食者，也是“寒武纪大爆发”最具代表性的明星动物之一。我国云南距今 5.2 亿年前的寒武纪澄江生物群中已发现至少 4 种已确定的奇虾类化石，包括帚状奇虾和赫德虾类奇虾化石等。澄江生物群中多种类型奇虾化石的发现表明，大型捕食型奇虾类动物在寒武纪大爆发的早期已经高度多样化，证实了类似现代海洋的复杂食物链和生态系统在寒武纪大爆发时期就已经形成。

云南工蕨
Zosterophyllum yunnanicum

裸蕨植物

最古老的陆生维管植物是裸蕨植物，顾名思义，它是一种植物体裸露、无叶子覆盖的早期蕨类植物。与后来出现的高等植物和现代植物相比，裸蕨植物的形态特征与组织器官显得十分简单和原始。其中有库克逊蕨，地球上最早出现的陆生维管植物，高约 10 厘米，没有枝叶，也没有根系，只是一种简单的分叉结构，主茎顶上有孢子囊，孢子囊中生有孢子，孢子会随风飘到其他地方后继续生长。还有工蕨类植物——一个相当自然的类群，它们不同于其他同时代植物的最大特点是，在枝轴顶部组成穗状的侧生孢子囊，它们大都呈肾形，基部有短柄，并有沿着前缘切线开裂以扩散孢子的细胞加厚带。后来一部分工蕨类植物演变成了现代石松类的远祖。

鱼龙

鱼龙的祖先来自陆地，却在中生代重返海洋后成为海上霸主，是一大批返回海洋生活的爬行动物中的佼佼者。鱼龙一般分布于古西特提斯海域的挪威、德国和阿尔卑斯山地区，古东特提斯海域的印度、泰国、日本，以及我国的西藏、四川、贵州、湖北、广西、安徽等地和西太平洋的加拿大、美国。

中华龙鸟

中华龙鸟属（*Sinosauropteryx*）生存于距今 1.4 亿年的早白垩世，1996 年发现于中国辽宁省西部北票热河生物群。开始人们以为它是一种原始鸟类，定名“中华龙鸟”，后经科学家证实为一种小型肉食性恐龙，属于兽脚亚目美颌龙科中华龙鸟属。骨架高 1 米左右，前肢粗短，爪钩锐利，后腿较长，适宜奔跑，全身披覆着原始羽毛。中华龙鸟体表的原始羽毛很可能是鸟类羽毛的前身，其没有飞翔功能，但主要起着保护皮肤和维持体温的功能。长期以来，人们对于鸟类是不是恐龙的后裔一直存在不同的看法，中华龙鸟的发现为鸟类起源研究提供了重要的化石证据。

辽宁古果

辽宁古果是一种早期被子植物，于 1996 年在北票市黄半吉沟组下部地层中发现。其植株纤细，主侧枝呈倒“人”字形，在貌似蕨类植物的枝条上，螺旋状排列着四十几枚类似豆荚的果实，每个果实中都包藏着 2 至 4 粒米粒般大小的种子。这些特征显示了它们在早期被子植物中的原始性。另外，其茎枝细弱，叶子细而深裂，根部发育，只具有几个简单的侧根，反映了其水生性质。

the earth, about 10 cm high, without branches, leaves or roots, but a simple bifurcated structure. There is a sporangium on the top of the main stem, and there are spores in the sporangium. The spores will continue to grow after floating to other places with the wind. There is also *Zosterophyllum*, a quite natural group. They are different from other contemporary plants in that they form spikes of lateral sporangia at the top of the branch axis. Most of them are kidney-shaped, with short stalks at the base and thickened bands of cells that split along the tangent line of the front edge to diffuse spores. Later, some of the *Zosterophyllum* became the forefathers of modern *Lycopodium*.

Ichthyosauria

The ancestors of *Ichthyosauria* came from the land, but after returning to the sea in the Mesozoic, they became the overlord of the sea. They are the leaders of a large number of reptiles returning to the sea. *Ichthyosauria* is generally distributed in the ancient West Tethys (Norway, Germany and the Alps), the ancient East Tethys (India, Thailand, Japan, and Tibet, Sichuan, Guizhou, Hubei, Guangxi, Anhui in China), and Canada, the United States in the Western Pacific Ocean.

贵州鱼龙
Guizhouichthyosaurus

Sinosauropteryx prima

Sinosauropteryx lived in the early Cretaceous 140 million years ago, and was found in Beipiao Jehol biota in the Western Liaoning Province of China in 1996. At first, people thought it was a primitive bird, named "*Sinosauropteryx*" (Chinese Dragon Bird), and then it was confirmed by scientists as a small carnivorous dinosaur. It belongs to the order Theropoda, the family Compsognathidae, the genus *Sinosauropteryx*. The skeleton is about 1 m in size, with short and thick forelegs, sharp claws, long hind legs, suitable for running, and covered with primitive feathers. The primitive feathers on the body surface of *Sinosauropteryx* are probably the predecessor of bird feathers. It has no flying function, but it mainly protects the skin and maintains the body temperature. For a long time, there have been different opinions on whether birds are descended from dinosaurs or not. The discovery of *Sinosauropteryx* provides important fossil evidence for the study of bird origin.

中华龙鸟
Sinosauropteryx prima

辽宁古果
Archaefructus liaoningensis

Archaefructus liaoningensis

Archaefructus liaoningensis is an early angiosperm, which was found in the lower stratum of Huangbanjigou formation in Beipiao City in 1996. The plant is thin, and the main and lateral branches are inverted herringbone. On the branches that look like ferns, there are more than 40 podlike fruits arranged spirally. Each fruit contains 2 to 4 seeds of the size of rice grains. These characteristics show their primitiveness in early angiosperms. In addition, its stems and branches are thin, leaves are thin and deep split, roots are developed, only a few simple lateral roots, reflecting the aquatic nature.

6）综合类馆

中国科学院沈阳应用生态研究所东北生物标本馆

基本信息

隶属于：中国科学院沈阳应用生态研究所
地址：中国辽宁省沈阳市文化路 72 号
网址：http://ifp-cas.cn
建馆年份：1954 年
收藏类群：维管植物、苔藓、地衣、高等真菌以及昆虫
标本藏量：62.2 万号
馆长：袁海生，研究员
联系人：曹伟，研究员
电子邮箱：caowei@iae.ac.cn
联系电话：+86 24 83970334

标本馆简介

东北生物标本馆成立于 1954 年，其前身是中国科学院林业土壤研究所林业（室）标本室，由中国科学院沈阳应用生态研究所奠基人——著名的植物学家刘慎谔先生领导创建。1999 年经中国科学院批准，在原标本室基础上成立"东北生物标本馆"。2006 年研究所进入中国科学院三期创新工程后，东北生物标本馆正式成为独立的所级科研支撑机构。目前，东北生物标本馆是东北地区标本保藏量最大、门类齐全的综合性生物标本馆，标本保藏量位居全国标本馆前列。标本馆建筑面积 1880 平方米，现保藏有维管植物、苔藓、地衣、高等真菌以及昆虫等标本 62.2 万份。现有科研和支撑人员 19 人，主要开展生物分类、生物系统与进化、多样性调查与保护、生物资源开发利用等方面的研究工作。作为研究所重要的科研支撑部门，积极争取、参与和支撑国家级项目申报与执行，建馆以来 40 余项科研成果获国家及省部级奖励，出版专著 80 余部，发表论文千余篇。近年来标本馆依托"全国中小学生研学实践教育基地"，辽宁省、沈阳市科普教育基地，充分发挥在自然生态领域科研和科普资源的优势，开展了丰富多彩的科普活动。东北生物标本馆坚持以东北地区为主，面向全国，辐射东北亚毗邻国家的区域定位，广泛收集、保藏和利用具有标本馆特色的生物

沈阳应用生态研究所东北生物标本馆
Northeast Biological Herbaria, Institute of Applied Ecology

6) Comprehensive Collection/Museum

Northeast Biological Herbaria (IFP), Institute of Applied Ecology, CAS

Basic Information

Affiliated to: Institute of Applied Ecology, CAS
Address: No.72 Wenhua Road, Shenyang City, Liaoning Province, P. R. China
Website: http://ifp-cas.cn
Established in: 1954
Species groups of collection: Vascular plants, bryophytes, lichens, higher fungi and insects
Collection quantity: 622,000
Curator: Dr. YUAN Haisheng, Professor
Contacts: Dr. CAO Wei, Professor
E-mail: caowei@iae.ac.cn
Tel: +86 24 83970334

Introduction

The Northeast Biological Herbaria (IFP) was founded in 1954. Its predecessor is the herbarium of the Institute of Forestry and Soil, CAS. It was founded under the leadership of the famous botanist Prof. LIU Shen-e, who is the founder of the institute. In 1999, with the approval of CAS, the "Northeast Biological Herbaria" was established based on the original Herbarium. Since 2006, the Institute entered the third phase of the Innovation Project of CAS, and the Northeast Biological Herbaria officially became an independent research support department. IFP is a comprehensive collection with the largest specimens in Northeast China, and the number of specimens ranks at the forefront of herbarias in China. The building area of the herbaria is 1,880 m^2. There are 620,000 specimens of vascular plants, bryophytes, lichens, higher fungi and insects. At present, there are 19 scientific research and supporting personnel who are mainly engaged in classification, systematics and evolution, diversity investigation and protection, biological resource development and utilization, etc. As an important scientific research support department of the institute, it actively strives for, participates in and supports the application and implementation of national projects. Since its establishment, more than 40 scientific research achievements have been awarded by the national, provincial and ministerial level, more than 80 monographs have been published, and more than 1000 papers have been published. In recent years, relying on the "National Primary and Secondary School Students' Research and Practice Education Base", Liaoning Province, Shenyang City and other popular science education bases, the herbaria take the advantages of scientific research and popular science resources in the field of natural ecology to carried out a variety of popular science activities. The herbaria will adheres to the regional positioning of focusing on the northeast region, facing the whole country and radiating the neighboring countries in Northeast Asia, extensively collects, preserves and utilizes the biological species resources, serves scientific research and

标本馆内景
Inside view of the Herbaria

科普走廊
Popular science corridor

物种资源，服务科学研究、辅助政府决策，逐步打造东北亚生物资源保藏、研究、科普研学等具有多重社会服务功能的重要基地。

收藏概况

东北生物标本馆现保藏有维管植物、苔藓、地衣、高等真菌、昆虫等标本 62.2 万份，另保存微生物菌株 2 万余株。

建馆初期维管植物标本主要来自刘慎谔先生于 1950-1953 年采集的东北地区标本 2000 多种，10 000 余号。在他的领导下，先后对大小兴安岭、长白山、千山、辽东山地及东北平原等进行了全面的植物调查和标本采集工作。20 世纪 90 年代之后逐渐开展植物资源的县域调查和区系研究工作，目前共保藏东北地区包括木本和草本植物在内的维管植物标本 33 万份，近 3100 种，其中包括 368 个新物种的模式标本 415 份，基本覆盖了东北地区所有的木本植物和 90% 以上的草本植物。

苔藓植物研究起始于 1957 年，在内蒙古东部、大小兴安岭、长白山、东北平原和沿海岛屿进行标本采集，20 世纪 70 年代后研究重点开始转向全国。建有“东北亚苔藓研究中心”，是东北地区最大的苔藓植物标本馆，在全国位居前三位，标本采集地域覆盖最广。目前保藏有全国 34 个省区市的苔藓标本 12.5 万份，包括 115 个新种的模式标本，以及采集和交换的东北亚国家如日本、韩国、俄罗斯等的标本。

高等真菌的分类学研究始于 1956 年，主要开展东北地区伞菌、多孔菌、锈菌等类群的研究工作。2000 年以后，重点开展木材腐朽菌和革菌类群的分类和系统学研究。对中国 34 个省区市的 240 余个自然保护区、森林公园和林场等进行广泛调查及标本采集，此外还包括欧洲、南北美洲、亚洲、非洲、大洋洲的 20 余个国家和地区的高等真菌标本。目前保藏标本 4 万余份，包括 200 个真菌新物种的模式标本和国外交换模式标本 65 份。

地衣研究始于 1962 年，标本采集地主要覆盖黑龙江、吉林、辽宁和内蒙古东北部，目前保藏标本 3 万份。

东北生物标本馆昆虫分类学研究始于建所之初。昆虫馆的收藏主要分为两大部分，一部分是从东北生物标本馆建立初期到 1993 年以前采集的东北三省的林业害虫标本，约 5 万号标本；从 1993 年开始，标本馆与奥地利维也纳自然历史博物馆开展国际合作，从全国 20 余个省级行政区的 800 余个水生鞘翅目栖息地采集水生鞘翅目标本 2 万余号，含模式标本 567 号。

特色收藏

维管植物

东北生物标本馆保藏的维管植物模式标本共 368 种 415 份，包括蕨类植物 5 科 6 份，裸子植物 1 科 12 份，被子植物 41 科 397 份，其中有许多采集自 20 世纪 20 年代的模式标本，如兴安杨（*Populus hsinganica*）、楔叶东北杨（*Populus girinensis* var. *ivaschkevitchii*）等。其中，兴安杨标本采集于 1928 年内蒙古扎兰屯，由王战先生定名。圆叶鼠李（*Rhamnus globosa* = *Rhamnus viridifolia*）模式标本是中华人民共和国成立初期东北生物标本馆奠基人刘慎谔先生在东北采集的首批标本之一，采集于 1950 年 5 月辽宁大连大赫山。

此外本馆也保藏有许多珍稀濒危物种标本，如人参、双蕊兰等。人参（*Panax ginseng*）为第三纪孑遗植物，珍贵的中药材，以“东北三宝”之首驰名中外，在我国药用历史悠久。此标本为 4 年生植株，俗称“灯台子”。近年来由于人为干扰严重，已很难见到野生人参。

苔藓

光藓（*Schistostega pennata*） 中国苔藓学奠基人陈邦杰先生曾在 1963 年编撰《中国藓类植物属志》时预言，光藓在东北有可能分布。1997 年标本馆曹同研究员等在长白山暗针叶林下分化的岩石缝内发现了该

assists the government's decision-making, and gradually builds an important base with multiple social service functions such as biological resource conservation, research and popular science in Northeast Asia.

Overview of Collections

There are 622,000 specimens of vascular plants, bryophytes, lichens, higher fungi, insects and so on, and more than 20,000 microbial strains are preserved in the Northeast Biological Herbaria.

In the early stage of the establishment of the herbaria, more than 10,000 specimens of 2,000 vascular plant species were preserved, and these specimens were collected from the Northeast of China between 1950 and 1953 by Prof. LIU Shen-e. After that, he has carried out a comprehensive plant survey for the Greater and the Lesser Khingan Mts., Changbai Mts., Qianshan Mts., Liaodong mountains and the Northeast Plain. Since the 1990s, investigation for counties and floristic research of plants have been carried out gradually. At present, there are 330,000 vascular plant specimens (around 3,100 species) of woody and herbaceous plants of Northeast China, including 415 type specimens of 368 new species. It covers all woody plants and more than 90% of herbaceous plants of Northeast China.

Bryophyte research began in 1957, specimen collection mainly conducted in the eastern part of Inner Mongolia, the Greater and the Lesser Khingan Mts., Changbai Mts., the Northeast Plain and coastal islands. After the 1970s, the research emphasis began to turn to the whole country. It has established the "Northeast Asia Bryophyte Research Center", which is the largest bryophyte herbarium in Northeast China, ranking the top three in the country, with the widest collection area. At present, there are 125,000 bryophyte specimens from 34 provinces, autonomous regions and cities in China, including the type specimens of 115 new species, as well as specimens collected and exchanged from Northeast Asian countries such as Japan, Republic of Korea, Russia, etc.

The taxonomy of higher fungi began in 1956, which mainly focused on Agaricomycetes, Polypores and rust fungi in Northeast China. After 2000, the herbaria focused on the classification and systematics of wood-inhabiting fungi. More than 240 nature reserves, forest parks and forest farms in 34 provinces, autonomous regions and cities of China were extensively investigated. In addition, more than 20 countries and regions in Europe, North and South America, Asia, Africa and Oceania were also included in the collection of higher fungi. At present, more than 40,000 specimens have been preserved, including the 200 type specimens of new species and 65 type specimens from foreign exchange.

Lichen's research began in 1962. Lichen specimens were collected from Heilongjiang, Jilin, Liaoning and the northeast of Inner Mongolia. At present, 30,000 specimens are preserved.

The research on insect taxonomy in Northeast Biological Herbaria has been carried out at the beginning of the institute. The collection of insects include two parts: about 50,000 specimens are the forest pest collection which collected before 1993; the other part is the water beetles which collected in more than 800 localities, 20 provincial districts since the cooperation was established between the Institute of Applied Ecology, Shenyang and the Natural History Museum, Vienna, Austria in 1993, and now more than 20,000 specimens of water beetles including 567 type specimens are preserved.

Special Collections

Vascular plant

A total of 415 type specimens of 368 species of vascular bundle plants are preserved in the Northeast Biological Herbaria, including ferns 6 duplicates in 5 families, gymnosperms 12 duplicates in 1 family and angiosperms 397 duplicates in 41 families. Among them, there are many specimens collected from the 1920s, such as *Populus hsinganica*, *Populus girinensis* var. *ivaschkevitchii*, etc. The type specimen of *Populus hsinganica*, collected in Zhalantun, Inner Mongolia in 1928, was identified by Prof. WANG Zhan. The type specimen of *Rhamnus viridifolia*, collected and identified by Prof. LIU Shen-e in May 1950 in Daheshan, Dalian, Liaoning Province. This was one of the earliest specimens collected by him in the initial stage of the People's Republic of China.

In addition, there are many rare and endangered species, such as *Panax ginseng* (ginseng), *Diplandrorchis sinica* and so on. *Panax ginseng* is a kind of rare and endangered species and relic plant of the Tertiary period. It is also a kind of famous and precious traditional Chinese medicine as the first of "Three treasures in Northeast China". It has a long history of medicinal use in China. This was a 4-year old specimen, commonly called as "lamp stand". Due to serious human disturbance, it is hard to see wild *Panax ginseng* in recent decades.

Bryophytes

Schistostega pennata: Prof. CHEN Panchieh, the founder of Chinese bryology, predicted that *Schistostega pennata*

兴安杨模式标本
Type specimen of *Populus hsinganica*

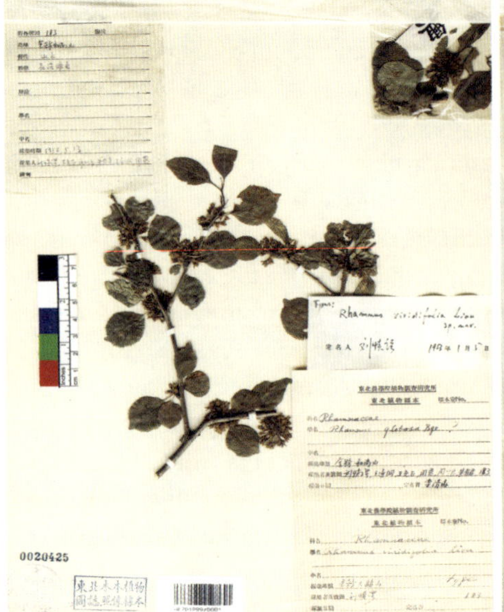
圆叶鼠李模式标本
Type specimen of *Rhamnus viridifolia*

人参标本
Specimen of *Panax ginseng*

光藓群落及其显微结构
Community and microscopic structures of *Schistostega pennata*

科模式种——光藓。该种生态位窄，需特殊生境，在我国属于极危物种。

无齿藓属（*Pseudochorisodontium*） 原为曲尾藓属（*Dicranum*）中的一个亚属，Brotherus 于 1924 年所描述的该亚属的特征范围并不唯一，因为一些种类与曲尾藓属其他亚属无法区分。1999 年标本馆高谦研究员等发现该亚属有些种具有共同的特殊的孢子体特征，如无蒴齿或蒴齿高度退化，这一特征和其他特征使其提升为一个属。无齿藓属均分布在喜马拉雅地区和中国西南高山地区，该属含 6 种，有 5 种为中国特有种，其中错那无齿藓（*P. conanenum*）、多枝无齿藓（*P. ramosum*）是濒危物种，瘤叶无齿藓（*P. mamillosum*）在中国是易危物种。

高等真菌

发现世界上最大的真菌 2010 年，在对海南省一个自然保护区进行野外调查中，东北生物标本馆科研人员发现了世界上最大的真菌，该菌生长在直径约 1 米的倒木下部二十余年，长 10.85 米，估计重量为 400-500 千克，是迄今为止世界上已发现的具有最大子实体的真菌。该真菌为我国新物种，戴玉成研究员将其命名为 *Fomitiporia ellipsoidea*。该真菌在我国福建省也有分布，在海南省发现的这个个体为其中最大的一个，经 BBC 采访报道后，受到了世界媒体广泛的关注。

桑黄的命名 桑黄作为重要的药用真菌，在我国具有悠久的应用历史，但对于其拉丁学名和分类地位一直以来都有争议。标本馆科研人员通过对东北生物标本馆馆藏标本的系统研究，将生长在桑树上的桑黄确定为新属和新种 *Sanghuangporus sanghuang*。

may distribute in Northeast China in 1963 when he edited *Genera Muscorum Sinicorum*. Prof. CAO Tong in the herbarium finally found the new distribution of species *Schistostega pennata* in the fracture of differentiated rock under dark coniferous forest in Changbai Mts in 1997. The niche of this species is narrow, and it needs special habitat, so it is critically endangered in China.

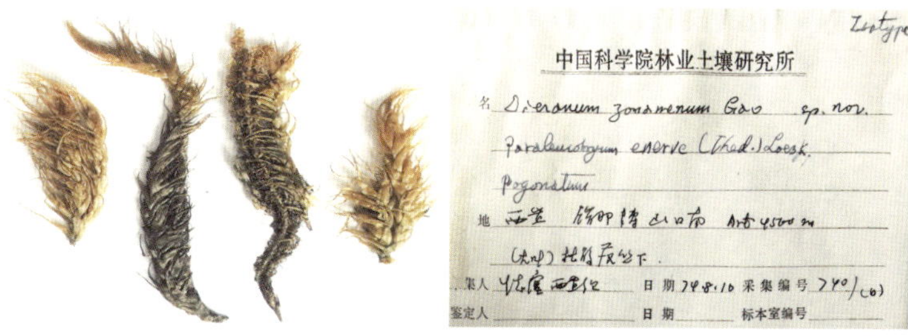

错那无齿藓标本及采集标签

Specimen and collection label of *Pseudochorisodontium conanenum*

The subgenus *Pseudochorisodontium* as circumscribed by Brotherus in 1924 was not exclusive, because it includes taxa that are not distinguishable from other subgenera of *Dicranum*. During the study of Chinese *Dicranum* species in 1999, Prof. GAO Chien found that a group of species share a peculiar feature of sporophyte structure in which the peristome teeth are absent or highly reduced. This feature in combination with other characters gives the group generic recognition. The genus *Pseudochorisodontium*, is found to be restricted to Himalayan regions and the southwestern high plateau of China. This genus includes six species, among which five species are endemic to China. *P. conanenum* and *P. ramosum* are endangered, while *P. mamillosum* is vulnerable in China.

Higher fungi

Discovery of the world's largest fungus: During a field survey for a natural reserve in Hainan Province in 2010, researchers of the Northeast Biological Herbaria discovered the world's largest fungus, which grows in the lower part of the fallen trunk with a diameter of about 1 m for more than 20 years. The fungus grew with a length of 10.85 m and an estimated weight of 400-500 kg. It is the largest fungus with a single fruiting body in the world. The fungus is a new species in China and named by Prof. DAI Yucheng as *Fomitiporia ellipsoidea*. The fungus is also distributed in Fujian Province of China. The individual found in Hainan Province is the largest one. After being interviewed and reported by the BBC, it has been widely concerned by the world media.

海南发现的世界最大真菌的担子果

The basidiocarps of the largest fungal species of the world found from Hainan Province, China

桑黄的子实体

Fruiting body of *Sanghuangporus sanghuang*

Name the Sanghuang species of China: as important medicinal fungi, Sanghuang have a long history of application in China, but their scientific names and taxonomic status have always been controversial. Through the systematic study of the specimens from Northeast Biological Herbaria, the researchers named the *Sanghuang* sp. growing on mulberry trees as a new genus and species: *Sanghuangporus sanghuang*.

Insect

Water beetle is the unique collection in the Northeast Biological Herbaria. There are 60 specimens of *Amphizoa sinica*, a rare insect species in China, in the collections. The family Amphizoidae is a primitive group, which belongs to the suborder Adephaga. Both larvae and adults are sub-aquatic, properly originated in the Triassic. Only five species were

国家珍稀昆虫——
中华两栖甲
Rare Chinese insect species:
Amphizoa sinica

昆虫

东北生物标本馆昆虫馆的特色收藏为水生鞘翅目，其中包括国家珍稀昆虫物种——中华两栖甲（*Amphizoa sinica*）标本 60 号。两栖甲科是肉食亚目的原始类群，幼虫、成虫均半水生，可能起源于三叠纪，属于活化石。该科全世界共有 5 个种，2 个种分布在中国，其余 3 种均分布在北美洲。东北生物标本馆的中华两栖甲标本多数采集于 20 世纪 80 年代，有关中华两栖甲的最近一次记录是沈阳应用生态研究所的姬兰柱研究员于 2000 年发现自朝鲜境内。在此后的近 20 年间，许多学者曾到过吉林长白山去考察中华两栖甲的生活现状，但均未发现其踪迹。

蛛溪泥甲属（*Ancyronyx*）是德国昆虫学家 Erichson 在 1847 年报道的，主要分布于东南亚。东北生物标本馆的边冬菊副研究员于 2012 年首次报告采自中国江西云居山的一个新种云居蛛甲（*Ancyronyx yunju*），系该属在中国的首次报道，亦是我国目前唯一的一种蛛溪泥甲。这个属的昆虫具六条修长的足，身体很小，因此看起来酷似蜘蛛，以至于在江西云居山第一次采集的标本差点被当成蜘蛛放掉。

多毛溪泥甲属（*Dryopomorphus*）隶属于多毛溪泥甲亚科（Larainae），主要分布于亚洲东部，目前已知 16 种。Jäch 和 Kodado 曾于 1995 年记录过中国的 *Dryopomorphus* 属，但未见具体的种类信息。2018 年，边冬菊副研究员首次对采自中国的两种多毛溪泥甲属物种进行了正式报道，模式标本均保存于东北生物标本馆。截至目前，多毛溪泥甲亚科在我国有分布记录的仅有 4 种。

中国科学院西北高原生物研究所青藏高原生物标本馆

基本信息

隶属于：中国科学院西北高原生物研究所
地址：中国青海省西宁市新宁路 23 号
建馆年份：1962 年
收藏类群：维管植物和脊椎动物
标本藏量：约 55.5 万份
馆长 / 联系人：陈世龙，研究员

青藏高原生物标本馆
Qinghai-Tibet Plateau Museum of Biology

植物标本库
Plant collection

recorded in this family, in which two species occurred in China and three species occurred in North America. Most specimens of *A. sinica* were collected in the 1980s, and the nearest record about *A. sinica* was in Democratic People's Republic of Korea in 2000 by Prof. JI Lanzhu. In the following 20 years, there is no report of this species though many researchers had visited Changbai Mountains to investigate.

云居蛛甲
Ancyronyx yunju

海纳多毛溪泥甲
Dryopomorphus heineri

The genus *Ancyronyx* was erected by Erichson in 1847, mainly occur in Southeast Asia. The only species of the genus in China, *Ancyronyx yunju*, was reported by Associate Prof. BIAN Dongju. And it is the first report of the genus in China. It looks quite similar to the spider, so the collector almost thought it was a spider and discard it when the first specimen was collected in Yunju Mountains, Jiangxi Provinces.

The genus *Dryopomorphus* belonging to the subfamily Larainae which mainly occur in the East of Aisa, and 16 known species were recorded at present. Chinese *Dryopomorphus* was mentioned in 1995 by Jäch & Kodada, but the authors did not give detailed species information. In 2018, Associate Prof. BIAN Dongju reported and described two new Chinese *Dryopomorphus* species, and the type specimens of which deposited in the Northeast Biological Herbaria. Presently, only four species of subfamily Larainae are recorded from China.

Qinghai-Tibet Plateau Museum of Biology (QTPMB), Northwest Institute of Plateau Biology, CAS

Basic Information

Affiliated to: Northwest Institute of Plateau Biology, CAS
Address: No.23, Xinning Road, Xining, Qinghai Province, P. R. China
Established in: 1962
Species groups of collection: Vascular plants and vertebrate
Collection quantity: 555,000
Curator/Contacts: Dr. CHEN Shilong, Professor
E-mail: herbarium@nwipb.cas.cn
Tel: +86 971 6110067

皮张库
Skin collection

头骨库
Skull collection

电子邮箱：herbarium@nwipb.cas.cn
联系电话：+86 971 6110067

标本馆简介

青藏高原生物标本馆是目前收藏、保存青藏高原这一特殊区域生物标本数量最多、种类最全、馆藏最丰富、标本采集覆盖青藏高原范围最广的标本馆。完成26万多号动植物标本信息化建设与共享。为青藏高原生物学研究、国家公园建设、青藏高原生物多样性及生态环境保护等提供重要技术支撑服务。作为科普教育基地，是宣传野生动植物保护和了解青藏高原动植物的窗口。

历史沿革

1962年建立动物标本室和植物标本室，后经研究所整合，于1999年改名为青藏高原生物标本馆。2002年标本馆改建和扩建，2012年重新设计布展约300平方米的科普展厅，陈列展示青藏高原独特、珍稀和濒危的动物形态标本以及特有的资源植物标本近300件。

收藏概况

标本采集和收藏历史、馆藏标本量 经过近60多年来研究所老、中、青三代科学家不畏艰险、流血流汗甚至以生命为代价的不懈努力，青藏高原生物标本馆生物标本收藏量达55万份左右。

标本类群覆盖范围 植物标本包括蕨类植物、裸子植物、被子植物共计38万号；动物标本16.25万号，包括昆虫11万号，哺乳类0.7万号，鸟类1万号，两栖爬行类0.85万号，鱼类2.7万号。

采集地覆盖范围 标本主要采集自青海和西藏，还有采自四川西部、云南西北部、甘肃南部、新疆、内蒙古、陕西、宁夏等地区的标本。基本涉及青藏高原的不同区域和各种动植物栖息地类型。

收藏特点 主要收藏青藏高原的维管植物及脊椎动物。

特色收藏

作为主要以收藏青藏高原动植物标本为目标的标本馆，无论从数量、种类、覆盖范围及馆藏特点上，青藏高原生物标本馆都已成为全国乃至世界范围内名副其实的青藏高原生物标本馆。

馆藏资源植物标本
Collection of the resource plant specimens

唐古特大黄
Rheum tanguticum

植物科学画——红花绿绒蒿
Botanical scientific illustration:
Meconopsis punicea

Introduction

As the only museum for the collection of Qinghai-Tibet Plateau, Qinghai-Tibet Plateau Museum of Biology (QTPMB) has the most amounts of biological specimens of this special area in the world. The Museum is becoming a multifunctional installation for scientific research, collection, scientific knowledge popularization and academic ex-changes, and also is cited by Worldwide Famous Specimen Museum Index. As a leading museum of China, QTPMB has established long-term cooperative relationships with quite some world-known museums and research institutions. QTPMB has been designated as a national-level (as well as provincial and city levels) base for popularizing sciences among the youth.

History

Herbarium and Zoological specimens' room were found independently in 1962, and were combined to one museum by the institute, which is named Qinghai-Tibet Plateau Museum of Biology in 1999. The warehouse of the museum was rebuilt in 2002. An exhibition hall of about 300 m^2 was put into use for scientific popularization in 2012. The exhibition hall was refitting, and about 300 unique, rare and endangered animal specimens and unique resource plant specimens of the Qinghai-Tibet Plateau were shown in it.

Overview of Collections

The history and the number of specimen collection. After about 60 years, with the effects of many scientists, the amount of specimens of Qinghai-Tibet Plateau Museum of Biology reached about 550,000.

The class and area of specimen. Plant specimens of about 380,000 included Pteridophyte, Gymnospermae, Angiospermae. For animal specimen in details: insect 110,000, mammal 7,000, bird 10,000, amphibian and reptile 8,500, fish 27,000.

Collection area. Mainly in Qinghai and Tibet of China, and also in Western Sichuan, Northwestern Yunnan, Southern Gansu, Xinjiang, Inner Mongolia, Shaanxi and Ningxia. The collection areas included most of the areas and different habitat type of plant and animal in the Qinghai-Tibet Plateau.

Characteristic of Collection. Focus on Vascular plants and vertebrates in the Qinghai-Tibet Plateau.

Special Collections

As a collection of animal and plant specimens of the Qinghai-Tibet Plateau, QTPMB has become a national and even worldwide biological specimen museum in terms of quantity, species, coverage and collection characteristics.

Plant specimens are those of endemic plants, cushion plants and other plants adapted to high altitude and cold in the Qinghai-Tibet Plateau, which are distributed in a narrow area. And the museum also had lots of specimen of the plants which is used as Tibetan and traditional Chinese medicine, such as *Rheum tanguticum*, *Rhodiola* spp., *Meconopsis* spp., *Fritillaria* spp, and so on. Invertebrates include Orthoptera of Insecta, among which Acridoidea is the main group studied by academician YIN Xiangchu, and plateau wingless-type is a unique species in Qinghai-Tibet Plateau. There are also vertebrate specimens peculiar or treasured to the plateau: Schizothoracinae and genus *Triplophysa* peculiar to the Qinghai-Tibet Plateau; amphibian and reptile specimens: *Rana kukunoris*, *Phrynocephalus* spp., *Laudakia* spp., *Thermophis baileyi* and *Gloydius strauchii*, etc.; birds: *Montifringilla* spp., *Tetraogallus* spp., *Aquila chrysaetos*, *Bubo bubo*, *Gypaetus barbatus*, *Grus nigricollis*, *Crossoptilon* spp. and other birds endemic to Qinghai-Tibet Plateau; Mammals: *Procapra przewalskii*, *Bos mutus*, *Pantholops hodgsonii*, *Procapra picticaudata*, *Pseudois nayaur*, *Ovis ammon*, *Cervus albirostris*,

红景天属植物生境照
Genus *Rhodiola* in the field

藏屹蝗
Oreoptygonotus tibetanus

裂腹鱼亚科 - 青海湖裸鲤
Schizothoracinae: *Gymnocypris przewalskii*

植物标本为狭域性分布的青藏高原特有植物、垫状植物及其他适应高海拔、寒冷的植物，收藏有大量汉藏药材标本，如唐古特大黄、红景天、绿绒蒿、贝母等。无脊椎动物有昆虫纲直翅目标本，其中蝗总科是印象初院士主要研究的类群，高原缺翅型是青藏高原的特有种类。还有高原特有或珍稀的脊椎动物标本：青藏高特有的裂腹鱼亚科和高原鳅属的鱼类；两栖爬行类收藏了高原林蛙、沙蜥、岩蜥、温泉蛇、高原蝮蛇等标本；鸟类包括青藏高原特有类群雪雀、雪鸡及金雕、雕鸮、胡兀鹫、黑颈鹤、马鸡等标本。标本馆还收藏大量具有青藏高原代表性动物的皮张标本和头骨标本，如普氏原羚、野牦牛、藏羚羊、藏原羚、岩羊、盘羊、白唇鹿、马麝、棕熊、雪豹、荒漠猫等，还有兔形目鼠兔科和啮齿目的鼢鼠等标本。

中国科学院新疆生态与地理研究所标本馆

基本信息

隶属于：中国科学院新疆生态与地理研究所

地址：中国新疆乌鲁木齐北京南路 818 号

建馆年份：1961 年

收藏类群：动物、植物、昆虫、土壤、矿石等

标本藏量：10.6 万号

馆长：冯缨，研究员

联系人：曹秋梅

电子邮箱 / E-mail：657478217@qq.com

联系电话 / Tel：+86 991 7885482

标本馆简介

中国科学院新疆生态与地理研究所标本馆（国际代码 XJBI），是一个具有内陆荒漠干旱区区域特色、涉及生态学和地理学两大学科，集合相关基础研究、标本保藏与科普教育基地的多功能科学博物馆。自成立至今已有约 60 年的历史，馆藏生物标本 10 万余号，为合理开发利用新疆特色资源，以及开展珍稀濒危生物的保护生物学研究提供了极为重要的基础信息和科学依据，是我国西北地区馆藏标本种类、数量最多的综合性标本馆之一，在国际、国内均具有典型的区域代表性。

标本馆长期致力于新疆生物标本 (含动物、植物、昆虫、土壤、矿石等) 的采集与分类研究，收集新疆各类生物资源的信息，更新和完善现有生物资源数据库，坚持干旱区生物区系研究，增进国内外学者的交流，

食肉类和有蹄类动物头骨
Skulls of carnivores and ungulates

雪豹标本
The specimen of Snow Leopard (*Uncia uncia*)

新疆生态与地理研究所标本馆
Specimen Museum of Xinjiang Institute of Ecology and Geography

Moschus chrysogaster, *Ursus arctos*, *Uncia uncia*, *Felis bieti*, the family Ochotonidae of order Lagomorpha and *Myospalax* spp. of the order Rodentia and so on.

Specimen Museum of Xinjiang Institute of Ecology and Geography, CAS

Basic Information

Affiliated to: Xinjiang Institute of Ecology and Geography, CAS
Address: No.818 South Beijing Road, Urumqi, Xinjiang, P. R. China
Established in: 1961
Species groups of collection: Animals, plants, insects, soil, ores, etc.
Collection quantity: 106,000
Curator: Dr. FENG Ying, Professor
Contacts: CAO Qiumei
E-mail: 657478217@qq.com
Tel: +86 991 7885482

Introduction

The Specimen Museum of Xinjiang Institute of Ecology and Geography (international code: XJBI), CAS is a multifunctional scientific and specimen preservation museum, and it has the inland arid regional features, relates to two main disciplines of ecology and geography, gathers the related fundamental researches, and is a popular science educational base. Its history is about 60 years. The museum has collected more than 100 thousand specimens. It provides very important basic information and scientific bases for rationally exploiting and utilizing the special resources in Xinjiang and carrying out the researches on the conservation biology of rare and endangered species, is one of the comprehensive specimen museums with the most species and number of collected specimens in northwest China, and is of the typical regional representativeness in China and even in the world.

XJBI has long been committed to the collection and classification of biological samples (including animals, plants, insects, soil, minerals, etc.) in Xinjiang, collecting information on all kinds of biological resources in Xinjiang, updating and improving the construction of the existing database of biological resources, persisting in the study of biota in arid areas, promoting exchanges between scholars, accumulating and providing background data for biodiversity research in Xinjiang. XJBI plays an important role in the sustainable development of arid areas.

XJBI has established a "Basic Geographic Information, Resource and Environmental Information System of Xinjiang" which includes information on the ecological environment, the social economy, plant resources, animal resources, soil resources and tourism resources.

On the basis of vouchered specimens housed at the XJBI, the following monographs on Xinjiang have been published: *Xinjiang Vegetation Utilization* (name in Chinese only), *Xinjiang Medicinal Flora* (name in Chinese only),

《新疆植物志》
Flora Xinjiangensis

一系列有关新疆生物资源的著作
A series of works on biological resources in Xinjiang

为新疆的生物多样性研究积累和提供本底数据，在干旱区区域可持续发展中发挥重要作用。

标本馆建有“新疆基础地理信息与资源环境信息系统”，包括生态环境、社会经济、植物资源、动物资源、土壤资源和旅游资源数据库。

以馆藏凭证标本为基础，先后出版了：《新疆植被及其开发利用》、《新疆药用植物志》、《新疆植物志》、《新疆经济植物及其利用》、《新疆盐生植物》、《中亚植物资源及其利用》、《新疆沙漠和改造利用》、《新疆鱼类志》、《中国天山自然地理》、《罗布泊科学考察与研究》、《新疆动物研究》、《新疆土壤及土地资源研究》、《塔克拉玛干沙漠及其周边环境考察路线指南》和《新疆土壤与改造利用》等专著。

收藏概况

采集的植物标本覆盖了中国所有干旱区，包括新疆全区，西藏部分地区，阿尔金山等地。馆藏标本还包括新疆的野生动物、土壤和昆虫。是中国科学院标本馆中唯一具有完整干旱区荒漠特色的标本馆。

馆藏植物标本 7 万号，脊椎动物标本 1 万号，昆虫标本 2 万余号，土壤（包括沙）与矿物标本近 2000 号。模式标本 97 号。当今世界上仅存的两份新疆植物特有种盐桦标本现藏于标本馆。由刘慎谔先生于 1931 年 7 月 30 日采于新疆迪化白杨沟的拟百里香标本是标本馆收集的最为久远的标本。

特色收藏

小沙冬青

小沙冬青是亚洲中部戈壁荒漠区特有的常绿豆科阔叶灌木之一。

胡杨

胡杨是生活在沙漠中的唯一的乔木树种，它自始至终见证了中国西北干旱区走向荒漠化的过程。虽然胡杨已退缩至沙漠河岸地带，但仍然是被称为“死亡之海”的沙漠的生命之魂。

馆藏最早标本——拟百里香，
1931 年 7 月 30 日采集
The oldest specimen collected by XJBI: *Betula platyphylla*, collected on July 30, 1931

Flora Xinjiangensis, *Xinjiang Economic Plants Utilization* (name in Chinese only), *Xinjiang Halophytes* (name in Chinese only), *Plant Resources and Utilization in Central Asia*, *Desert and Its Transformation and Utilization in Xinjiang* (name in Chinese only), *Xinjiang Fish Records* (name in Chinese only), *Physical Geography of the Tianshan Mountains in China*, *Scientific investigation and research of Lop Nur* (name in Chinese only), *Research of Xinjiang Animals* (name in Chinese only), *Research on Soil and Land Resources in Xinjiang* (name in Chinese only), *Guide to the Investigation Route of Taklimakan Desert and Its Surrounding Environment and Transformation* and *Utilization of Soil in Xinjiang.*

Overview of Collections

The collecting area covers all the arid areas of China, including the whole region of Xinjiang, parts of Xizang, Altun Mountain and other places. The collection also includes wild animals, soil and insect from Xinjiang. It is the only CAS herbarium that includes a complete set of desert characteristics.

There are 70,000 plant specimens, 10000 vertebrate specimens, more than 20,000 insect specimens, and nearly 2,000 soil (including sand) and mineral specimens. There are also 97 type specimens. The only two specimens of *Betula halophila* endemic to Xinjiang in the world are stored in XJBI. Mr. LIU Shen-e collected *Betula platyphylla* specimens in Baiyanggou, Dihua, Xinjiang, on July 30, 1931, which is the oldest specimen collected by the XJBI.

Special Collections

Ammopiptanthus nanus

Ammopiptanthus nanus is one of the endemic evergreen broad-leaved shrubs in the Gobi desert area of Central Asia.

小沙冬青
Ammopiptanthus nanus

Populus euphratica

Populus euphratica experiences the desertification of the arid area in northwest China all along, and they are still the soul of life in the sand desert named as the "sea of death" although the forests have shrunk to the riparian zone in the desert.

Saussurea involucrata

Saussurea involucrate is commonly distributed in the zones of 4000-5000 m a. s. l. on the southern and northern slopes of the Tianshan Mountains, Altay Mountain and Kunlun Mountain in Xinjiang, China, and is listed as a protective herb species at class III. *S. involucrata* is a kind of excellent Chinese herbal medicine.

胡杨
Populus euphratica

Xinjiang Wild Fruit Forest

The Xinjiang Wild Fruit Forest is the only natural gene pool of economic forest resources in China, also an important component of the gene pool of wild apples

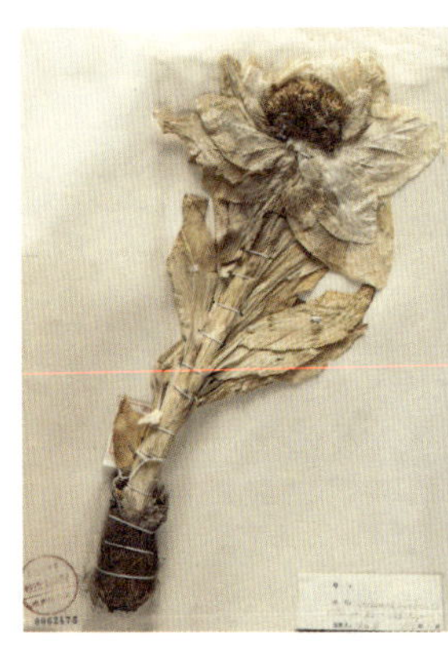

雪莲（天山雪莲）
Saussurea involucrata

野杏
Armeniaca vulgaris

新疆野果林
Xinjiang Wild Fruit Forest

雪莲（天山雪莲）

雪莲常分布于中国新疆天山、阿尔泰山及昆仑山南北山坡，海拔4000-5000米的地带，被列为国家三级保护植物，是一种优良的中药材。

新疆野果林

新疆野果林是我国经济林资源中唯一天然的基因库，也是世界野苹果基因库的重要组成部分，具有重要的科研和保护价值。我国仅新疆有天然分布，均被列为中国濒危二级保护植物。

新疆野果林有6个类型：野苹果林、野杏林、野核桃林、野扁桃林、野李群系、野樱桃群系。已划为自然保护区的有3处：新源野苹果保护区、巩留野核桃保护区、裕民野巴旦杏（或野扁桃）保护区。树种多为第三纪阔叶林的孑遗，主要分布于伊犁谷地及塔城盆地。

普氏野马

普氏野马被列入中国国家一级保护动物；2015年濒危物种红色名录中被列为濒危。

蒙古野驴

分布于中亚及西亚各国，在中国分布于内蒙古、甘肃和新疆。为中国一级保护动物，被IUCN列为濒危。

野骆驼

新疆有世界上仅存的纯基因野骆驼种群，具有极高的科学研究和保护价值。野骆驼为国家一级保护动物。

雪豹

雪豹是亚洲高山高原地区最具代表性的物种，国家一级保护动物，也是国际濒危动物。

扁吻鱼（新疆大头鱼）

是中国的特产鱼类，也是世界裂腹鱼中的珍贵物种，起源于3亿年前，有着古鱼类活化石之称。仅一属一种，其在世界上的分布仅存于塔里木河水系，属国家一级保护动物。

in the world, and it is of important scientific and conservational values. There is a natural distribution of wild fruit forests only in Xinjiang, China, where the plants are all listed as the endangered plants at class II in China.

In the Xinjiang Wild Fruit Forest, there are 6 forest types, including the forests of *Malus sieversii*, *Armeniaca vulgaris*, *Juglans regia*, *Amygdalus ledebouriana* and the populations of *Liwa tarnina* and *Amygdalus nana*, in which 3 places have been delimited as the nature reserves. They are the Xinyuan *Malus sieversii* Nature Reserve, Gongliu *Juglans regia* Nature Reserve and Yumin *Amygdalus ledebouriana* Nature Reserve. The tree species in these forests are mostly the relicts of Tertiary broad-leaved forests, and they are mainly distributed in the Ili Valley and the Tacheng Basin.

Equus ferus

Equus ferus is listed as the first-class protected animal in China, and it is also listed in the Red List of Threatened Species - endangered animal.

普氏野马
Equus ferus

Equus hemionus

E. hemionus inhabits in the countries in Central Asia and West Asia and in Inner Mongolia, Gansu and Xinjiang of China. *E. hemionus* is listed as a protected animal species at class I in China, and it is in imminent danger in IUCN.

蒙古野驴
Equus hemionus

Camelus bactrianus ferus

Xinjiang has the only pure gene *Camelus bactrianus ferus* population in the world, which has high scientific research and protection value. *Camelus bactrianus ferus* is class I protected animals in China.

Uncia uncia

Uncia uncia is the most representative species in the Asian alpine regions and plateaus, and the national protective

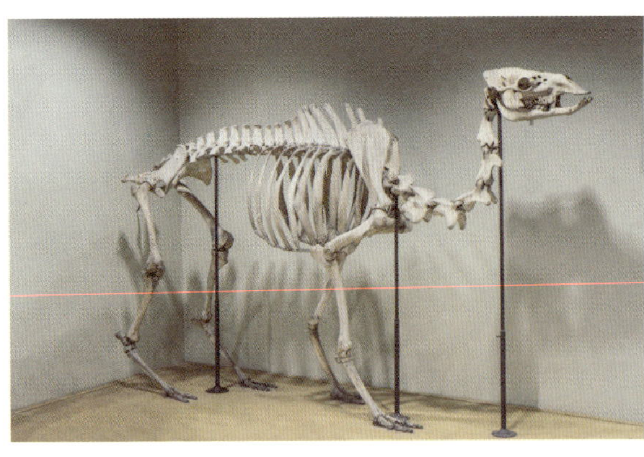

野骆驼
Camelus bactrianus ferus

雪豹
Uncia uncia

新疆北鲵（娃娃鱼、水四脚蛇）

新疆北鲵是距今 3 亿 -4 亿年前最原始的有尾两栖动物物种，被列入《中国濒危动物红皮书》，濒危等级为“极危”。

阿波罗绢蝶

阿波罗绢蝶栖息于海拔 750-2000 米的高山区，耐寒性强，常常生活在雪线上下，是国家二级保护动物。

扁吻鱼标本及其生活环境
Specimen and living environment of *Aspiorhynchus laticeps*

新疆北鲵
Ranodon sibiricus

animals at class I. And it is listed as Endangered animal.

Aspiorhynchus laticeps

Aspiorhynchus laticeps is an endemic species in China, also a precious species in Schizothoracinae in the world. It originated before 3 hundred million years and is called a living fossil of ancient fishes. There is only one genus and one species, *A. laticeps* inhabits only in the Tarim River system, *A. laticeps* is listed in the protected animals of China at class I.

Ranodon sibiricus

Ranodon sibiricus is the most primitive amphibian species that originated before 3-4 hundred million years, and it is listed in the "China Red Data Book of Endangered Animals" at the "extremely endangered level".

Parnassius apollo

The butterflies of *P. apollo* inhabit in the alpine zones 750-2000 m a. s. l. They have strong cold endurance and generally inhabit around the nival line. It is listed in the protected animals of China at class II.

阿波罗绢蝶及其生活环境
Specimen and living environment of *Parnassius apollo*

4

近年来工作进展和成果

Progress and achievements in recent years

标本馆的工作主要包括三方面。第一是以标本为中心的收藏、整理、鉴定、数字化及管理技术研发等战略生物资源收集和保藏工作；第二是依托馆藏标本，在支撑科学研究、服务国家战略计划、服务国家重大需求、支撑专业人才培养和服务社会需求方面所做的科研支撑工作;第三是面向社会大众的科学普及工作。

4.1 战略生物资源收集和保藏

战略生物资源的收集与保藏是生物标本馆的重要职能之一。2014-2019 年，配合国家重大研究计划，各馆开展针对性考察和采集，共组织国内外考察 1400 余次，采集标本 317.3 万号，整理制作 166.9 万号。国内考察主要集中在生物多样性研究热点地区和以往考察薄弱的地区，如青藏高原、西南山地等。各馆也根据自身定位和特色进行采集，如东北生物标本馆对东北地区动植物的采集，海洋生物标本馆和南海海洋生物标本馆对我国近海及其他海域的考察和采集等。国外考察则主要集中于“一带一路”沿线国家和地区，包括中亚、东南亚、非洲、印度洋和西太平洋等地。

不断的考察和采集积累了大量标本，使各馆馆藏量稳步增加，标本总量从 2013 年底的 1774.0 万号持续、平稳地增加到 2019 年的 2126.9 万号，增长率达 19.9%。同时，定名标本和标本数字化量也稳步增长，定名标本增长了 10%，目前已占馆藏标本的 54.8%，数字化标本占 47.6%。

2013-2019 年标本总量及各类群标本量（单位：万号）

Total amount and specimen amount of various groups in 2013-2019 (×10k)

类别 Category	2013	2014	2015	2016	2017	2018	2019*
植物 Plants	683.3	691.4	700.7	705.7	718.4	738.3	761.4
动物 Animals	999.7	1040.0	1081.1	1126.7	1169.7	1201.4	1232.2
菌物 Fungi	51.3	52.2	54.2	54.9	55.8	56.4	63.6
化石 Fossils	39.7	40.4	40.8	41.6	42.0	42.3	69.7
总量 Total	1774.0	1824.0	1876.8	1928.9	1985.9	2038.4	2126.9

* 庐山植物园标本馆于 2019 年底加入，植物标本量增加 19 万号。

* Herbarium of Lushan Botanical Garden was added at the end of 2019, with an increase of 190,000 plant specimens.

The work of the Herbarium/Collection mainly includes three aspects. First, the collection and preservation of strategic biological resources such as collection, sorting, identification, digitization and management technology research and development centered on specimens; second, the scientific research support in supporting scientific research, serving the national strategic plan, serving the major needs of the country, supporting the cultivation of professional talents and serving the social needs relying on specimens in the collection; third, facing the public scientific popularization.

4.1 Collection and preservation of strategic biological resources

The collection and preservation of strategic biological resources are the important functions of biological Collection. From 2014 to 2019, in line with the national major research plan, each Collection has carried out targeted investigation and collection, organized more than 1400 domestic and international investigations, collected 3.173 million specimens and made 1.669 million. The domestic researches are mainly focused on the hotspot areas of biodiversity research and the areas with weak researches in the past, such as the Qinghai Tibet Plateau, the southwest mountains, etc. Each Collection also collects according to its own location and characteristics, such as the collection of animals and plants in the northeast by the Northeast Biological Herbaria, the investigation and collection of China's offshore and other sea areas by the Marine Biological Museum and Marine Biodiversity Collections of South China Sea. Foreign studies mainly focus on the country and region along the B&R, including Central Asia, Southeast Asia, Africa, India ocean and the Western Pacific.

A large number of specimens have been accumulated through continuous investigation and collection, which has steadily increased the collection of each Collection. The total number of specimens has increased continuously and steadily from 17.74 million at the end of 2013 to 21.269 million in 2019, with a growth rate of 19.9%. At the same time, the number of classified specimens and specimen digitization has also increased steadily. The number of classified specimens has increased by 10%, now accounts for 54.8% of the collection specimens, and the number of digitated specimens accounts for 47.6%.

Investigations and collections of important areas in China

After the founding of the People's Republic of China, with the establishment of various institutes of CAS, investigation, collection and research of various biological resources in China have been started. For example, the Institute of Zoology, the Institute of Botany, the Institute of Microbiology and other units have successively organized the biological resources investigations of Yunnan, Xinjiang, Tibet, Mount Qomolangma, Mount Tomur, areas along the South-to-North Water Diversion Project, Wuyi Mountain, Namjagbarwa Mountain, Shennongjia, Xiaowutai Mountain, Qinling, Daba Mountain, Kunlun Mountain, Hoh Xil, Hengduan Mountain, Wuling Mountain, Mt. Shiwanda, Three Gorges Reservoir area, Qinling, Hainan Island, etc. These early large-scale scientific research obtained many important biological specimen resources, which constitute the majority of various biological collections, provide a large number of research materials, support a large number of basic research work in the field of biology in China, and lay a solid foundation for the development of life science in China, especially support the "Three Flora/Fauna" representing the most comprehensive understanding of biological resources in China.

In recent years, with the support of key projects of NSFC and special projects of basic work of the Ministry of science and technology, all biological collections have closely focused on strategic conservation objectives, national ecological civilization construction, national ecological security and national economic construction, focused on biodiversity hotspots and rare taxa in Herbarium, actively expanded field investigation areas and enriched investigation and collection taxa.

Qinghai-Tibet Plateau

The Qinghai-Tibet Plateau has a special geographical environment, and its poor survival pressure has created a unique composition of biological species. This area has always been the focus of researchers' attention. In recent years, the state has launched the project of "The Second Comprehensive Scientific Investigation of the Qinghai-Tibet Plateau", with the active participation of relevant research institutes and various biological collections, which has obtained a large number of biological specimens.

2014-2019 年定名标本及数字化标本量（单位：万号）

The number of classified specimens and digitated specimens in 2014-2019 (×10k)

类别 Category	2014	2015	2016	2017	2018	2019
定名标本量 Classified specimens	1058.5	1068.7	1070.9	1107.6	1146.1	1165.0
数字化标本量 Digitated specimens	876.8	876.4	906.4	946.4	971.6	1012.1

国内重要地区考察和采集

中华人民共和国成立后，随着中国科学院各个研究所的建立，陆续开始了对我国各个生物类群的资源调查、采集和研究，如动物研究所、植物研究所、微生物研究所等单位先后组织了对云南、新疆、西藏、珠穆朗玛峰、托木尔峰、南水北调工程沿线地区、武夷山、南迦巴瓦峰、神农架、小五台山、秦岭、大巴山、昆仑山、可可西里、横断山区、武陵山区、十万大山、三峡库区、秦岭、海南岛等区域的生物资源考察。这些早期大型科考获得的许多重要的生物标本资源，构成了现今各个生物标本馆馆藏的大部分，提供了大量的研究材料，支撑了中国生物领域的大量基础性研究工作，为中国生命科学的发展奠定了坚实的基础，特别是支撑了代表我国对生物资源最全面认识的“三志”的编研。

近年来，在国家自然科学基金重点项目和科技部基础性工作专项等项目资助下，各生物标本馆紧紧围绕战略保藏目标，并关注国家生态文明建设、国门生态安全、国民经济建设，重点关注生物多样性热点区域及标本库稀缺类群，积极拓展野外考察区域、丰富考察采集类群。

青藏高原

青藏高原有着特殊的地理环境，其恶劣的生存压力也造就了独一无二的生物物种组成。这一地区历来是研究者关注的热点。近年来，国家启动了“第二次青藏高原综合科学考察”项目，相关研究所及各生物标本馆积极参与，获得了大量生物标本。

西北高原生物研究所青藏高原生物标本馆作为唯一坐落于青藏高原的生物标本馆，担负着保藏青藏高原生物的责任。近年来标本馆开展的战略生物资源调查收集工作主要有：虎耳草属植物调查，青海灌丛植物调查，三江源植物旗舰物种野外考察，“第二次青藏高原综合科学考察”青海和西藏植物多样性调查，“第二次青藏高原综合科学考察”中国北方内陆盐碱地植物种质资源调查，祁连山国家公园植物本底调查及大型真菌调查，澜沧江源区植物多样性调查，西藏植物种质资源采集等。

成都生物研究所两栖爬行动物标本馆也参与了“第二次青藏高原综合科学考察”项目。标本馆以青藏高原极具代表性的爬行类物种温泉蛇 (*Thermophis baileyi*) 和四川温泉蛇 (*Thermophis zhaoermii*) 为研究对象，分别于 2018 年和 2019 年对这两个物种开展专项调查研究。2018 年在西藏调查 17 个点，采集了 13 个点的温泉蛇样品，共 62 号标本。2019 年在四川甘孜藏族自治州调查 9 个点，采集了 3 个点的四川温泉蛇样品，共 9 号标本。针对这类特殊而濒危的高原蛇类的生存状况进行了本底调查，为日后的科学研究与物种保护工作奠定了重要的基础。

成都生物研究所两栖爬行动物标本馆还参与了“藏东南动物资源综合考察与重要类群资源评估”项目。2014-2019 年，标本馆与研究团队共同圆满完成子课题“两栖爬行类资源考察与重要类群资源评估”。该项

As the only biological collection located on the Qinghai-Tibet Plateau, the Qinghai-Tibet Plateau Museum of Biology of Northwest Institute of Plateau Biology is responsible for the preservation of Qinghai-Tibet Plateau organisms. In recent years, the museum did the collections as follows: Survey about the resources of Genus *Saxifraga*, Survey about vegetation structure of shrubs in Qinghai, flagship species of plant in Three-river source region, plant diversity investigate in Qinghai and Xizang as part of The Second Comprehensive Scientific Investigation of the Qinghai-Tibet Plateau, investigation on plant germplasm resources of in saline-alkali land in Northern China, the background of plant and fungi of Qilian Mountain National Park, Survey of plant diversity in Lancang-Mekong river source region, Germplasm resources collection in Xizang and so on.

The Herpetological Museum of Chengdu Institute of Biology participated in the project of "The Second Comprehensive Scientific Investigation of the Qinghai-Tibet Plateau", taking the representative reptile species of the Qinghai Tibet Plateau, *Thermophis baileyi* and *Thermophis zhaoermii*, as the research objects, and held special investigation and research on these two species in 2018 and 2019 respectively. In 2018, 17 sites were investigated in Tibet, and 62 specimens of *T. baileyi* from 13 localities were collected. In 2019, 9 sites were investigated in Ganzi Prefecture, Sichuan Province, and 9 specimens of *T. zhaoermii* from 3 localities were collected. Such background investigations conducted especially for these special and endangered snakes accumulated massive information on basic data, which will strongly facilitate future study and conservation.

The Herpetological Museum of Chengdu Institute of Biology also participated in the project of "The Comprehensive Investigation of Animal Resources and Evaluation of Important Group Resources in Southeastern Tibet". From 2014 to 2019, the museum participated in the project of "Comprehensive Investigation of Animal Resources and Evaluation of Important Group Resources in Southeastern Tibet", together with the research team, the subproject "Investigation of Amphibian and Reptile Resource and Evaluation of Important Group Resource" was successfully completed. This project recorded 47 amphibians and 56 reptiles in Southeast Tibet. The project has obtained more than 12,000 amphibian and reptile specimen data, including 4,659 submitted by us, and all of which are vouchered specimens in the museum.

The Fungarium of Institute of Microbiology conducted continuous surveys of Fungi in the Qinghai-Tibet Plateau. With the large-scaled field investigations were performed in the Qinghai-Tibet Plateau from 2008 to 2016, the excursions covered 155 cities and counties in six provinces, and 24 of them had never been explored. More than 21,000 specimens belonging to 322 genera in 143 families of 50 orders were collected and added to the Fungarium. Currently, more than 60,000 specimens of 5,228 species from the area, including more than 400 type specimens are preserved in the Fungarium, which updated the knowledge of fungal diversity and provided critical materials for systematic studies.

Xinjiang

Xinjiang is located in the hinterland of the Eurasian continent, bordering many countries. It is an important passage of the ancient Silk Road in history. Now it is the only place to pass the second "Eurasian Continental Bridge". Its strategic position is very important. It is also the provincial administrative region with the largest land area in China, with a variety of geographical environments and rich natural resources.

As the only collection of CAS with complete desert characteristics in arid areas, XJBI is responsible for the preservation of biological and other natural resources specimens in Xinjiang. With the support of the "Special Fund for the Operation of Strategic Biological Resources Science and Technology support system" of the Bureau of Science and Technology for Development, Chinese Academy of Sciences, biological specimens were collected in the field, and taxonomic research was carried out. A large number of field scientific data have been accumulated for the distribution and characteristics of desert wildlife resources in Xinjiang. Every year, more than 7000 biological specimens such as animals, plants, soil and insects are collected, and the collected specimens are classified, identified, preserved and digitized, so that they can be shared on the Internet. Since 2003, XJBI has participated in the project of the Chinese Virtual Herbarium (CVH) system. More than 80,000 copies of XJBI specimen data have been uploaded. Since the launch of the CVH website in 2006, XJBI has received a total of more than 500,000 visits every year. XJBI also participated in the second comprehensive scientific investigation of the Qinghai-Tibet Plateau, organized the scientific examination of biodiversity in Tianshan and Pamirs Plateau, and all the specimens collected were preserved in the herbarium of Xinjiang Institute of ecology and geography.

South China Sea

Since ancient times, the South China Sea has been the main channel for exchanges between the East and the west,

水生生物博物馆参加“第二次青藏高原综合科学考察”项目
The Museum of Hydrobiological Sciences participated in “The Second Comprehensive Scientific Investigation of the Qinghai-Tibet Plateau” project

东北生物标本馆参加“第二次青藏高原综合科学考察”项目
The Northeast Biological Herbaria participated in “The Second Comprehensive Scientific Investigation of the Qinghai-Tibet Plateau” project

西藏萨迦县的温泉蛇
Thermophis baileyi from Sa’gya, Tibet

四川新龙县的四川温泉蛇
Thermophis zhaoermii from Xinlong County, Sichuan Province

青藏高原的大型真菌和地衣
Macrofungi and lichens from Qinghai-Tibet Plateau

and its strategic position is very important. It is self-evident to conduct a comprehensive investigation and collection of plant resources in the South China Sea Islands. However, the islands in the South China Sea are far away from the mainland, and the distribution of the islands is extremely scattered. In addition, the wind and waves in the adjacent waters are strong, and there is a lack of fresh water on the islands, so the field work is extremely difficult.

In recent years, by the special projects of basic work of the Ministry of Science and Technology and the CAS pilot project, the Herbarium of South China Botanical Garden has investigated the plant resources of Xisha Islands, Zhongsha Islands, Nansha Islands and other South China Sea Islands for many times. More than 6,000 plant specimens were collected, and the *Flora of South China Sea Islands* was published.

新疆荒漠植物标本的数字化
Digitization work of desert plant specimens of Xinjiang

天山生物多样性科考
Biodiversity investigation in Tianshan

荒漠生物多样性科考
Biodiversity investigation of desert

Other areas

In addition, according to their own collection characteristics, and in combination with scientific research projects, each collection actively expanded the collection range of specimens and went to other areas such as northeast China and Yunnan Province to collect specimens.

The Fungarium organized fungal investigations to the Hengduan Mountains (1973-1980), the Tomur Peak (1977-1978), the Namcha Barwa Peak (1982-1983), Shennongjia (1984), the Xiaowutai Mountains (1990), the Qinling Mountains (1991-1992), the Daba Mountains (1994) in China, and to the King George Island (Antarctica, 1983-1984, 1994-1995) and Murmansk (Arctic, 1995). Many monographs have been published base on the specimens housed in the Fungarium which provides a scientific basis for revealing the background of fungi in China and the world. Recently, supported by CAS, the National Natural Science Foundation and Ministry of Science and Technology of China for fundamental Research, the Fungarium organized successively the field investigations in tropical regions (1997-2000) and northwest of China (2003-2004), Hainan Province (2007-2009), the Karst regions in Southwestern China (2014-2018), the Greater and the Lesser Khingan Mountains (2014-2018), and the Svalbard Archipelago (Norway, 2002), the East Siberia (Russia, 2004-2005), and the Shetland Archipelago (Antarctica, 2004-2005). A great number of collections were accumulated, and comprehensive systematic studies were carried out. New species were published continuously, and the knowledge of fungal diversity increased significantly. And the research results have been highly concerned by researchers at home and abroad.

The collection center of Shanghai Entomological Museum is East China, based on East China and facing the whole country. In recent years, the museum has been paid more attention to the area of the B&R. In the past 5 years, 49 field

研究共记录藏东南地区两栖动物 47 种，爬行动物 56 种。项目团队共获得两栖爬行动物标本数据 1.2 万余条，其中本馆提交标本数据 4659 条，均来自馆藏标本。

微生物研究所菌物标本馆 2008-2016 年针对青藏高原进行连续考察，涵盖 6 省区的 155 个市县，其中 24 个县为首次针对菌物资源进行考察，共采集菌物标本 2.1 万余份，涵盖 50 目、143 科、322 属，是针对青藏高原菌物资源开展的规模最大的科学考察活动。目前菌物标本馆馆藏青藏高原地区标本 6 万余份（含模式标本 400 余份），涉及 5228 个物种，为研究和保护青藏高原地区菌物多样性奠定了坚实的基础。

新疆

新疆地处亚欧大陆腹地，与多个国家接壤，在历史上是古丝绸之路的重要通道，现在是第二座“亚欧大陆桥”的必经之地，战略位置十分重要，也是中国陆地面积第一大的省级行政区，具有多样的地理环境和丰富的自然资源。

作为中国科学院唯一具有完整干旱区荒漠特色的标本馆，新疆生态与地理研究所标本馆担负着保藏新疆地区生物和其他自然资源标本的重任。在中国科学院科技促进发展局“战略生物资源科技支撑体系运行专项经费”的资助下，标本馆每年进行生物标本采集和以区域特色为重点的馆藏建设工作，野外采集生物标本并进行分类学研究。对新疆荒漠野生生物资源的分布及其特点，积累了大量的野外科学数据。每年新增采集动物、植物、土壤、昆虫等生物标本 7000 余号，并将采集标本进行分类鉴定、保藏以及标本数字化，使之可在网上共享，2003 年以来，参加科技部项目中国数字植物标本馆（Chinese Virtual Herbarium, CVH）共建，先后上传 XJBI 标本数据 8 万余份。CVH 网站自 2006 年开通上线以来，XJBI 年度访问量累计超过 50 万次。此外还参与第二次青藏高原综合科学考察，组织前往天山和帕米尔高原进行生物多样性科考，所采集的标本均保存于新疆生态与地理研究所标本馆。

南海

南海岛屿植物调查
Plant Survey of islands in the South China Sea

自古以来，南海便是东西方交流的主要通道，战略地位十分重要，对南海诸岛的植物资源进行全面调查与收集具有不言而喻的重要性。但由于南海诸岛远离大陆，岛屿之间的分布极为分散，加之邻近海域风大浪急，岛上缺少淡水，野外工作极为艰难。

近年来，结合华南植物园承担的科技部基础性工作专项与中科院先导专项，华南植物园标本馆多次赴西沙群岛、中沙群岛、南沙群岛及其他南海岛屿进行植物资源的调查与收集，共收藏南海岛屿植物标本 6000 余号，并整理出版了《中国南海诸岛植物志》。

其他地区

此外，各馆还根据自身收藏特色，结合承担科研项目，积极扩展标本收藏范围，赴国内其他地区如东北、云南等地进行标本采集。

东北草地植被调查
Investigation of grassland vegetation in Northeast China

对云南高黎贡山水生昆虫进行调查和标本采集
Investigation and specimen collection of aquatic insects in Gaoligong Mts. of Yunnan Province

collections have been organized, covering Qinghai, Jilin, Zhejiang, Shanghai, Guangxi, Guangdong, Hainan and other provinces, with a total of 190,000 specimens collected. In order to play an important role in the protection of the ecological environment and the utilization of resources, the collection of specimens is mainly based on wetland and soil environment, while taking advantage of the collection of dominant groups and some rare species.

在天山、珠峰等地区的野外考察
Fungal surveys at Tianshan Mountains, Mount Qomolangma and so on

LBG mostly depends on the special key project of basic work which is supported by the National Ministry of Science and Technology, "The Integrated Scientific Expedition of the Biodiversity of Luoxiao Mountains Region", CAS key deployment project "East China-Jiangxi Native Plant Inventory and Protection" and other projects. More than 40,000 specimens were collected in Jiangxi, Hunan and Hubei provinces. All the collected specimens were already identified, organized and digitized in the past 5 years. Expedition of the Luoxiao Mountains region has been found to be an area with abundant biodiversity in China, with more than 4,000 species during the comprehensive survey, 1 new plant species, and more than 30 new distribution groups have been identified and incorporated in the plant database of Luoxiao Mountains region.

Countries and regions along the Belt and Road routes

Overseas cooperative research mainly focuses on the countries and regions along the B&R, including Central Asia, Southeast Asia, Africa, India ocean and the Western Pacific. Since "The Belt and Road Initiative" was proposed in 2013, CAS has invested a lot of manpower and resources in developing science and technology exchanges and cooperation with the countries and regions along the route. Depending on various scientific and technological cooperation projects of China, the biological collections vigorously implement the strategy of "going out", and constantly strengthen cooperation and exchanges with other countries in biodiversity, species protection and pest control.

In recent years, the Collection of National Zoological Museum of China has paid attention to the construction of cooperative relations with neighboring countries, and has visited Mongolia, Vietnam, Tajikistan and other countries many times for cooperative research and joint expedition, which not only accumulated valuable specimen resources for

2010 年海南昌江霸王岭地衣考察（左二魏江春院士）
Lichen survey in the Bawangling Mountains, Changjiang, Hainan Province in 2010 (Academician Wei Jiang-Chun, No. 2 from left)

上海九段沙湿地考察
Investigation on Jiuduansha Wetland in Shanghai

在罗霄山脉地区开展植物资源调查
Investigation of plant resources in Luoxiao Mountains region

菌物标本馆早期组织了横断山区（1973-1980 年）、托木尔峰（1977-1978 年）、南迦巴瓦峰（1982-1983 年）、神农架（1984 年）、小五台山（1990 年）、秦岭（1991-1992 年）、大巴山（1994 年），以及南极乔治王岛（1983-1984 年，1994-1995 年）和北极摩尔曼斯克（1995 年）等地区的资源考察。基于对馆藏标本和考察获得的材料的研究，出版了大批专著，为揭示我国及世界菌物本底提供了科学依据。近年来，在国家自然科学基金重点项目和科技部基础性工作专项等项目资助下，开展对我国热带地区（1997-2000 年）、西北地区（2003-2004 年）、海南省（2007-2009 年）、西南喀斯特地区（2014-2018 年）和东北大小兴安岭地区（2014-2018 年），以及挪威斯瓦尔巴德群岛（2002 年）、俄罗斯东西伯利亚地区（2004-2005 年）和南极设得兰群岛（2004-2005 年）等地菌物资源的大规模考察，收集大量标本，发现并报道了大量新物种，大幅提升了人们对我国及世界菌物多样性的认识，研究成果获得国内外同行高度关注。

上海昆虫博物馆的采集重心是华东地区，立足华东，面向全国。近些年的工作尤为注重“一带一路”沿途省份的标本采集。近 5 年来组织野外采集 49 次，采集范围包括青海、吉林、浙江、上海、广西、广东、海南等省份，累计采集标本约 19 万号。标本采集以湿地和土壤环境为主，同时兼顾收集优势类群及部分稀有种类，以期能在生态环境保护和资源利用方面起到作用。

庐山植物园标本馆依托科技部基础性工作重点专项“罗霄山脉地区生物多样性综合科学考察”、中国科学院重点部署项目“华东 - 江西本土植物清查与保护”及其他项目，在江西、湖南和湖北三省采集到标本

赴蒙古国合作研究和联合科考
Visit Mongolia for cooperative research and joint scientific expedition

赴塔吉克斯坦合作研究和联合科考
Visit Tajikistan for cooperative research and joint scientific expedition

赴哈萨克斯坦合作研究和联合科考
Visit Kazakhstan for cooperative research and joint scientific expedition

赴吉尔吉斯斯坦合作研究和联合科考
Visit Kyrgyzstan for cooperative research and joint scientific expedition

赴乌兹别克斯坦合作研究和联合科考
Visit Uzbekistan for cooperative research and joint scientific expedition

在塔吉克斯坦开展标本制作技术培训
Technical training on specimen preparing in Tajikistan

共计 4 万多份，完成了过往 5 年间所有采集标本的鉴定、整理和数字化工作。罗霄山脉地区是我国生物多样性非常丰富的地区，分布着 4000 多种高等植物，此次全面调查发现植物新种 1 种，新分布类群 30 多种，极大地完善了罗霄山脉地区植物资料。

“一带一路”沿线国家和地区

国外合作研究主要集中于“一带一路”沿线国家和地区，包括中亚、东南亚、非洲、印度洋和西太平洋等地。自“一带一路”倡议于 2013 年提出以来，中国科学院投入大量人力物力，与沿线国家和地区开展科技交流合作。各生物标本馆也依托国家各项科技合作项目，大力实施“走出去”战略，不断在生物多样性、物种保护和有害生物防治方面与其他国家加强合作交流。

国家动物博物馆标本馆近年来注重构建与周边国家的合作关系，并多次前往蒙古国、越南、塔吉克斯坦等国进行合作研究和联合科学考察，不但为本地区生物多样性研究积累了宝贵的标本资源，而且加深了双方科研人员的了解，建立了深厚的友谊。还通过国际合作推动了中亚青年人才培养与专业技术培训，拓展了合作空间，推动了后续一系列重大合作项目的联合申报。该成果被中国科学院国际合作局推荐参加“一带一路”倡议实施 5 周年成果展示，并被《中国科学院院刊》报道。

仅分布于越南南部与老挝的中介角蟾
Megophrys intermedia, known only from Southern Vietnam and Laos

成都生物研究所两栖爬行动物标本馆注重通过国际交流与合作，开展战略生物资源的采集、保存与利用，先后与越南国家自然博物馆、俄罗斯科学院动物学研究所与巴西圣保罗大学动物博物馆等国外多家科研院所、博物馆建立了长期合作伙伴关系。例如，从 2015 年至 2019 年，先后 4 次访问越南，开展了大量的联合野外爬行动物科学考察工作。其间对于部分类群如水游蛇科（Natricidae）的后棱蛇属（*Opisthotropis*）与异纹蛇属（*Pararhabdophis*）进行了分类学研究，对水游蛇科 3 属进行了厘定，已在国际期刊上发表 SCI 论文两篇。在国外访问期间，课题组人员还到访了国际知名的多家博物馆或标本馆，如越南科学技术院生态与生物研究所，开展了大量合作研究。

标本馆工作人员访问越南国家自然博物馆
Staff of the museum visited the Vietnam National Museum of Nature

标本馆工作人员出访俄罗斯
Staff of the museum visited Russia

biodiversity research in the region, but also deepened the understanding of researchers from both sides and established profound friendship. Through international cooperation, it has also promoted the training of young talents and professional and technical training in Central Asia, expanded cooperation space, and promoted the joint application of a series of major cooperation projects in the future. This achievement was recommended to the 5th Anniversary Achievements of the B&R Initiative and reported by the *Bulletin of the Chinese Academy of Sciences*.

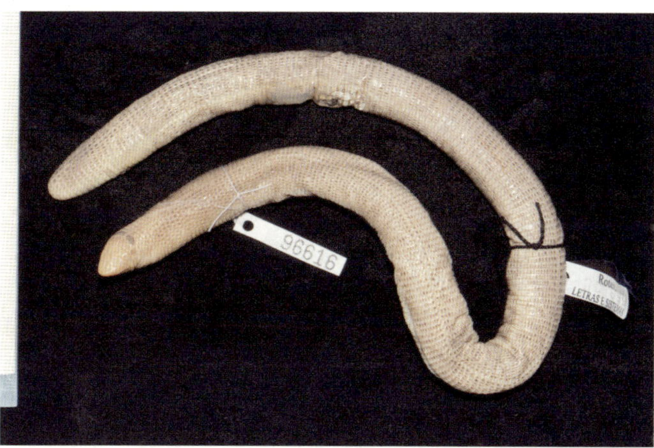

交换标本 *Leposternon microcephalum*，仅分布于南美洲，隶属于蚓蜥科（Amphisbaenidae），对研究爬行动物的适应性演化与性状演化具有重要价值

Specimen obtained via exchange, Smallhead Worm Lizard, *Leposternon microcephalum* (family: Amphisbaenidae) endemic to South America, is of great value in the study of adaptive evolution

The Herpetological Museum of Chengdu Institute of Biology pays attention to international collaboration, especially on collecting, preserving, and utilizing strategic resources. The museum has established long-term cooperative partnership with many renowned institutions or museums, including Vietnam National Museum of Nature (Vietnam), Zoological Institute Russian Academy of Sciences (Russia), and Museum of Zoology, University of São Paulo (Brazil). In Vietnam, for example, researchers were designated to conduct 4 joint herpetofauna surveys from 2015 to 2019. Based on these cooperative works, researchers published two SCI articles in international journals, which include taxonomic revisions of *Opisthotropis* and *Pararhabdophis* within the snake family Natricidae. During these field trips, researches had visited many famous herpetological collections, e.g., the Institute of Ecology and Biological Resources, Vietnamese Academy of Science and Technology, and conducted a lot of cooperative research.

Furthermore, the Herpetological Museum of Chengdu Institute of Biology is forging ahead in an innovative and enterprising spirit. Not only international collaborations, the museum also extends the herpetological collections of strategic resources by exchanging specimens. In 2019, the museum exchanged specimens with the Museum of Zoology, University of São Paulo, three reptilian specimens from its collections were exchanged for 19 specimens. All the 19 Brazilian specimens are endemic to South America, and they belong to 18 families, 16 of them were known out of China. In addition, a total of 23 tissue samples were also received, some of which were used in ongoing studies. These important herpetological materials remarkably increased the collections of foreign amphibians and reptiles preserved in the museum, even for China. And they provides important research materials for researchers visiting the museum, and they tremendously enriched the intellectual merit and international influence of this museum.

In the past five years, relying on MHBS, eight researchers of IHB, CAS have conducted about 30 individual or group visits to Kenya, Ethiopia, Vietnam, Malaysia, Laos, Myanmar, India, Tajikistan, Czech Republic, France, Germany, Japan, USA, Canada, Australia, etc. for cooperative investigating on zoobenthos, fish, amphibians and reptiles, cooperating to cultivate graduate students, undertaking academic exchanges, etc. These cooperation enriched the collection of MHBS and the B&R countries and increased the communication between the museum and the countries and regions along the B&R, strengthening the relationship with relevant institutions and experts of these countries and regions, and laying foundation of further cooperation via B&R. Related research projects include: East African freshwater fish diversity and evolution research; Investigation of the Mekong River Fish Diversity Project; Cooperative studies on the taxonomic classification of the fish in *Cirrhinus* distributed in Bangladesh; Survey and Systematic Taxonomy of Cave Fishes in North-Central Vietnam; Fish diversity and important fish biology in northern Laos; Joint field survey and data collection on the status of fish diversity conservation in Tajikistan; Establishment of a database on fish diversity in the surrounding areas of China; Collaborative research on fish invasion, climate change, and river health environmental assessment; Studies on the interaction between invasive fish and freshwater mollusks; Fishway evaluation of the Elbe River and Rhine River in Germany and Netherlands; Deep-sea fish joint taxonomy studies; Research and exchange of standards related to aquatic organisms resources protection.

The Indian Ocean is closely related to the South China Sea and the East Indian Ocean area is important to the Maritime Silk Road. From 2011 to 2019, the Marine Biodiversity Collections of South China Sea has participated Indian Ocean voyages relying on scientific research vessels from the South China Sea Institute of Oceanology of CAS, and the

与肯尼亚国家博物馆同行交流
Communication with peer of the National Museums of Kenya

东印度洋海洋学考察
East Indian Ocean Oceanographic Survey

此外，成都生物研究所两栖爬行动物标本馆在与国际知名博物馆进行合作期间，进一步扩大战略生物资源馆藏量与国外标本保有量，2019 年与圣保罗大学动物博物馆进行了部分馆藏标本交换，换出标本 3 号，换入标本 19 号，均为南美洲特有种，隶属于 18 科，其中 16 科在我国无分布。此外还取得了组织样品 23 份，其中部分珍贵样品已经投入科学研究。这些研究材料扩展了标本馆馆藏甚至我国境内保存的国外两栖爬行动物标本量，为到馆访问的科研人员提供了重要的研究材料，为标本馆的进一步国际合作、提高国际影响力、发挥出更大的学术价值奠定了重要的基础。

依托水生生物博物馆，近 5 年来，水生生物研究所 8 位研究人员约 30 次率组或单独赴肯尼亚、埃塞俄比亚、越南、马来西亚、老挝、缅甸、印度、塔吉克斯坦、捷克、法国、德国、日本、美国、加拿大、澳大利亚，开展底栖生物、鱼类、两栖爬行类资源合作调查，合作培养研究生，开展学术交流等。增加了博物馆与“一带一路”沿线国家和地区之间的交流，还加强了与这些国家和地区相关机构和专家的联系，为下一步“一带一路”合作打下了基础。相关研究包括：东非淡水鱼类多样性与进化研究；湄公河流域鱼类多样性项目调查；孟加拉国鲮属鱼类分类整理合作研究；越南中北部洞穴鱼类调查与系统分类学研究；老挝北部鱼类多样性与重要鱼类生物学研究；塔吉克斯坦鱼类多样性保护现状及联合实地调查和资料收集；建立我国周边地区鱼类多样性数据库；鱼类入侵、气候变化和河流健康环境评价等方面的合作研究；鱼类入侵种和淡水贝类相互作用关系的研究；在德国、荷兰开展易北河和莱茵河鱼道过鱼效果评估；深海鱼类联合分类研究；水生生物资源保护相关标准制定研究与交流等。

浮游生物连续分层采样网作业
Multinet plankton continuous stratified sampling network

印度洋与南海密切相关，东印度洋航段是海上丝绸之路重要路段。2011-2019 年，南海海洋生物标本馆依托中国科学院南海海洋研究所科考船开展印度洋综合考察航行，累计航程超过 80 000 海里，共采集海洋生物标本 14 769 号，包括许多珍贵的海洋生物标本，如海蛇标本等。开展合作研究的区域包括印度洋、斯里兰卡和阿拉伯海等地。通过相关数据和样品分析，获得了宝贵的现场数据资料，为印度洋海域的生物地球化学循环过程等研究提供了重要的观测证据，为我国“一带一路”建设提供了科技支撑。

从 2015 年起，植物研究所标本馆参与了中国科学院东南亚生物

赴缅甸北部联合科考
Visit Northern Myanmar for cooperative scientific expedition

东南亚中心第一次中缅联合考察
The first China-Myanmar joint expedition with CAS-SEABRI

cumulative range is more than 80,000 nautical miles. A total of 14,769 marine biological specimens have been collected, including many precious marine life specimens, such as sea snake specimens, etc. The regions for cooperative research include the Indian Ocean, Sri Lanka and the Arabian Sea. Through the relevant data and sample analysis, valuable field data were obtained, which provided important observational evidence for research on the biogeochemical cycle in the Indian Ocean and provided scientific and technological support for the B&R Initiative construction.

Since 2015, PE has participated in the scientific investigations of the Southeast Asia Biodiversity Research Institute of CAS. Southeast Asia is very rich in biodiversity and with a high proportion of plant endemic species, including four global hotspots of biodiversity. Under the leadership of the Southeast Asia Biodiversity Research Institute of CAS, the researchers carried out more than a dozen cooperative scientific investigations. These joint research works discovered one newly recorded order and two families and eight new species of angiosperms from Myanmar. Two special issues of plant diversity of southeastern Asia were published in *Phytokeys*.

The plants from Southeast Asia are important for the study of the global tropical flora. XTBG is located in the southwest of China and is not far away from the international borders. This area has tropical environments that are rare in China. HITBC opens its doors toward Southeast Asia. Most of the Southeast Asian specimens are obtained by exchange with other herbaria. Recently, XTBG has been taking part in the development of the B&R Initiative and has established overseas branches or offices in Myanmar and Laos, especially the Southeast Asia Biodiversity Research Institute, CAS (CAS-SEABRI) in Myanmar. At the same time, based on the cooperation with CAS-SEABRI, a series of cooperative research work has been carried out by HITBC and other institutes of Southeast Asia.

To be more influential internationally, serve for the B&R Initiative, the KUN also sign cooperation agreements with herbaria from Central and Southeastern Asian countries. KUN actively strengthens cooperation with countries along the B&R, advances communication, expands the space for international inspection and specimen exchange, and seeks a new situation of mutual benefit and win-win. In 2019, cooperation agreements were signed between KUN and the herbaria of Kyrgyzstan, Uzbekistan, Kazakhstan and Tajikistan, respectively. This is also the first step to implement the advocate to jointly build the "Center for Biodiversity Research and Protection in Central Asia" by the five Central Asian countries (Uzbekistan, Kazakhstan, Tajikistan, Kyrgyzstan, Turkmenistan) and China (Kunming Institute of Botany, etc.) (referred to as "5 + 1").

Recently, several joint expedition groups of KUN have been to the eastern United States, Uzbekistan, Georgia, Kyrgyzstan, Kazakhstan and Tajikistan in central Asia and Georgia in West Asia for cooperative botanical research and field works. Specimens and related molecular materials were collected, and hundreds of plants (by seeds) were introduced. Special attention is paid to the collection and conservation of endemic and important economic plants in Central Asia.

The cooperative biological resources investigations of the Herbarium of Chengdu Institute of Biology. was mainly concentrated in the Qinghai-Tibet Plateau and its surrounding areas. In the winter of 2017, we conducted a plant diversity survey along the gradient from 200 m to 3000 m in the Pokhara area in the central and western part of Nepal. In recent years, we have a lot of important cooperation research with Southeast Asian regions, such as Vietnam Fern Diversity Survey (2014-2016), Vietnam Asteraceae Diversity Survey (2018), Thailand Fern Diversity Survey (2018-2019), Northern Myanmar plant diversity (2009), Myanmar Asteraceae Diversity Survey (2019), and Laos Asteraceae Diversity Survey

多样性中心的科学考察工作。东南亚生物多样性非常丰富，植物特有种类比例非常高，具有四个全球生物多样性热点地区。科研人员在中国科学院东南亚生物多样性中心的组织和领导下，先后开展了十多次联合科学考察，发现缅甸被子植物 1 个新记录目 2 个新记录科和 8 个新种；在国际植物分类学杂志 *Phytokeys*，组织了 2 期东南亚植物多样性研究专刊。

东南亚热带地区的植物对研究全球热带植物区系等有着重要的意义。西双版纳热带植物园标本馆地处中国少有的热带地区，立足边疆，面向东南亚，通过标本交换等方式获得一批珍贵的东南亚热带国家的标本资源。近年来，随着“一带一路”的推广，西双版纳热带植物园标本馆在东南亚缅甸、老挝等国设立海外机构“东南亚生物多样性研究中心”（简称“东南亚中心”），同时开展了一系列合作研究工作。

同时，昆明植物研究所标本馆也与“一带一路”沿线的中亚、东南亚等国家和地区的标本馆开展合作，深化交流，拓展国际考察和标本交换空间，寻求互利共赢新局面。例如，2019 年分别与吉尔吉斯斯坦、乌兹别克斯坦、哈萨克斯坦、塔吉克斯坦 4 个国家标本馆之间签署合作协议。这也是贯彻落实未来由中亚五国（乌兹别克斯坦、哈萨克斯坦、塔吉克斯坦、吉尔吉斯斯坦、土库曼斯坦）和中国（昆明植物研究所等）（简称“5+1”）合作共同建设“中亚生物多样性研究和保护中心”倡议内容之一。

此外，近年来昆明植物研究所标本馆分别赴美国东部，中亚的乌兹别克斯坦、吉尔吉斯斯坦、哈萨克斯坦、塔吉克斯坦，以及西亚的格鲁吉亚开展合作研究和联合考察，采集植物标本以及相关分子材料，引种（种子）数百种。尤其重点关注中亚特有和重要经济植物的采集保存。

成都生物研究所植物标本馆的战略生物资源的合作考察主要集中在青藏高原及其周边地区，如 2017 年冬季前往尼泊尔，在尼泊尔中西部博卡拉一线，从低海拔（200 米）至中高海拔（3000 米）沿海拔梯度进

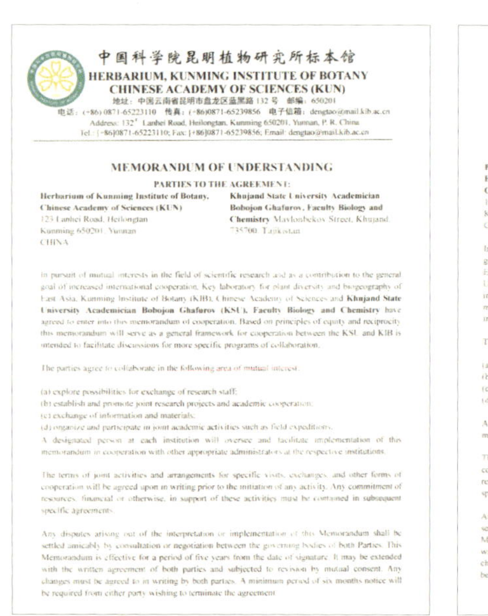

中国科学院昆明植物研究所标本馆
HERBARIUM, KUNMING INSTITUTE OF BOTANY
CHINESE ACADEMY OF SCIENCES (KUN)
地址：中国云南省昆明市盘龙区蓝黑路 132 号 邮编：650201
电话：(+86) 0871-65223110 传真：(+86)0871-65239856 电子信箱：dengtao@mail.kib.ac.cn
Address: 132 Lanhei Road, Heilongtan, Kunming 650201, Yunnan, P. R. China
Tel.: [+86]0871-65223110; Fax: [+86]0871-65239856; Email: dengtao@mail.kib.ac.cn

MEMORANDUM OF UNDERSTANDING

PARTIES TO THE AGREEMENT:

Herbarium of Kunming Institute of Botany, Chinese Academy of Sciences (KUN)
123 Lanhei Road, Heilongtan
Kunming 650201, Yunnan
CHINA

Khujand State University Academician Bobojon Ghafurov, Faculty Biology and Chemistry Mavlonbekov Street, Khujand 735700, Tajikistan

In pursuit of mutual interests in the field of scientific research and as a contribution to the general goal of increased international cooperation, Key laboratory for plant diversity and biogeography of East Asia, Kunming Institute of Botany (KIB), Chinese Academy of Sciences and **Khujand State University Academician Bobojon Ghafurov (KSU), Faculty Biology and Chemistry** have agreed to enter into this memorandum of cooperation. Based on principles of equity and reciprocity this memorandum will serve as a general framework for cooperation between the KSU and KIB is intended to facilitate discussions for more specific programs of collaboration.

The parties agree to collaborate in the following area of mutual interest.

(a) explore possibilities for exchange of research staff;
(b) establish and promote joint research projects and academic cooperation;
(c) exchange of information and materials;
(d) organize and participate in joint academic activities such as field expeditions.

A designated person at each institution will oversee and facilitate implementation of this memorandum in cooperation with other appropriate administrators at the respective institutions.

The terms of joint activities and arrangements for specific visits, exchanges, and other forms of cooperation will be agreed upon in writing prior to the initiation of any activity. Any commitment of resources, financial or otherwise, in support of these activities must be contained in subsequent specific agreements.

Any disputes arising out of the interpretation or implementation of this Memorandum shall be settled amicably by consultation or negotiation between the governing bodies of both Parties. This Memorandum is effective for a period of five years from the date of signature. It may be extended with the written agreement of both parties and subjected to revision by mutual consent. Any changes must be agreed to in writing by both parties. A minimum period of six months notice will be required from either party wishing to terminate the agreement.

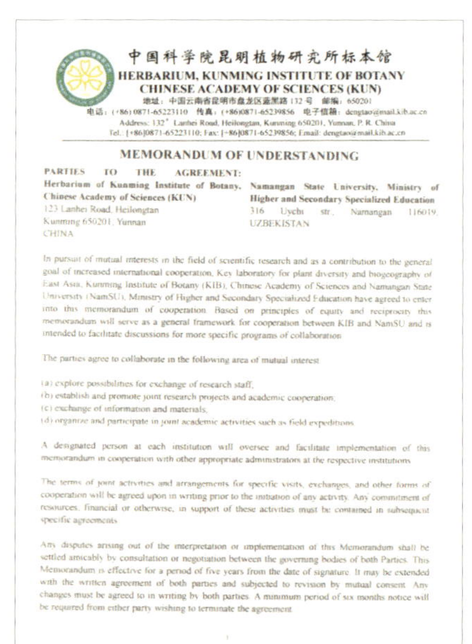

中国科学院昆明植物研究所标本馆
HERBARIUM, KUNMING INSTITUTE OF BOTANY
CHINESE ACADEMY OF SCIENCES (KUN)
地址：中国云南省昆明市盘龙区蓝黑路 132 号 邮编：650201
电话：(+86) 0871-65223110 传真：(+86)0871-65239856 电子信箱：dengtao@mail.kib.ac.cn
Address: 132 Lanhei Road, Heilongtan, Kunming 650201, Yunnan, P. R. China
Tel.: [+86]0871-65223110; Fax: [+86]0871-65239856; Email: dengtao@mail.kib.ac.cn

MEMORANDUM OF UNDERSTANDING

PARTIES TO THE AGREEMENT:

Herbarium of Kunming Institute of Botany, Chinese Academy of Sciences (KUN)
123 Lanhei Road, Heilongtan
Kunming 650201, Yunnan
CHINA

Namangan State University, Ministry of Higher and Secondary Specialized Education
316 Uychi str., Namangan 116019,
UZBEKISTAN

In pursuit of mutual interests in the field of scientific research and as a contribution to the general goal of increased international cooperation, Key laboratory for plant diversity and biogeography of East Asia, Kunming Institute of Botany (KIB), Chinese Academy of Sciences and Namangan State University (NamSU), Ministry of Higher and Secondary Specialized Education have agreed to enter into this memorandum of cooperation. Based on principles of equity and reciprocity this memorandum will serve as a general framework for cooperation between KIB and NamSU and is intended to facilitate discussions for more specific programs of collaboration.

The parties agree to collaborate in the following area of mutual interest

(a) explore possibilities for exchange of research staff,
(b) establish and promote joint research projects and academic cooperation;
(c) exchange of information and materials,
(d) organize and participate in joint academic activities such as field expeditions

A designated person at each institution will oversee and facilitate implementation of this memorandum in cooperation with other appropriate administrators at the respective institutions

The terms of joint activities and arrangements for specific visits, exchanges, and other forms of cooperation will be agreed upon in writing prior to the initiation of any activity. Any commitment of resources, financial or otherwise, in support of these activities must be contained in subsequent specific agreements

Any disputes arising out of the interpretation or implementation of this Memorandum shall be settled amicably by consultation or negotiation between the governing bodies of both Parties. This Memorandum is effective for a period of five years from the date of signature. It may be extended with the written agreement of both parties and subjected to revision by mutual consent. Any changes must be agreed to in writing by both parties. A minimum period of six months notice will be required from either party wishing to terminate the agreement

中国科学院昆明植物研究所标本馆
HERBARIUM, KUNMING INSTITUTE OF BOTANY
CHINESE ACADEMY OF SCIENCES (KUN)
地址：中国云南省昆明市盘龙区蓝黑路 132 号 邮编：650201
电话：(+86) 0871-65223110 传真：(+86)0871-65239856 电子信箱：dengtao@mail.kib.ac.cn
Address: 132 Lanhei Road, Heilongtan, Kunming 650201, Yunnan, P. R. China
Tel.: [+86]0871-65223110; Fax: [+86]0871-65239856; Email: dengtao@mail.kib.ac.cn

Manager
Herbarium of Kunming Institute of Botany,
Chinese Academy of Sciences
Date:

Dean of faculty
Mirzobakhodurova Sh.R.
Khujand State University
Academician Bobojon Ghafurov,
Faculty Biology and Chemistry
Date

中国科学院昆明植物研究所标本馆
HERBARIUM, KUNMING INSTITUTE OF BOTANY
CHINESE ACADEMY OF SCIENCES (KUN)
地址：中国云南省昆明市盘龙区蓝黑路 132 号 邮编：650201
电话：(+86) 0871-65223110 传真：(+86)0871-65239856 电子信箱：dengtao@mail.kib.ac.cn
Address: 132 Lanhei Road, Heilongtan, Kunming 650201, Yunnan, P. R. China
Tel.: [+86]0871-65223110; Fax: [+86]0871-65239856; Email: dengtao@mail.kib.ac.cn

Manager
Herbarium of Kunming Institute of Botany,
Chinese Academy of Sciences
Date:

Rector
Namangan State University,
Ministry of Higher and Secondary
Specialized Education
Date

KUN 与其他国家的合作协议
Cooperation agreements between KUN and other countries

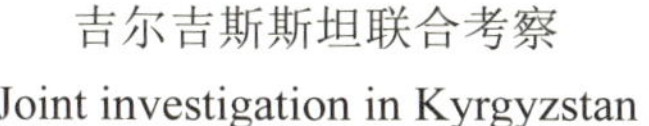

吉尔吉斯斯坦联合考察
Joint investigation in Kyrgyzstan

赴北美东部联合考察
Visited the eastern part of North America for joint expedition

北马其顿联合考察采集真菌标本
Joint fungal investigation in northern Macedonia

赴肯尼亚开展合作研究和联合野外考察
Visit Kenya for cooperative research and joint scientific expedition

(2018). By strengthening cooperation with Southeast Asian countries such as Nepal, Myanmar, Vietnam, Thailand, and Laos in the areas of plant diversity research and species resource protection, it serves the B&R Initiative.

With the support of various projects, the researchers of the Northeast Biological Herbaria have extensively carried out the comprehensive collection, identification and preservation of biological resources in Northeast China. In addition, we selected the national key areas to collect and study the key biological groups. Besides, under the agreement of international cooperation, several joint investigations have been carried out in the B&R countries, such as Russia, Japan, Republic of Korea, Vietnam, Thailand and Kenya. In the past five years, we have organized 105 joint field investigations and 337 person-times. In order to further continuous field investigation will improve the understanding of the local biodiversity and play an important role in the systematic study of various biological groups and the promotion of global ecological environment protection.

The implementation of the B&R Initiative has created many conditions and opportunities for scientific research through international cooperation projects. XJBI has also expanded the study of biodiversity to five Central Asian countries by taking advantage of geographical advantages. For example, XJBI carried out the Central Asia cooperation project "Desert Plant Diversity and Digitization Research in Uzbekistan", visit the National Herbarium of Institute of Botany, Academy of Sciences of Uzbekistan (TASH), and carried out the digitization of desert plant specimens; participated in the environmental investigation of Gwadar Port in Pakistan; visit Tajikistan for the investigation of desert plant resources; cooperated with Kazakhstan to open cooperative project of investigation on the survival status of wild

访问乌兹别克斯坦科学院植物研究所国家标本馆
Visit the National Herbarium of Institute of Botany, Academy of Sciences of Uzbekistan (TASH)

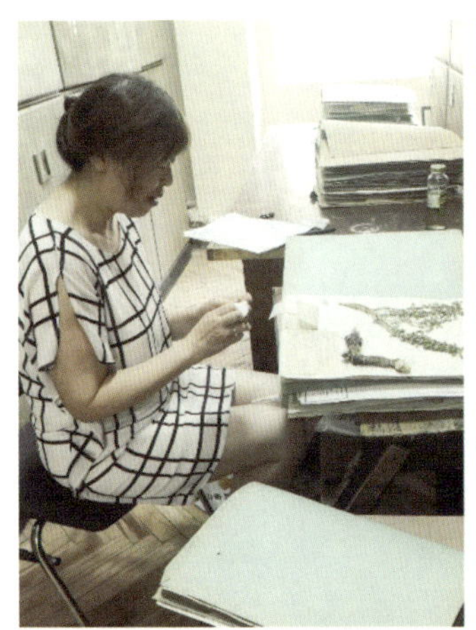

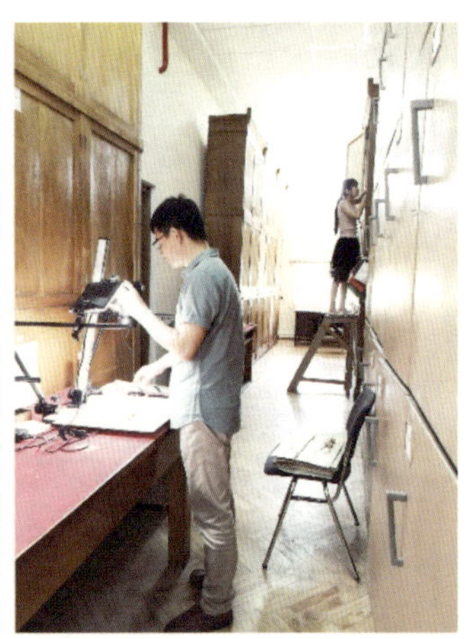

与乌兹别克斯坦科学院植物研究所国家标本馆合作开展荒漠植物标本的数字化工作
Cooperation with TASH on the digitization of desert plant specimens

行了联合植物多样性调查。标本馆近年来还与越南合作开展了蕨类植物多样性调查（2014-2016 年）及菊科植物多样性调查（2018 年），与泰国合作开展了蕨类植物多样性调查（2018-2019 年），与缅甸北部合作开展了植物多样性调查（2009 年）及与缅甸合作开展了菊科植物多样性调查（2019 年）、与老挝合作开展了菊科植物多样性调查（2018 年）等。通过加强与尼泊尔、缅甸、越南、泰国和老挝等东南亚国家在植物多样性研究、物种资源保护等领域的合作研究，服务于国家“一带一路”倡议。

东北生物标本馆也在各类项目的支持下，立足东北，广泛开展东北地区各类生物资源的全面采集、鉴定及保藏工作，同时选择全国重点区域，对重点类群生物资源进行标本收集与研究。此外，广泛开展国际合作，在俄罗斯、日本、韩国等东北亚国家，以及越南、泰国、肯尼亚等“一带一路”沿线国家进行考察。近 5 年来，共组织联合野外考察 105 次 337 人次。持续的野外调查和标本收集将为进一步认识当地的生物多样性，在全球尺度下系统研究各生物类群，以及促进全球生态环境保护发挥十分重要的作用。

国家“一带一路”倡议提出以来，通过国际合作项目为科学研究创造了不少条件和机遇。新疆生态与地理研究所标本馆利用地缘优势，也将生物资源多样性研究拓展到中亚五国。例如，开展了中亚合作项目“乌兹别克斯坦荒漠植物多样性与数字化研究”，前往乌兹别克斯坦科学院植物研究所国家标本馆（TASH），进行荒漠植物标本的数字化工作；参与巴基斯坦瓜达尔港生态环境考察；赴塔吉克斯坦进行荒漠植物资源调查；与哈萨克斯坦合作开展野苹果生存现状调查合作项目。依托这些项目，完成中亚植物多样性的编目；构建了“中亚生物多样性数据库”，为中亚地区生物多样性保护与可持续管理提供了重要数据支撑。还开展科技基础资源调查专项——“一带一路”国家传统草药品种本底整理及数据库建设，这将有利于“一带一路”国家传统医药体系的互联互通，形成政治互信、经济融合、文化包容的“一带一路”传统医药体系利益共同体。此外还帮助合作单位申请中国科学院国际人才计划（PIFI），培养国际合作团队，形成了良好的合作渠道。

古脊椎动物与古人类研究所标本中心以研究所为依托，广泛收集国外标本。特别是近年来加大了与“一带一路”沿线国家的交流合作，通过开展联合科考和合作研究等形式，填补馆藏空白，丰富藏品类型。2018 年，

apple. Based on these projects, the catalogue of plant diversity in Central Asia has been completed; the "Central Asia Biodiversity Database" has been constructed, which provides important data support for biodiversity conservation and sustainable management in Central Asia. They also carry out the special investigation on basic resources of science and technology. Traditional Herbal Medicine Background Survey and Database Construction of B & R countries. This will benefit the interconnection of the traditional medicine system of the B&R countries and form the interest community of traditional medicine system with mutual political and trust, economic integration and cultural inclusion. In addition, it also helped cooperative units to apply for the CAS President's International Fellowship Initiative (PIFI), cultivate international cooperation teams and form good cooperation channels.

Collection Center of IVPP collected a large number of specimens from abroad based on IVPP. Especially in recent years, communication and cooperation with the B&R countries has been strengthened. Specimens from these countries have filled in the blank of collection and enriched collection types through the joint expedition and cooperative research. In 2018, NI Xijun's group of IVPP and the Government College University Faisalabad and other institutions formed the Sino-Pakistan Paleontological Expedition Team, which carried out the first joint scientific expedition in Pakistan. The team collected fossils and rock samples, did stratigraphic comparison work in Punjab Province and other places. Verified original information of several classical fossil sites published by western researchers in early years, corrected some key erroneous records, but also found several new fossil sites and stratums. It provides valuable specimens for further research on paleontology cooperation, and exploration of the evolution and response mechanism of Cenozoic vertebrates under the background of the disappearance of the Eastern Tethys Ocean and uplift of the Qinghai-Tibet Plateau. In 2019, NI Xijun's group went to Tajikistan to carry out joint expedition there, conducted bio-stratigraphic investigation and field excavation

中 - 巴古生物学联合考察队
Sino-Pakistan Paleontological Expedition Team

联合考察队发现的鱼化石
Fish fossils found by the joint expedition team

《泛喜马拉雅植物志》尼泊尔考察
Botanical survey in Nepal for *Flora of Pan-Himalaya*

中 - 非联合考察队在坦桑尼亚的梅鲁山考察
Sino-Africa Joint investigation team visit the Mt. Meru in Tanzania

研究所倪喜军课题组与巴基斯坦费萨拉巴德政府学院大学等机构组成中 - 巴古生物学联合考察队，在巴基斯坦开展了首次联合科考。考察队在旁遮普省等地开展了古生物化石和岩石样品采集、地层对比等工作。不仅重新核实了早期西方学者发表的多个经典化石地点的原始坐标和岩石属性等信息，纠正了一些关键的错误记录，还发现了多个新的化石地点和层位，为双方进一步开展古生物学合作研究、探讨东特提斯海消亡与青藏高原隆升背景下的新生代脊椎动物演化与响应机制提供了珍贵的标本材料。2019 年，倪喜军课题组还赴塔吉克斯坦开展了联合科学考察工作，共同进行了生物地层学考察和古生物化石野外发掘工作，并建立了友好、密切的合作关系。

泛喜马拉雅地区覆盖喜马拉雅山、阿富汗境内的兴都 - 库什山的东北部、喀喇昆仑山和横断山等，涉及 7 个国家。这里生物多样性不仅丰富度高，而且特有性强，涵盖保护国际（CI）确定的全球 34 个生物多样性热点地区中的 3 个。

中国科学院植物研究所标本馆一直致力于植物标本的广泛合作收集，增加馆藏标本的类型、数量、质量和代表性，力争做到国内标本物种和地域全覆盖、国外标本科级和属级全覆盖。近年来，标本馆加强与周边国家和地区合作，特别是与全球生物多样性热点地区的合作研究，如东南亚（越南、印度尼西亚、缅甸）、泛喜马拉雅地区（印度、尼泊尔）和非洲（肯尼亚）；平均每年与全球十余个国家交换的植物标本达 3000 余份。中国科学院植物研究所 2010 年牵头国内几十家科研院所与大专院校，启动了泛喜马拉雅地区植物多样性调查和《泛喜马拉雅植物志》的编撰。通过该项目于 2013-2019 年与泛喜马拉雅地区 7 个“一带一路”沿线国家开展多项合作研究，出版《泛喜马拉雅植物志》7 卷册和总名录。

热带东非沿海地区森林生物多样性非常丰富，植物特有种类比例非常高，是世界上 34 个生物多样性热点地区之一。武汉植物园从 2009 年起承担了非洲植物资源调查的相关项目，前往非洲开展合作研究成为武汉植物园标本馆的一项重要任务。武汉植物园作为中 - 非联合研究中心的挂靠单位，一直主导着在非洲开展植物资源调查方面的研究。自 2014 年中 - 非联合研究中心成立至今，标本馆与肯尼亚、坦桑尼亚、埃塞俄比亚、马达加斯加等东非国家开展了大量合作研究和联合野外考察工作，不但丰富了标本馆对非洲植物的标本收藏，还支撑了相关著作，如《非洲常见植物野外识别手册——肯尼亚山册》和《肯尼亚植物志》的编研。

从 2016 年起，植物研究所标本馆也参与了中国科学院中 - 非联合研究中心的联合科学考察工作。科研人员克服重重困难，在非洲的肯尼亚及马达加斯加地区开展了十多次联合科学考察，使馆藏标本的属、种以及分子材料样品的覆盖面均得到了提高。

中 - 非联合考察队登上非洲最高山乞力马扎罗山
Sino-Africa Joint investigation team reached the summit of Mt. Kilimanjaro

安第斯山脉是世界上最长的山脉，素有“南美洲脊梁”之称，亚马孙热带雨林是世界上生物多样性最丰富的地区，南美还是栽培植物的主要起源中心。自 2008 年开始，华南植物园标本馆在中国科学院对外合作项目的支持下多次对亚马孙流域上游热带雨林生物多样性和安第斯山脉高山植被情况进行合作调查，为今后开展全球变化对该地区生物多样性影响研究奠定基础。

菌物标本馆除了对我国热带地区、西北地区、海南省、青藏高原、西南喀斯特地区和东北大小兴安岭等地区进行考察外，还赴南极乔治王岛、北极摩尔曼斯克、挪威斯瓦尔巴德群岛、俄罗斯东西伯利亚地区和南极设得兰群岛

中 - 非联合考察队在非洲第二高山肯尼亚山考察
Sino-Africa Joint investigation team visit the Mt. Kenya

中 - 非联合考察队在肯尼亚的塔伊塔丘陵考察
Sino-Africa Joint investigation team visit the Taita Hills in Kenya

of fossils, and established a friendly and close cooperative relationship.

The Pan-Himalayan regions include the Himalayas, the northeast of Mt. Hindu-Kush in Afghanistan, Karakorum, and Hengduan Mountains in seven countries. The biodiversity of Pan-Himalayan regions has unique and rich biota, covering three of the 34 biodiversity hotspots in the world identified by the Conservation International (CI).

PE has long term plan for extensive cooperative collections of plant specimens, aiming to increase the resource types, quantity, quality and representativeness of the specimens and try to achieve full coverage of native flora and regions, and family and genus of world flora. In recent years, PE has strengthened cooperation with surrounding countries and regions, especially in the research of global biodiversity hotspots, such as Southeast Asia (Vietnam, Indonesia, Myanmar), pan Himalaya (India, Nepal) and Africa (Kenya). More than 3,000 plant specimens from more than 10 countries around the world are exchanged each year. In 2010, PE started the pan Himalayan plant diversity survey and the compilation of *Flora of Pan-Himalaya* cooperated by dozens of domestic research institutes. The project currently has carried out many cooperative researches with 7 countries along the B&R in the Pan-Himalayan region during the period of 2013-2019 and published 8 volumes of *Flora of Pan-Himalaya*.

The plant biodiversity in tropical coastal areas of East Africa is very rich, and the proportion of endemic species is very high. It is one of the 34 hotspots of biodiversity in the world. Since 2009, Wuhan Botanical Garden has undertaken projects related to the investigation of African plant resources. To carry out cooperative research in Africa has become an important task for the Herbarium of Wuhan Botanical Garden. Wuhan Botanical Garden, as the affiliated unit of "Sino-

亚马孙雨林联合野外考察
Joint field trip in Amazon rainforest

秘鲁联合野外考察
Joint field trip in Peru

在西太平洋卡罗琳海山采集的部分深海生物标本
Some deep sea biological specimens collected in the Caroline seamount of the Western Pacific Ocean

等地进行菌物资源的大规模合作考察。基于对馆藏标本和考察获得的材料的研究，发现并报道了大量新物种，出版了大批专著，为揭示我国及世界菌物本底提供了科学依据，大幅提升了对我国及世界菌物多样性的认识。

此外，我国在极地海域和深海等领域也有所收获。例如，海洋生物标本馆在2010-2019年近10年间，依托科学院先导A、B类专项，科技部基础专项，科技部基础调查专项，国家重点基础研究发展计划（973计划），以及国家自然科学基金等项目，分别在南极、北极、西太平洋以及中国沿海对各门类海洋生物进行调查，发表1个新科，13个新属，266个新种。

4.2 科研支撑

生物标本是开展许多生物学研究的基础，特别是对物种、进化、物种多样性等这些生物学中极其重要的概念的阐释和不断深入理解，都离不开生物标本。对全国以及世界其他地区的不同时期的生物标本进行收集，并长久妥善地保存，是生物标本馆最重要的工作。而这些标本支撑了大量的生物学领域的科学研究，发挥着巨大作用。

1）经典分类及物种资源研究

标本和标本馆对我国生物物种资源研究发挥了重要支撑作用。生物标本馆是我国摸清生物资源家底、有效保护和利用生物资源，以及防止有害生物入侵的坚强后盾。《中国植物志》、《中国动物志》、《中国孢子植物志》、《泛喜马拉雅植物志》、《中国药用植物志》、《中国大百科全书》（第三版）等重要志书的编研均是国家为了摸清家底、有效保护和利用生物资源及防止有害生物入侵而相继启动的一系列重大工程，是功在当代利在千秋的伟大事业。而各个标本馆作为坚强后盾，承担了大量的工作，完成了繁重的任务。80%以上的物种标本信息依托中国科学院生物标本馆体系，各馆也为这些重大工程提供了卓有成效的服务，保障了这些计划的顺利执行和志书的完成。

例如，植物研究所标本馆主持的由15国参与的重大国际合作项目“泛喜马拉雅地区综合考察和植物志

Africa Joint Research Center, Chinese Academy of Sciences" (SAJOREC), has been leading the research of plant resources investigation in Africa. Since the establishment of the SAJOREC in 2014, the herbarium has carried out a large number of cooperative research and joint field investigations with Kenya, Tanzania, Ethiopia, Madagascar and other East African countries, not only enriching the herbarium's collection of African plants, but also supporting the compilation of relevant works, such as *Field book of African Common Plants: Kenyan Mountain* and *Flora of Kenya*.

Since 2016, PE has also participated in the scientific investigation of the Sino-Africa Joint Research Center of CAS. The researchers overcame many difficulties and carried out more than 10 joint scientific investigations in Kenya and Madagascar. These collections greatly improved the coverage of genera, species, and molecular material samples in PE.

部分极地生物标本

Some polar biological specimens

The Andes is the longest mountain range in the world, known as the "backbone of South America". The Amazon rainforest is the most biodiversity-rich area in the world, and South America is also the main origin center of cultivated plants. Since 2008, supported by the CAS foreign cooperation project, the Herbarium of South China Botanical Garden has carried out cooperative investigation on the biodiversity of tropical rainforest in the upper reaches of the Amazon basin and the alpine vegetation in the Andes many times, which will lay a foundation for future research on the impact of global change on biodiversity in the region.

The Fungarium organized fungal investigations to the tropical areas and Northwest of China, and Hainan, Qinghai Tibet-Plateau, karst areas in Southwest and the Greater and the Lesser Khingan Mountains in Northeast of China. In addition, a large-scale cooperative investigation was carried out in King George Island in Antarctica, Murmansk in the Arctic, Svalbard Islands in Norway, eastern Siberia in Russia and Shetland Islands in Antarctica. Based on the research on the collected specimens and materials obtained from the investigation, a large number of new species were found and reported, and many monographs were published, which provided a scientific basis for revealing the background of fungi in China and the world, and greatly improved the understanding of the diversity of fungi in China and the world.

In addition, China has also gained in polar waters and the deep sea. For example, in the past 10 years of 2010-2019, relying on the projects of class A and class B projects, basic projects of the Ministry of science and technology, basic investigation projects of the Ministry of science and technology, 973 projects and National Natural Science Foundation of China, the Marine Biological Museum carried many investigations in the Antarctic, Arctic, Western Pacific and coastal areas of China, and published a New family, 13 new genera and 266 new species.

4.2 Scientific research support

The biological specimen is the basis of many biological researches, especially for the interpretation and continuous in-depth understanding of the most important concepts in biology, such as species, evolution and species diversity. It is the most important work for the biological collections to collect and keep the specimens of different periods in China and other parts of the world for a long time. These specimens support a large number of scientific researches in the field of biology and play a huge role.

1) Classical classification and species resources research

Specimens and collections play an important supporting role in the study of biological species resources in China. The biological collections are the strong backing for China to find out the background of biological resources, effectively protect and utilize biological resources, and prevent the invasion of harmful organisms. The compilation and research of important works, such as the *Flora Reipublicae Popularis Sinicae*, *Fauna Sinica*, *Flora Cryptogamarum Sinicarum*, *Flora of Pan-Himalaya*, *Flora of Chinese Medicinal Plants*, *Encyclopedia of China* (Third Edition), is a series of major projects launched by the state in order to find out the background, effectively protect and utilize biological resources and prevent the invasion of harmful organisms. It is a great cause of great contribution in the future. As a strong backing, each collection has undertaken a lot of work and completed a heavy task. More than 80% of the information of species and specimens relies on the biological collection system of CAS, and each collection also provides effective services for these major projects, ensuring the smooth implementation of these plans and the completion of the works.

新近出版的《泛喜马拉雅植物志》部分卷册
Some newly published volumes of *Flora of Pan-Himalaya*

编研”，现已出版4本《泛喜马拉雅植物志》，该项目集中体现了传统方法与生物学的最新发展和手段的紧密结合——大规模野外调查、形态性状和DNA分析等，揭示了这一重要地区的植物物种多样性格局和生物资源的丰富度。

在分类学逐渐衰落、研究力量极度萎缩的情况下，各馆坚持不断克服困难，通过各种方式进行研究，如聘请退休老专家或邀请其他单位相关类群研究者前来鉴定标本。2014-2019年共鉴定标本近120万号，使得馆藏定名标本量有了显著提升，标本鉴定率也能够维持在相对合理的水平。此外，还支撑了5600余篇论文和近230部专著的发表，以及多项专利的申请和行业标准的制定。

“三志”编研

生物志书的编研是对地球上生物种类的认识和整理，是生物多样性保护、生物安全及生物资源可持续利用的基础。我国生物资源丰富，在国家自然科学基金委员会及有关部门的支持下，国家陆续开展了各类生物资源志书的编研，其中最重要的就是“三志”。“三志”及相关志书的编研，是对生物物种名录在国家尺度的编制，极大地提高了对中国生物区系的认知水平和生物分类学的发展速度。

《中国植物志》

《中国植物志》从20世纪20年代开始酝酿，历经四代科学家薪火相传、费时八十余载研究编纂完成，中国植物从此有了权威而全面的“户口簿”。《中国植物志》自1959年出版第一卷始，历时45年，于2004年10月全部完成80卷126册，包含国产及归化的种子植物和蕨类植物，分别归属于301科3408属31 142种。全书有5000多万字，图版9000余幅，对每种植物的科、属、种的中外文献，植物形态特征，国内外的分布及其生态环境，科学研究价值及经济用途，系统分类的讨论等进行论述，是世界上已出版植物志中种类最丰富的科学著作，为合理开发利用植物资源提供了极为重要的基础信息和科学依据，对陆地生态系统研究将起到重大促进作用，对科研和经济建设都有重要的价值。

首先提出编写《中国植物志》并进行开拓性研究的中国学者是胡先骕。他在1919年就亲自到野外调查考察和采集植物标本，在同行的共同努力下，采集到近百万份植物标本，开始创建中国人自己的标本馆。1934年，他在中国植物学会第一届年会上，提出编写《中国植物志》的倡议。

《中国植物志》的出版为全世界植物学家瞩目，自1989年开始，中美合作编写英文版《中国植物志》，即 *Flora Reipublicae Popularis Sinicae*。前后历时20多年编撰，2013年9月《中国植物志》英文修订版全部出版，共50卷，是目前世界上规模最大的高水平英文版植物志。这是中国植物学界又一里程碑式的成果。

《中国植物志》是新中国生物学领域的重大学术成就。这主要体现在两个方面，一是包含极其丰富的第一手资料。中华人民共和国成立后，由中国科学院及其下属研究所为主体，开始地质性植物学野外调查考察与标本采集工作，在1964-2000年，还曾组织10余次大型的全国性野外调查考察。采集到的植物标本总计达1700万份，这在世界采集史上是罕见的。可以说，《中国植物志》是全部由中国植物学工作者在紧密结合自身野外调查与考察的基础上完成的，含有大量新信息、新内容，有很高的科学价值。二是为植物学各分

For example, under the auspices of the Herbarium of Institute of Botany, the major international cooperation project "Comprehensive Investigation and Compilation and Research of Flora in the Pan-Himalayan Region" in which 15 countries participate, has published 4 volumes of *Flora of Pan-Himalaya*. The project focuses on the close combination of the latest development of traditional methods and biology, such as large-scale field investigation, morphological characteristics and DNA analysis, etc., and reveals the diversity of plant species and the abundance of biological resources in this important area.

With the gradual decline of taxology and the extremely shrinking of research power, all collections have continued to overcome difficulties, such as hiring retired experts or inviting researchers from other units to identify specimens. In 2014-2019, nearly 1.2 million specimens were identified, which has significantly improved the number of identified specimens in the collection, and the specimen identification rate can also be maintained at a relatively reasonable level. In addition, it has supported the publication of more than 5,600 papers and nearly 230 monographs, as well as the application of many patents and the formulation of industry standards.

The compilation and research of "Three Flora/Fauna"

The compilation and research of Flora and Fauna is the understanding and arrangement of the biological species on the earth, and is the basis of biodiversity protection, biosafety and sustainable utilization of biological resources. China is rich in biological resources. With the support of NSFC and relevant departments, the country has successively carried out the compilation and research of various kinds of biological resources records, the most important of which is "Three Flora/Fauna". The compilation and research of "Three Flora/Fauna" and related books is the compilation of the list of biological species at the national level, which greatly improves the cognitive level of China's biota and the development speed of biological taxonomy.

Flora Reipublicae Popularis Sinicae

Flora Reipublicae Popularis Sinicae has been brewing since the 1920s. After four generations of scientists and more than 80 years of research and compilation, the *Flora Reipublicae Popularis Sinicae* has an authoritative and comprehensive household register. *Flora Reipublicae Popularis Sinicae* has been published for 45 years since the first volume was published in 1959. In October 2004, it completed 80 volumes of 126 volumes, including domestic and domesticated seed plants and ferns, which belong to 31,142 species of 3,408 genera, 301 families, respectively. The book has more than 50 million words and more than 9,000 plates. It discusses the Chinese and foreign literatures on the families, genera and species, plant morphological characteristics, distribution and ecological environment at home and abroad, scientific research value and economic use, and systematic classification of each plant. It is the most abundant scientific work in the world's published flora, which provides great importance for the rational development and utilization of plant resources. The essential basic information and scientific basis will play an important role in promoting the research of terrestrial ecosystem and have important value in scientific research and economic construction.

The first Chinese scholar to write *Flora Reipublicae Popularis Sinicae* and carry out pioneering research is HU Hsen-Hsu. In 1919, he personally went to the field to investigate and collect plant specimens. With the joint efforts of his peers, he collected nearly one million plant specimens and began to establish his own herbarium. In 1934, he put forward the proposal of compiling *Flora Reipublicae Popularis Sinicae* at the first annual meeting of the Chinese society of Botany.

The publication of *Flora Reipublicae Popularis Sinicae* has attracted the attention of botanists all over the world. Since 1989, China and the United States have cooperated in the preparation of *Flora Reipublicae Popularis Sinicae* in English. It took more than 20 years to compile. In September 2013, the revised English version of *Flora Reipublicae Popularis Sinicae* was published in 50 volumes. It is the largest high-level English version of flora in the world. This is another milestone achievement in the field of botany in China.

《中国植物志》
Flora Reipublicae Popularis Sinicae

Flora Reipublicae Popularis Sinicae is a great academic achievement in the field of biology in P. R. China. This is mainly reflected in two aspects. First,

支学科研究提供重要支持。为解决某些科、属的系统位置或分类问题，在编写过程中曾结合形态学、解剖学、孢粉学、细胞生物学或其他植物学分支学科进行研究，发表大量相关专著和论文，获得国家自然科学奖多项奖励。

《中国植物志》摸清了中国植物资源的家底，奠定了我国植物学发展的基石，为合理开发利用植物资源提供了重要基础信息和科学依据，对陆地生态系统研究将起到重要促进作用，对国家和全球的可持续发展将做出重要贡献并产生深远影响。此外，在宣传和普及植物科学知识、提高公众对生物多样性的认识等方面，也发挥了重要作用。

《中国动物志》

《中国动物志》
Fauna Sinica

《中国动物志》的编研是为摸清我国动物资源家底而进行的一项跨世纪系统工程，是我国历史上首次进行的动物物种资源研究方面的创新工程。1956 年，国务院将编著《中国动物志》的任务列入我国科学技术发展远景规划。20 世纪 70 年代以来，为使中国动物学立足于国际前沿，中国动物志编辑委员会组织了 104 个单位的 600 余位动物学家，立志编写出一套高水平的《中国动物志》来，为中国动物学的发展做出更大贡献。1962 年中国动物志编辑委员会正式成立。1963 年，中国科学院将《中国动物志》列为院重点科研项目。目前《中国动物志》已出版 161 卷，全套志书共 11 471.4 万字，共记载我国 8528 属 38 161 种动物。其中，昆虫纲已出版 68 卷，共包括昆虫 12 目 25 000 余种。基本涵盖了我国昆虫的主要类群。膜翅目蚁科、蜉蝣目、蜚蠊目、蜻蜓目等，由于研究起步较晚，现在正在编研阶段，还未出版见书。

由于在 20 世纪 50 年代及 60 年代前期建立了较雄厚的基础，在中国动物志编辑委员会和全体编研人员的共同努力下，目前已经出版《中国动物图谱》27 卷、《中国经济昆虫志》55 卷、《中国经济动物志》12 卷和《中国动物志》161 卷。迄今为止，《中国动物志》共获国家、省部级奖 25 项；2001 年，《中国经济昆虫志》荣获国家自然科学奖二等奖。

《中国动物志》是反映我国动物分类区系研究工作成果的系列专著，是进一步研究物种多样性、探讨物种演化和系统发育的奠基石，是动物资源开发利用、有害物种控制、濒危物种保护的理论根据。

《中国孢子植物志》

《中国孢子植物志》自 1987 年起由中国科学出版社分卷出版，共 150 万字，分为《中国海藻志》、《中国淡水藻志》、《中国真菌志》、《中国地衣志》及《中国苔藓志》。各志根据分类单位的种数情况，又分为若干卷册，计划出版 131 卷，现已出版 86 卷。其中记述了我国孢子植物物种的形态、解剖、生理、生化、生态、地理分布及其与人类关系等方面的综合信息，并附有根据中国标本绘制的形态图、参考文献和索引。全书是在生物系统学原理和方法的指导下对中国孢子植物进行系统分类的研究成果，是物种保护的重要依据，也是我国生物资源开发利用、科学研究与教学的重要参考文献。

《中国苔藓志》
Flora Bryophytorum Sinicorum

it contains extremely rich first-hand information. After the founding of P. R. China, with CAS and its subordinate research institutes as the main body, geological botany field investigation and specimen collection began. In 1964-2000, more than 10 large-scale national field investigations were organized. A total of 17 million plant specimens have been collected, which is rare in the history of world collection. It can be said that *Flora Reipublicae Popularis Sinicae* was completed by Chinese botanists based on their own field investigation and investigation. It contains a lot of new information and content and has a high scientific value. The second is to provide important support for the research of various branches of Botany. In order to solve the systematic position or classification problems of some families and genera, in the process of compilation, this work has studied in combination with morphology, anatomy, palynology, cell biology or other branches of botany, published a large number of related monographs and papers, and won many awards of the National Natural Science Award.

Flora Reipublicae Popularis Sinicae has found out the background of Chinese plant resources, lay the foundation stone for the development of Botany in China, provide important basic information and scientific basis for the rational development and utilization of plant resources, play an important role in promoting the research of terrestrial ecosystem, make important contributions to the sustainable development of the country and the world and have a profound impact. In addition, it has also played an important role in publicizing and popularizing plant science knowledge and raising public awareness of biodiversity.

Fauna Sinica

The compilation and research of *Fauna Sinica* is a cross-century systematic project to find out the background of animal resources in China, and it is the first innovative project in the research of animal species resources in the history of China. In 1956, the State Council listed the task of compiling the *Fauna Sinica* in the long-term plan for the development of science and technology in China. Since the 1970s, in order to make Chinese zoology stand on the international front, the editorial board of *Fauna Sinica* has organized more than 600 zoologists from 104 units to write a set of high-level *Fauna Sinica* to make a greater contribution to the development of Chinese zoology. In 1962, the Editorial Committee of *Fauna Sinica* was formally established. In 1963, CAS listed the *Fauna Sinica* as a key scientific research project. At present, 161 volumes of *Fauna Sinica* have been published, with a total of 114.714 million characters, recording 38,161 species of animals belonging to 8,528 genera in China. Among them, 68 volumes of Insecta have been published, including more than 25,000 species of 12 orders of insects. It basically covers the main groups of insects in China. Hymenoptera Formicidae, Ephemeroptera, Blattaria, Odonata and so on, due to the late start of the research, are now in the stage of compilation and research, and have not been published.

Due to the solid foundation established in the 1950s and early 1960s, with the joint efforts of the Editorial Committee of *Fauna Sinica* and all the editors and researchers, 27 volumes of *Animal Atlas of China*, 55 volumes of *Economic Insect Fauna of China*, 12 volumes of *Economic Animal Chronicles of China* and 161 volumes of *Fauna Sinica* have been published. So far, *Fauna Sinica* has won 25 national and provincial awards; in 2001, *Economic Insect Fauna of China* won the second prize of the National Natural Science Award.

The *Fauna Sinica* is a series of monographs reflecting the achievements of the research on animal taxonomy and fauna in China. It is the cornerstone of further study on species diversity, species evolution and phylogeny, and the theoretical basis for the development and utilization of animal resources, the control of harmful species and the protection of endangered species.

Flora Cryptogamarum Sinicarum

Since 1987, *Flora Cryptogamarum Sinicarum* has been published in volumes by China Science Press, with a total of 1.5 million words. It is divided into five parts: *Flora Algarum Marinarum Sinicarum*, *Flora Algarum Sinicarum Aquae Dulcis*, *Flora Fungorum Sinicorum*, *Flora Lichenum Sinicorum* and *Flora Bryophytorum Sinicorum*. According to the number of classification units, the chronicles are divided into several volumes. 131 volumes are planned to be published, and 86 volumes have been published. It describes the comprehensive information about the morphology, anatomy, physiology, biochemistry, ecology, geographical distribution and the relationship with human beings of spore plant species in China, with the morphological map, reference and index drawn based on Chinese specimens. Under the guidance of the principles and methods of biosystematics, the book is the research achievement of systematic classification of Chinese sporophytes, the important basis of species protection, and the important reference for the development and utilization of biological resources, scientific research and teaching in China.

其他生物学研究著作

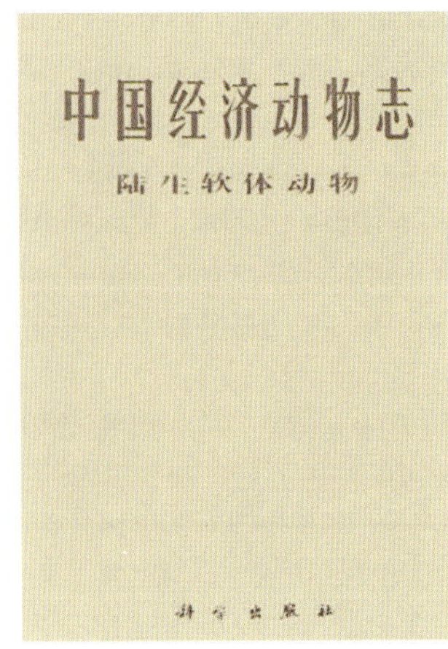

《中国经济动物志》
Economic Fauna of China

《中国经济昆虫志》
Economic Insect Fauna of China

在“三志”的基础上，根据对我国生物标本馆，特别是中国科学院系统的各个生物标本馆的标本的研究，陆续出版了大量各类全国性或地方性的生物资源志书。

在已出版的161卷《中国动物志》、12卷《中国经济动物志》和55卷《中国经济昆虫志》中，分别有53卷、7卷和39卷是中国科学院动物研究所学者主要依托国家动物博物馆馆藏标本完成的，标本馆为这三大类志书其他诸多卷册的编研提供了标本。此外，依托馆藏标本，所内外学者还编研了《西藏昆虫》(1981年、1982年)、《海南森林昆虫》(2002年)、《秦岭西段及甘南地区昆虫》(2005年)、《中国海洋蟹类》(1986年)、《中国土壤动物检索图鉴》(1998年)、《西南武陵山地区无脊椎动物》(1997年)、《金沙江流域鱼类》、《中国内陆鱼类物种与分布》、《中国鸟类分布名录》(1955年、1958年、1976年)、《中国鸟类系统检索（第三版)》(2002年)、《中国鸟类鸣声》(2017年)、《中国兽类彩色图谱》(2017年)、《中国哺乳动物多样性及地理分布》(2015年）等一批针对某一类群或某一地区的志书，以及《食用蜗牛养殖技术》(1990年)、《人工养蝎技术》(1999年)、《蜈蚣养殖技术》(2000年)、《蚯蚓养殖技术》(2009年）等实用技术手册。

西北高原生物所青藏高原生物标本馆参与并支撑“三志”编写，还相继编著了《世界蝗虫及其近缘种类分布目录》、*A Worldwide Monograph of Gentiana*、*A Worldwide Monograph of Swertia and Its Allies*、《青藏高原的蝗虫》、《青藏高原的鱼类》、《青海经济动物志》、《青藏高原药物图鉴》、《藏药志》、《青海经济植物志》、《青海植物志》、《青海植物检索表》、《青藏高原维管植物及其生态地理分布》、《喀喇昆仑和昆仑山的禾本科植物》、《昆仑植物志》、《西藏植物志》、《西藏哺乳类》和《西藏鸟类》等多部学术专著。《世界蝗虫及其近

《昆仑植物志》
Flora Kunlunica

《世界龙胆属专著》
A Worldwide Monograph of Gentiana

《獐牙菜属和近缘属的世界性分类》
A Worldwide Monograph of Swertia and Its Allies

Other biological resources works

On the basis of "Three Flora/Fauna", a large number of national or local works of biological resources have been published in succession according to the research on the specimens of China's biological collection, especially those of CAS.

Among the 161 volumes of *Fauna Sinica*, 12 volumes of *Economic Fauna of China* and 55 volumes of *Economic Insect Fauna of China*, 53 volumes, 7 volumes and 39 volumes are respectively completed by the scholars of the Institute of Zoology, who mainly rely on the specimens collected by the National Zoological Museum of China. The NZMC provides specimens for the compilation and research of other volumes of these works. In addition, relying on the specimens, scholars have also compiled and researched on a number of works for a certain group or region, such as *Insects of Tibet* (1981, 1982), *Hainan Forest Insects* (2002), *Insects in the Western Qinling Mountains and Gannan Region* (2005), *Marine Crabs in China* (1986), *Index Atlas of Soil Animals in China* (1998), *Invertebrates in the Southwest Wuling Mountain Region* (1997), *Fish in the Jinsha River Basin, Species and Distribution of Inland Fish in China, List of Bird Distribution in China* (1955, 1958, 1976), *Classification System Key of Chinese Birds* (Third Edition) (2002), *Song of Chinese Birds* (2017), *Color Atlas of Chinese Mammals* (2017), *Diversity and Geographical Distribution of Chinese Mammals* (2015), etc., as well as practical technical manuals such as *Edible Snail Culture Technology* (1990), *Scorpion Culture Technology* (1999), *Centipede Culture Technology* (2000), *Earthworm Culture Technology* (2009).

A Synonymic Catalogue of Grasshoppers and Their Allies of the World, A Worldwide Monograph of Gentiana, A Worldwide Monograph of Swertia and Its Allies, The Locusts of Qinghai-Tibet Plateau, The fishes of the Qinghai-Tibet Plateau, Economic Fauna of Qinghai, Illustrated Handbook for Medicine of Qinghai-Tibet Plateau, Tibetan Medicine Records, Economic Flora of Qinghai, Flora of Qinghai, Index *Florae Qinghaiensis, The Vascular Plants and Their Eco-geographical Distribution of the Qinghai-Tibet Plateau, The grasses of Karakorum and Kunlun Mountains, Flora Kunlunica, Flora of Tibet* and so on, *The Mammals of Xizang, The Avifauna of Xizang. A Synonymic Catalogue of Grasshoppers and Their Allies of the World* (English Vision) was published, which has recorded all the known 2,261 genus and 10,136 species of locusts in the world from 1758 to 1990 with 2,000,000 words. The book is the most complete monograph in the world up to now.

The Specimen Museum of Xinjiang Institute of Ecology and Geography also participated in the compilation and research of *Flora Kunlunica* and the second revision of *Flora Xinjiangensis* funded by NSFC. Relying on its collection, researchers are completing the compilation of the *List of Tianshan Plants* and the *Atlas of Shrub Plants in Xinjiang*.

The Herbarium of Wuhan Botanical Garden focuses on the collection and preservation of plant specimens in Central China, especially in Shennongjia area in Western Hubei. It has published 1-4 volumes of *Flora of Hubei*, which has made important contributions to the research of plant resources, flora, ecology and taxonomy in Central China, and won the science conference award of Hubei Province. Based on the collection of specimens, during the 9th Five Year Plan period, it undertook a number of projects supported by the National Natural Science Foundation and the Institute's special support for biological classification and flora, as well as scientific research projects of the Three Gorges Construction Committee, and published two monographs and more than 60 academic papers. During the 10th Five Year Plan period, It undertook the research work of national 976 research project, National Natural Science Foundation project, scientific research project of Three Gorges Construction Committee and directional research project of Chinese Academy of Sciences, published 5 monographs and more than 80 academic papers, among which Shennongjia plant investigation and research won the third prize of natural science of Chinese Academy of Sciences in 1989, Hubei flora research and Hubei flora compilation and research ever and won the second prize of natural science of Hubei Province in 2005. During the 11th Five Year Plan period, it participated in the publication of 5 monographs and more than 40 academic papers. During the 12th Five Year Plan period, he participated in publishing 2 monographs and nearly 50 academic papers.

《新疆植物志》
Flora Xinjiangensis

In 2008, based on the Northeast Biological Herbaria of studies on bryophytes in China, Institute of Applied Ecology won the second award of National Natural Science as the first completion unit. This was the first National Natural Science Award in the field of botany research in China. The research results have greatly enriched the flora of China and the world, provided an important scientific basis for further understanding of the composition, origin and systematic evolution of bryophytes in China, and contributed to the protection of rare bryophytes.

《湖北植物志》

Flora of Hubei

缘种类分布目录》，记录了1758-1990年所有已知的蝗虫类2261属、10 136种，是目前世界上最全面的同类专著。

新疆生态与地理研究所标本馆也参加了国家自然科学基金资助的《昆仑植物志》的编研以及《新疆植物志》的第2次修订工作。依托其馆藏标本，科研人员正在完成《天山植物名录》和《新疆灌木植物图谱》的编写。

武汉植物园标本馆重点收集保存了以华中地区为主，特别是以鄂西神农架地区为重点的植物标本。出版了《湖北植物志》1-4卷，为华中地区植物资源、区系、生态以及分类学研究做出了重要贡献，并荣获湖北省科学大会奖。依托馆藏标本，“九五”期间承担国家自然科学基金及院生物分类和区系特别支持项目与三峡建委科研项目数个，出版专著2本、发表学术论文60余篇。“十五”期间承担国家976研究项目、国家自然科学基金项目、三峡建委科研项目和中国科学院方向性研究项目研究工作，出版专著5本、发表学术论文80余篇，其中神农架植物考察研究曾获得1989年度中国科学院自然科学三等奖，湖北植物区系研究和湖北植物志编研曾获2005年度湖北省自然科学二等奖。“十一五”期间参与专著出版5本、发表学术论文40余篇。“十二五”期间参与出版专著2本，学术论文近50篇。

苔藓植物研究专著

Monographs on bryophytes research

依托东北生物标本馆，2008年中国科学院沈阳应用生态研究所作为第一完成单位的《中国苔藓植物研究》获得国家自然科学奖二等奖，是我国植物学研究领域第一个国家级自然科学奖励。研究成果极大丰富了世界和中国植物区系，为深入了解中国苔藓植物区系组成、起源、系统演化提供了重要科学依据，也为珍稀苔藓植物保护做出了贡献。

以昆明植物研究所标本馆馆藏标本为依托，研究所科研人员出版的重要著作有：《中国植物志》（总论、樟科、罂粟科、紫堇科、唇形科、杜鹃花科、豆科、葫芦科、漆树科、冬青科、天南星科、禾本科、竹亚科等部分卷册）、《中国苔藓志》（参编第1-2卷、主

Based on the KUN specimens, more than 70 important monographic works have been published by staff members of the institution: e.g. *Flora Reipublicae Popularis Sinicae* (General, Camphor, Poppy, Corydaceae, Labiatae, Rhododendron, Leguminous, Cucurbitaceae, Anacardiaceae, Hollyceae, Araceae, Gramineae, Bamboo Subfamily, etc.), *Flora Bryophytorum Sinicorum* (volume 1-2, editor 3-4), *Flora of Yunnan* (21 volumes in total), *Checklist of Seed Plant in Yunnanica* (Volume I and Volume II), *Flora of Tibet* (volumes 1-5), *Bryoflora of Tibet*, *Fungi of Tibet*, *Fungi of Hengduan Mountain*, *Plants in Yunnan* (Japanese version), *Chinese Vegetation*, *Natural Geography of China — Plant Geography (Part 1)*, *A Brief History of Plant Collection in Yunnan*, *Synthesis of Angiosperms in China*, etc. 32 scientific research achievements have received national, CAS, provincial and ministerial awards.

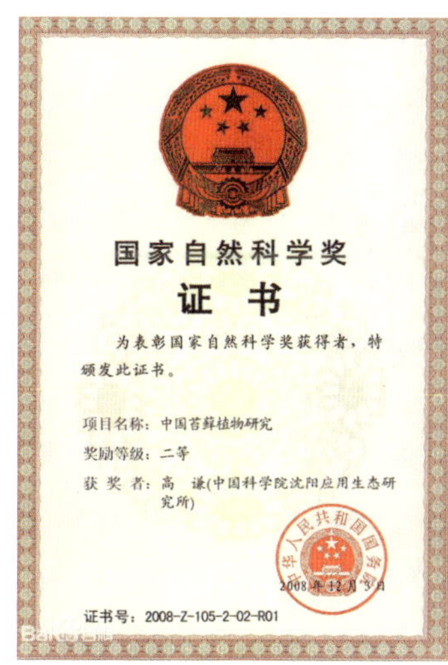
国家自然科学奖
证　书
为表彰国家自然科学奖获得者，特颁发此证书。
项目名称：中国苔藓植物研究
奖励等级：二等
获 奖 者：高　谦(中国科学院沈阳应用生态研究所)
2008年12月3日
证书号：2008-Z-105-2-02-R01

获得国家自然科学奖二等奖
The second award of National Natural Science

《云南植物志》
Flora of Yunnanica

戴芳澜院士与邓叔群院士
Academician Tai Fang-Lan and Academician Teng Shu-Chün

《中国的真菌》
Fungi of China

《中国真菌总汇》
Sylloge Fungorum Sinicorum

《中国真菌志》
Flora Fungorum Sinicorum

《中国菌物学 100 年》
100 Years of Mycology in China

《2018 世界菌物现状》
State of the World's Fungi – 2018

编第 3-4 卷）、《云南植物志》（共 21 卷）、《云南种子植物名录》（上、下册）、《西藏植物志》（1-5 卷）、《西藏苔藓植物志》、《西藏真菌志》、《横断山区真菌》、《云南の植物》（日文版）、《中国植被》、《中国自然地理——植物地理（上）》、《云南植物采集史略》、《中国被子植物科属综论》等 70 余卷册。基于本标本馆的标本，研究所有 32 项次科研成果获得国家、中国科学院、省部级奖励。

《中国的真菌》（1963 年）是邓叔群院士基于菌物标本馆 2.8 万余份馆藏标本撰写的菌物分类学巨著，描述了我国分布的菌物 2400 余种。戴芳澜院士的《中国真菌总汇》（1979 年）也继而出版，收录我国报道的 6800 余种菌物。两部巨著已成为研究我国菌物资源与分类的学者案头必备的典籍。依托菌物标本馆馆藏标本，目前已出版《中国真菌志》59 卷和《中国地衣志》3 卷，共描述菌物 92 科 898 属 8702 种。新版 *The Enumeration of Lichenized Fungi in China* 收录 2 门 9 纲 25 目 98 科 441 属 3114 种。这些专著是说明我国菌物多样性本底的重要科学文献。

菌物标本馆还参与了《中国菌物学 100 年》和《2018 世界菌物现状》的编写。《中国菌物学 100 年》对中国菌物的学科历史、重要人物、分类研究和菌物栽培及产业化等方面进行重点论述，是对中国 100 年来相关研究的小结或概述。由于中国菌物学研究的突出地位，《2018 世界菌物现状》将中国列为“聚焦国家”，重点介绍了我国的菌物学研究进展，提高了我国在世界菌物学界的学术地位。

东北生物标本馆已出版专著 27 部，包括主编 4 卷和参编 3 卷《中国苔藓志》，主编 2 卷和参编 2 卷英文版 *Moss Flora of China*；发表论文 180 篇，参与编写了“中国首批濒危苔藓植物红色名录简报”。这些分类研究工作也获得了大量科技奖励。

新分类单元的发表

作为生物学中最重要的概念之一，物种也是构建生物标本馆最重要的理论基础。新物种的诞生离不开标本馆，研究者通常需要将新发表物种的模式标本保存在重要的标本馆，或者从标本馆已有的馆藏中发现新的物种。新分类单元的发表，是生物标本馆对科研支撑作用最直接的体现。2014-2019 年，各馆系统厘定并发表了新种、属、科等新分类单元 800 余个，馆藏模式标本量增加到近 40 万号。

截至 2020 年 2 月，在分类学研究中，以成都生物研究所两栖爬行动物标本馆为依托参与发表并命名的两栖爬行类新种共计 187 个，其中两栖类 151 个，爬行类 36 个。此外，还建立了若干两栖爬行类的新属，修订了部分两栖爬行类物种的分类地位。这些分类学研究是中国两栖爬行动物系统建立的基础，以标本馆为依托的“中国两栖动物系统学研究”获 2014 年度国家自然科学奖二等奖。成都生物研究所两栖爬行动物标本馆团队近 5 年来的部分分类学研究成果如下。

高等真菌研究专著
Monographs on higher fungi

建立树蛙科 Rhacophoridae 一新属：广义树蛙属 *Rhacophorus sensu lato* 是树蛙科 Rhacophoridae 物种多样性最高的属之一，但由于新种发表较快，以

Fungi of China (1963), written by Prof. Teng Shu-Chün was based on an extensive examination of more than 28,000 specimens in the Fungarium, with more than 2,400 Chinese species described in this monograph. *Sylloge Fungorum Sinicorum* (1979), compiled by Prof. Tai Fang-Lan, recorded more than 6,800 species of China, being the most comprehensive record of Chinese fungi. These two great works have long been the essences of Chinese mycological studies. Hitherto, 59 volumes of *Flora Fungorum Sinicorum* and 3 volumes of *Flora Lichenum Sinicorum* have been published, and they recorded and described 8,702 fungal species (belong to 92 families and 898 genera) in China. Additionally, the new edition *The Enumeration of Lichenized Fungi in China* included 3,114 lichen species in China, belonging to 2 kingdoms, 9 classes, 25 orders, 98 families and 441 genera.

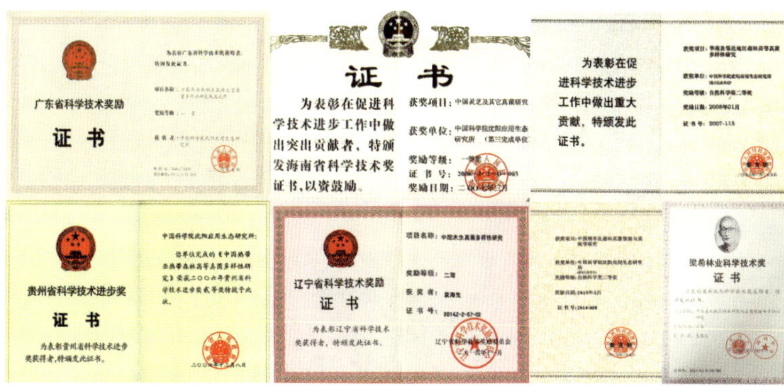

基于菌物标本馆馆藏的科技奖励

Based on the collection of Fungarium, awards of science and technology

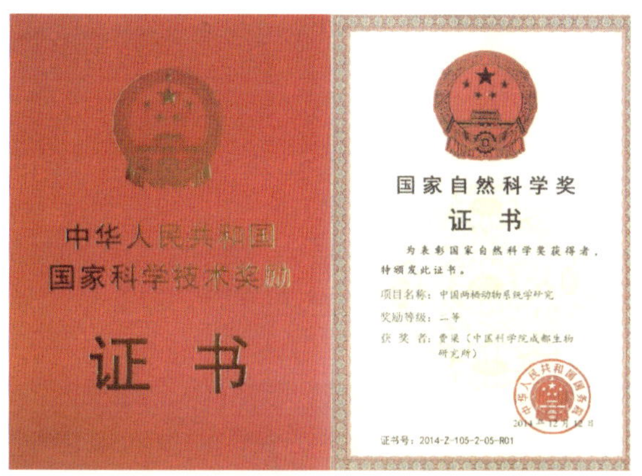

2014 年度国家自然科学奖获奖证书

Award certificate of National Natural Science Prize in 2014

The staff of the Fungarium participated in the complications of two important books, *100 Years of Mycology in China* and *The State of World's Fungi*-2018. In the former, the history of mycological studies in China was thoroughly reviewed, and future perspectives were discussed inspiringly. In the latter, China was selected as the focused country for her outstanding mycological research in the past several decades. The report briefly introduced the appreciating mycological research progress in China, which significantly improved the academic status of mycology research of China in the world.

Twenty-seven monographs have been published by the Northeast Biological Herbaria, including 4 volumes (edited) and 3 volumes (participated) in the compilation of the *Flora Bryophytorum Sinicorum* 2 volumes (edited) and 2 volumes (participated) in the compilation of the English version of *Moss Flora of China*; 180 papers and "Brief report on the red list of the first batch of endangered bryophytes in China" has been published. These classified research work also received a lot of science and technology awards.

Publication of new taxon

As one of the most important concepts in biology, species is also the most important theoretical basis for the construction of biological collections. The birth of new species can not be separated from the collections. Researchers usually need to keep the type specimens of newly published species in important collections or find new species from the existing specimens of the collections. The publication of the new taxon is the most direct embodiment of the supporting role of the biological collections in scientific research. In 2014-2019, more than 800 new taxa, such as new species, genera and families, have been systematically identified and published, and the number of type specimens has increased to nearly 400,000.

Up to Feb. 2020, based on the Herpetological Museum of Chengdu Institute of Biology, 187 new species of amphibians and reptiles was established (151 amphibians and 36 reptiles), as well as several new genera, and the taxonomic status of some species was also revised. These taxonomic studies are the basis of the systematics of amphibians and reptiles in China. Based on the museum, "The systematic study on amphibians of China" won the second prize of National Natural Science Prize in 2014. The researches of recent five years by the museum group are listed as follows.

Established a new genus of Rhacophoridae: Genus *Rhacophorus* is one of the most diverse genera of the family Rhacophoridae, and its taxonomy of genus *Rhacophorus sensu lato* faces major challenges because of rapidly described new species and complex interspecific relations. The researchers investigates the generic taxonomy within the genus

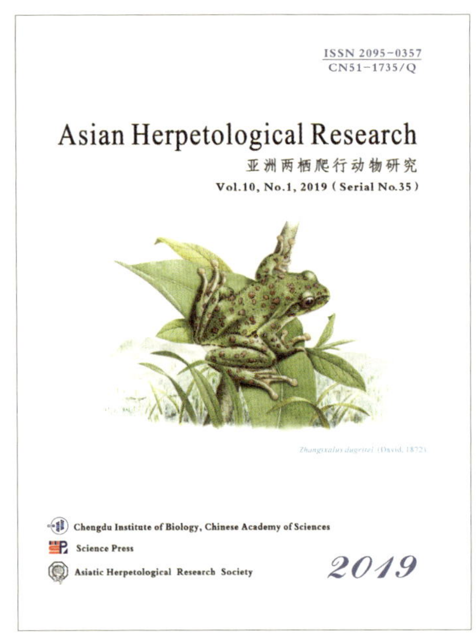

Asian Herpetological Research 10(1) 封面——新属张树蛙属 *Zhangixalus* 的模式种：宝兴树蛙 *Zhangixalus dugritei*

The cover of *Asian Herpetological Research* Vol. 10, No. 1. Type species of newly established genus *Zhangixalus*: *Zhangixalus dugritei*

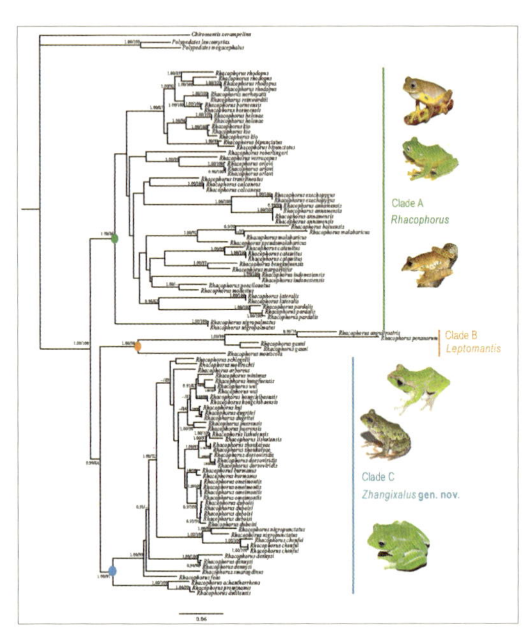

广义树蛙属 *Rhacophorus sensu lato* 系统发育树

Phylogenetic tree of *Rhacophorus sensu lato*

及复杂的种间关系，其分类面临着巨大挑战。研究人员根据分子和形态学数据研究了 *Rhacophorus* 属的分类地位，发现了高度分化的三支，这三支符合形态学特征和地理分布且有较高支持率，因此，研究人员认为这三支可划分为三个属：*Rhacophorus sensu stricto* (Clade A)、恢复至属级水平的 *Leptomantis* (Clade B)，以及新命名的属 *Zhangixalus* (Clade C)。

发表刘树蛙属 *Liuixalus* 新种 2 个　由于刘树蛙属物种体型小、形态相似性大及其隐秘的生活习性，该属的分类学与分子系统学研究一直面临着巨大挑战。研究使用线粒体 DNA 片段构建了该属的分子系统发生树，结果表明刘树蛙属内有 6 个共有遗传支系，其中二者为隐存种。基于分子系统学、比较形态学与声学分析，分别描述了产自我国广西大瑶山的金秀刘树蛙 *L. jinxiuensis* 与广西十万大山的十万大山刘树蛙 *L. shiwandashan*。

发表后棱蛇属 *Opisthotropis* 新种 1 个　后棱蛇属 *Opisthotropis* 物种多样性很高，然而该属的物种多样性仍有待进一步挖掘。基于湖南西部古丈县所采 1 雄 2 雌标本，依据分子系统学、比较形态学研究发现，该种群应系后棱蛇属一新种，将其描述并命名为赵氏后棱蛇 *Opisthotropis zhaoermii*，以纪念赵尔宓院士对中国及世界两栖爬行动物研究做出的卓越贡献。

发表过树蛇属 *Dendrelaphis* 新种 1 个　结合形态学与分子系统学数据，将采自云南省西双版纳傣族自治州勐仑的过树蛇标本订为新种——沃氏过树蛇(*Dendrelaphis vogeli*)。同时，依据形态学和分子系统学数据，将银山过树蛇（*D. ngansonensis*）暂时作为蓝绿过树蛇 *D. cyanochloris* 复合种，并报道蓝绿过树蛇分布于我国海南、云南和西藏。

国家动物博物馆标本馆为国内外学者开展系统发育、物种进化等学科研究工作提供了坚实的支撑。标本馆长期积累并妥善保藏的标本资源，为开展这些领域的科学研究提供了翔实可靠的实物资源和数据资源。依托馆藏标本，国内外科研人员已鉴定发表动物新种上万种，其中绝大多数是采自中国的标本。例如，科

十万大山刘树蛙 *Liuixalus shiwandashan*（左）
与金秀刘树蛙 *Liuixalus jinxiuensis*（右）
Liuixalus shiwandashan (left) and *Liuixalus jinxiuensis* (right)

新种赵氏后棱蛇模式标本生态照片
General and close-up views of *Opisthotropis zhaoermii* in life

Rhacophorus based on both molecular and morphological data. The results reveal three well-supported and highly diverged matrilines that correspond with morphological characteristics and geographic distribution. Accordingly, researchers consider these three lineages as distinct genera: *Rhacophorus sensu stricto* (Clade A), resurrected genus *Leptomantis* (Clade B), and a newly named genus *Zhangixalus* (Clade C).

Described two new species of the genus *Liuixalus*. Due to small body sizes, superficial similarities in morphologies, and obscure activity behaviors, the phylogeny and taxonomy of species in the genus *Liuixalus* were very troublesome. The museum constructed the matrilineal genealogy of the genus using mitochondrial DNA sequences. Analyses recovered six well-supported matrilines and revealed that two undescribed species. Based on the genealogical, morphological, and bioacoustic analysis, researchers described *L. jinxiuensis* from the type locality Mt. Dayao, Guangxi, China and *L. shiwandashan* from the type locality Mt. Shiwanda, Guangxi, China.

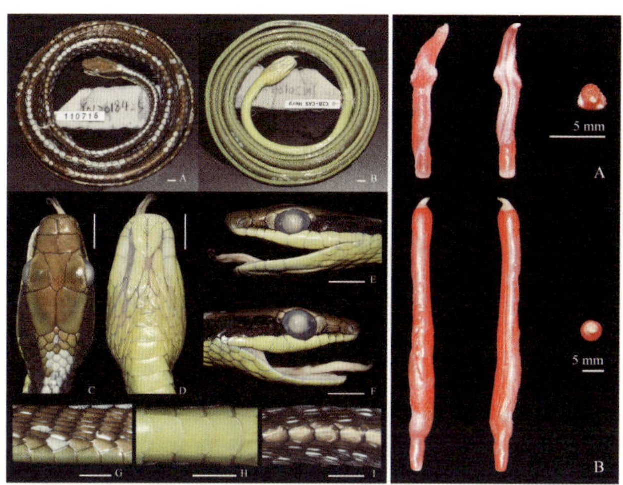

新种沃氏过树蛇（*Dendrelaphis vogeli*）正模标本照片（左）与半阴茎对比（右）
Photographs of the preserved holotype of *Dendrelaphis vogeli* (left) and morphological comparisons of the hemipenis (right)

Described a new species of the genus *Opisthotropis*. The genera *Opisthotropis* is one of the most speciose genus of Natricidae, and the biodiversity is still to be uncovered. Based on newly collected 1 male and 2 female specimens from Guzhang County, western Hunan Province, China, researchers name the population as a new member of *Opisthotropis* according tomolecular systematics and comparative morphology. The specific epithet, *zhaoermii*, is derived from the name of an internationally renowned Chinese herpetologist, Prof. ZHAO Er-mi, to honor his great contribution to herpetological research in China and the World.

Described a new species of the genus *Dendrelaphis*. A new species of the genus *Dendrelaphis* is described from Menglun, Xishuangbanna, southern Yunnan, China, based on molecular and morphological data. According to molecular and morphological data, *D. ngansonensis* is temporally considered to the *D. cyanochloris* complex. *D. cyanochloris* complex from Tibet, Yunnan and Hainan is also discussed.

The collection of the NZMC provides solid support for scholars at home and abroad to carry out the research work of phylogeny, species evolution and other disciplines. The specimen resources accumulated and properly preserved by the collection for a long time, provide detailed and reliable material resources and data resources for scientific research in these fields. Relying on the collection of specimens, researchers at home and abroad have identified and published tens of thousands of new animal species, most of which are collected from China. For example, on the basis of consulting, comparing and measuring the specimens collected by the NZMC, the researchers reported a new species

我国鸟类新种——四川短翅蝗莺
New species of birds in China: *Locustella chengi*

研人员在对国家动物博物馆馆藏标本进行查阅、比对和测量形态等研究的基础上，于 2015 年报道了在我国中部地区发现的一个鸟类新种——四川短翅蝗莺（*Locustella chengi*），这也是首个以中国鸟类学家来命名的鸟类；此外还在一块缅甸出土的琥珀中发现了古鸟类一种新属新种陈光琥珀鸟（*Elektorornis chenguangi*）。在此基础上，标本馆在对物种资源的认识、利用，以及推动生物多样性保护规则制定等方面，也提供了重要的服务。

依托昆明动物博物馆 90 万号馆藏标本，科研人员发现和描述鱼类新种 200 余种，哺乳类新种 8 种，昆虫新种 200 余种，两栖爬行类新种 10 种，螺类新种 5 种，鸟类新亚种 4 个，支撑项目数百项，包括国家重点基础研究发展计划（973 计划）、中国科学院先导专项、科技部重大基础专项、中国科学院重大专项等重大项目，为国家公园和各级自然保护区的建立、动物多样性本底调查、珍稀物种的监测保护等提供了第一手资料和基础支撑。为《中华人民共和国野生动物保护法》的修订，向云南省林业草原局提供了动物保护的专业意见和建议。在生物多样性编目与保护、动物资源调查、生物多样性信息库的建设、收集与保存等方面做出了突出的贡献。

从 20 世纪 50 年代开始，依托上海昆虫博物馆丰富的馆藏标本和研究者的努力，本馆对土壤六足动物、直翅目、等翅目、双翅目等类群的系统分类学研究不断深入。迄今为止，已发现新属、新种 500 余种，发表研究论文 300 余篇，出版专著 10 余部，并获得国家级奖项 2 项，中国科学院奖项 8 项，地方级奖项 2 项。近 5 年来，发表新种近百种，涵盖直翅目、蜚蠊目、革翅目、原尾纲。

依托南海海洋生物标本馆馆藏历史标本以及新增标本资源，科研人员对南海近岸典型生境的浮游生纤毛虫多样性及区系结构进行了分析总结，共发现了 39 个种，隶属于 17 个属、7 个科、4 个目，其中包括了 16个新种；对多种近岸生境的纤毛虫多样性进行分析，发现红树林湿地拥有最高的浮游生纤毛虫多样性。在我国率先开展深海介形亚纲、桡足亚纲怪水蚤目的多样性研究，相继报道南海新种 19 种。查清了牡蛎生物多样性最高的华南沿海牡蛎的种类与分布，共鉴定出 5 个属、11 种牡蛎，其中新种 2 种。明确了我国海域海龙科和鲽形目鱼类的种类组成，并发表 2 个新种。

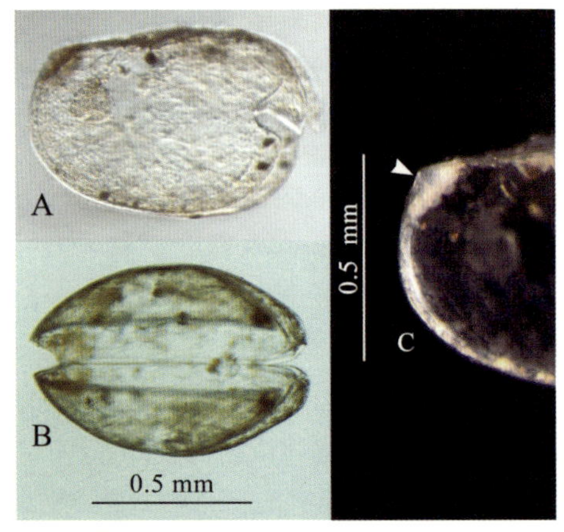

介形亚纲新种：南沙深海浮萤
Bathyconchoecia nanshaensis sp. nov.

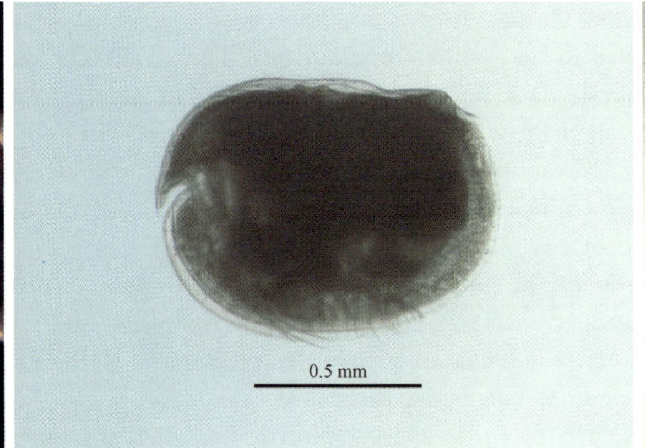

介形亚纲新种：缺刻深海浮萤
Bathyconchoecia incisa sp. nov.

桡足亚纲新种：陈氏怪水蚤
Cymbasoma cheni sp. nov.

of bird, *Locustella chengi*, which was found in Central China in 2015. And this is the first bird named after a Chinese ornithologist. In addition, a new genus and new species of ancient bird, *Elektorornis chenguangi*, was found in a piece of amber unearthed in Myanmar. On this basis, the collection also provides important services for the understanding and utilization of species resources, as well as promoting the formulation of biodiversity protection rules.

Relying on more than 900,000 specimens in KNHMZ, researchers have discovered and described more than 200 new species of fish, 8 new species of mammals, 200 new species of insects, 10 new species of amphibians and reptiles, 5 new species of snails, and 4 new subspecies of birds. The KNHMZ has also supported hundreds of projects, including some important projects such as 973 project, CAS pilot projects, Ministry of Science and Technology major foundation projects and CAS major projects. For the establishment of national parks and nature reserves at all levels, animal diversity baseline surveys and monitoring and protection of rare species, KNHMZ has provided first-hand information and basic support, and has also provided professional opinions and suggestions on animal protection to the Yunnan Forestry and Grassland Bureau for the revision of the "Wildlife Protection Law of the People's Republic of China", which has made outstanding contributions in biodiversity cataloging and protection, animal resource survey, and construction, collection and preservation of biodiversity information databases.

Since the 1950s, continuous systematic taxonomy researches on the soil hexapods, Orthoptera, Isoptera, Diptera, etc. have progressed by Shanghai Entomological Museum. So far, there have been more than 500 new genera and species discovered, over 300 research papers released, more than 10 monographs published, and also won 2 national prizes, 8 awards of C.A.S and 2 local prizes. In the past 5 years, nearly one hundred new species have been published, including Orthoptera, Blattella, Dermaptera and Protura.

Based on the collection of specimens in MBCSCS, the diversity and flora of planktonic ciliates in typical habitats near the South China Sea were reviewed, and 39 species were found which belonging to 17 genera, 7 families and 4 orders, and 16 new species were published. The diversity of ciliates in various coastal habitats was also analyzed. It was found that mangrove wetlands have the highest diversity of planktonic ciliates. In China, the collection took the lead in the research on the classification and diversity of Ostracoda and Copepoda and has reported 19 new species from the South China Sea. The species and distribution of oysters along the coast of South China with the highest oyster biodiversity were investigated, and a total of 5 genera and 11 species of oysters were identified, of which 2 new species were recorded. The species composition of the Syngnathidae and Pleuronectiformes from the China seas were clarified, and two new species of each taxon were reported.

The study of the taxonomy of marine organisms is a traditional dominant discipline of the Institute of Oceanology, which gather the famous marine biologists in China and even in the world, such as ZENG Chengkui, ZHANG Xi, ZHANG Fusui, LIU Ruiyu, ZHENG Shouyi, etc. Since 1950, relying on the specimens of the Marine Biological Museum, more than 140 monographs and 2,000 research papers have been edited and published on the systematics and floristic research of marine life in China, and 1 new subclass, 6 new families, 92 new genera and 1,770 new species of marine life have been published, and more than 40 achievement awards have been won at the provincial and ministerial level. It covers the fields of marine morphology, classification, floristics, systematics, biodiversity, biogeography, ecology, biooceanography and

鰈形目新种：南海舌鳎
Cynoglossus nanhaiensis sp. nov.

芍药属、党参属、冠唇花属等类群的分类研究专著
Taxonomic monographs of *Paeonia*, *Codonopsis*, and *Microtoena*

美丽桐科，被子植物一新科
Wightiaceae, a new family of angiosperm

肠蕨科，蕨类植物一新科
Diplaziopsidaceae, a new family of ferns

海洋生物分类系统学研究是海洋研究所的传统优势学科，集中了我国乃至世界著名的海洋生物学家，如曾呈奎、张玺、张福绥、刘瑞玉、郑守仪等。自1950年以来，依托海洋生物标本馆标本，已编辑出版了有关我国海洋生物分类系统学和区系研究的专著140余部，研究论文2000余篇，发表海洋生物新物种1个新亚纲、6个新科、92个新属、1770个新种，获省部级以上成果奖40余项。内容涉及海洋生物形态、分类、区系、系统学、生物多样性、生物地理学、生态学、生物海洋学等学科领域，基本摸清了我国近海海洋生物的物种、多样性构成、区系特点和资源状况，为我国海洋科学和社会经济发展做出了重大贡献。同时，也为国内外的海洋生物分类学和区系研究培养了大批高级研究人才。每年接待数位国际同行前来合作研究，促进了学术交流。奠定了我国海洋生物分类系统学和区系研究的基础，是我国海洋生物多样性研究的中心和策源地。

近十年来，在植物研究所标本馆的支撑下，标本馆科研人员在物种资源和生物多样性保护研究取得很大进展：①发表新分类群约380个，包括维管植物3个新科、5个新族、9个新属、约350个新种；②全面评估中国高等植物物种的濒危状况；③根据最新的研究成果，全面更新中国高等植物名录，出版了《中国生物物种名录》高等植物各类群分册；④在国家林业和草原局的领导下，修订和提出全国重点保护植物名录；⑤编撰出版《泛喜马拉雅植物志》第30、47、48(2)、48(3)等7卷册和总名录；⑥查明了我国外来入侵植物的种类，建立入侵植物的DNA条码数据库；⑦系统揭示了中国被子植物属演化的时空分布特征；⑧完成其他分类学志书和专著53部，发表论文约300篇，其中80篇发表在分类学和进化生物学主流(前30%)刊物中。

近十年来，依托成都生物研究所植物标本馆馆藏和采集标本，发表蕨类植物新科2个（牙蕨科Pteridryaceae和翼囊蕨科Didymochlaenaceae），新属4个（水龙骨科的非洲石韦属*Hovenkampia*、三叉蕨科的龙蕨属*Draconopteris*和马来蕨属*Malaifilix*、凤尾蕨科的美洲翠蕨属*Gastoniella*），新种182个。

依托华南植物园标本馆馆藏标本主要进行热带、亚热带重要类群的分类学研究，重点开展了禾本科竹亚科、姜科、杜鹃花科、爵床科、无患子科、菊科千里光族和毛茛科的研究。自2016年以来，发表禾本科竹亚科1新属、爵床科1新属、菊科2新属，发表新种30个、新变种2个、越南新记录种6个、中国新记录种2个，还有新组合、新名称、新异名若干个。

近5年来，昆明植物研究所标本馆发表植物及菌物的新属24个、新种151个以及新组合64个，被

新科——牙蕨科植物形态
Morphology of the new family: Pteridryaceae.

新科——翼囊蕨科植物形态
Morphology of the new family: Didymochlaenaceae.

other disciplines. It has basically found out the species, diversity composition, floristic characteristics and resource status of China's offshore marine organisms and has made significant contributions to China's Marine Science and social and economic development. At the same time, a large number of senior researchers have been trained for marine taxonomy and floristic research at home and abroad. Every year, the museum receives several international colleagues to cooperate and research, which promotes academic exchanges. The MBCSCS has laid the foundation for taxonomic systematics and floristic research of marine organisms in China, and it is the center and source of marine biodiversity research in China.

In last decade, with the support of PE researchers made exciting progress in species resources and conservation of biodiversity: ① 380 new taxa were published, including three new families, five new tribes, nine new genera and approximately 350 new species of vascular plants; ② all known species of higher plants in China were evaluated; ③ the checklist of higher plan in China was updated based on recent research and all volumes of higher plants of *Species Catalogue of China* were published; ④ the checklist of state key protected plants were revised under the leadership of National Forestry and Grassland Administration of China; ⑤ seven volumes (30、47、48(2)、48(3) , etc.) and species checklist of *Flora of Pan-Himalaya* were published; ⑥ The checklist of invasive plant species in China were documented and DNA barcode database of invasive plant species was established; ⑦ The spatial-temporal evolutionary patter of China angiosperm flora was discovered; ⑧ Fifty-three monographs and approximately 300 papers were published, out of which 80 were published in evolutionary biology or taxonomy mainstream (top 30%) journals.

During the last decade, two new fern families (Pteridryaceae and Didymochlaenaceae) of Polypodiales, four new genera (*Hovenkampia* in Polypodiaceae, *Draconopteris* and *Malaifilix* in Tectariaceae, and *Gastoniella* in Pteridaceae), and 182 new species have been reported based on the specimens deposited in CDBI.

Taxonomic studies of Herbarium of South China Botanical Garden focus on the important taxa in tropical and subtropical regions: Bambusoideae of Poaceae, Zingiberaceae, Ericaceae, Acanthaceae, Sapindaceae, *Senecioneae*

菊科新种——肇骞千里光

New species of Compositae: *Sencecio changii*

肯尼亚植物新种——近革叶马㼎儿

New species of plants from Kenya: *Zehneria subcoriacea*

Flora of Pan-Himalaya、《中国高等植物彩色图鉴》、《云南省生物物种名录》、《神农架植物志》和《湖北植物大全》等收录。标本馆还为所内外科学家的日常查阅、类群的经典分类及系统学研究以及植物资源利用的比对查阅提供了优质服务；已实现 100 多万份标本的数字化，为植物学工作者查阅 KUN 馆藏标本提供了极大的便利。

武汉植物园标本馆在非洲组织和开展植物资源合作研究过程中，发现了一些非洲植物新的类群，研究人员联合非洲的合作者，至今已正式命名和发表非洲植物新类群 11 个，其中在肯尼亚发现植物新种 9 个、新变种 1 个，在马达加斯加发现植物新种 1 个。包括肯尼亚的近革叶马㼎儿（*Zehneria subcoriacea*）、肯尼亚景天（*Sedum keniense*）、肯尼亚锡生藤（*Cissampelos keniensis*）、白花拟堇菜凤仙花（*Impatiens pseudoviola*）、有棱蒴莲（*Adenia angulosa*）、长花马㼎儿（*Zehneria longiflora*）、丹尼尔多穗兰（*Polystachya danielana*）、块茎马㼎儿（*Zehneria tuberifera*）、滨海马豆（*Croton kinondoensis*）、埃尔贡悦猴瓜（*Peponium elgonense*）和马达加斯加的拟天麻（*Gastrodia elatoides*）。

据不完全统计，全国的科研人员依托庐山植物园标本馆发表的植物新种达 500 种以上。全国多个省的植物资源调查，特别是在江西省各保护区开展的大量植物资源科学考察，有力推动了江西地区的生物多样性保护工作。

得益于微生物研究所菌物标本馆丰富的收藏，菌物学家在真菌系统学方面取得了重大进展，提升了我国菌物系统学研究的国际影响力。例如，基于表型和基因型相结合的综合分析，科研人员将石耳科地衣从

肯尼亚植物新种——肯尼亚景天
New species of plants from Kenya: *Sedum keniense*

肯尼亚植物新种——肯尼亚锡生藤
New species of plants from Kenya: *Cissampelos keniensis*

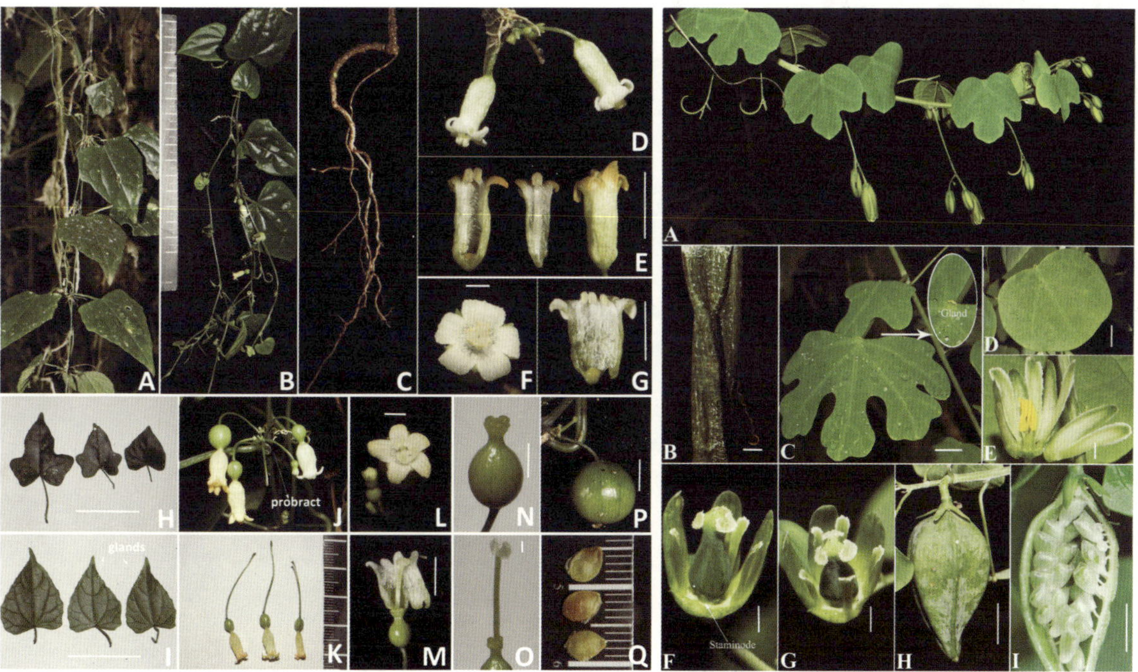

肯尼亚植物新种——长花马瓟儿
New species of plants from Kenya: *Zehneria longiflora*

肯尼亚植物新种——有棱蒴莲
New species of plants from Kenya: *Adenia angulosa*

肯尼亚植物新种——丹尼尔多穗兰
New species of plants from Kenya: *Polystachya danielana*

肯尼亚植物新种——块茎马㼎儿
New species of plants from Kenya: *Zehneria tuberifera*

肯尼亚植物新种——滨海马豆
New species of plants from Kenya: *Croton kinondoensis*

茶渍目中分出，另建立新目石耳目。同时，建立了子囊菌门肉座菌目中一个新科——茧壳菌科，以容纳一个单一的和独特的物种 *Cocoonihabitus sinensis*。

木生真菌是高等真菌中与陆地生态系统密切相关的重要类群，木生真菌分类和系统学也是东北生物标本馆的特色研究领域。目前标本馆是我国该类群真菌多样性和系统学研究的两个中心之一，近年来发表的新种数量、提交物种的分子序列数量等在国际上该类群研究的相关机构中也居于前列。东北生物标本馆菌物组深入挖掘馆藏标本的科研价值，发表新科 1 个，新属 17 个，新种 200 个，新组合种 59 个，新记录科、属、种 386 个。使中国木生真菌种类由 800 种增加到了 1400 多种，占全球的 60% 左右，位居世界第一。此外，在对苔藓植物研究中共发现新分类单位 3 新属，115 新种，中国新记录 3 科，28 属。

菌物中的小薄孔菌属于 1980 年由 Ryvarden 和 Johansen 建立，模式种为 *Polyporus semisupinus*，作为多孔菌的核心类群之一，目前该属在世界范围内约有 50 个种。*Antrodiella gypsea* 和 *A. thujae* 是生长在针叶树上的干基腐

肯尼亚植物新种——埃尔贡悦猴瓜
New species of plants from Kenya: *Peponium elgonense*

马达加斯加植物新种——拟天麻
New species of plants from Madagascar: *Gastrodia elatoides*

(Asteraceae) and Ranunculaceae. Since 2016, 1 new genus of Bambusoideae, 1 new genus of Acanthaceae, 2 new genera of Asteraceae, 30 new species, 2 new varieties, 6 new records of Vietnam, 2 new records of China, as well as several new combinations, new names and new synonyms have published.

Recent five years, KUN published 24 new genera, 151 new species and 64 new combinations of plants and fungi, which have been cited in *Flora of Pan-Himalaya*, *Atlas of Chinese Higher Plants*, *Checklist of the Yunnan province*, *Flora of Shennongjia* and *Plants of Hubei*. KUN provides high-quality services for the researches on taxonomy and systematics, and plant resource utilization by botanists both at home and abroad; more than one million specimens have been digitized, providing great convenience for botanists accessing the KUN collection.

In the process of organizing and carrying out the cooperative research on plant resources in Africa, the Herbarium of Wuhan Botanical Garden has found some new African flora. The researchers, together with their African collaborators, have officially named and published 11 new African flora, including 9 new species found in Kenya, 1 new variety and 1 new species found in Madagascar. These include *Zehneria subcoriacea*, *Sedum keniense*, *Cissampelos keniensis*, *Impatiens pseudoviola* Gilg var. *alba*, *Adenia angulosa*, *Zehneria longiflora*, *Polystachya danielana*, *Zehneria tuberifera*, *Croton kinondoensis*, *Peponium elgonense* from Kenya, and *Gastrodia elatoides* from Madagascar.

According to incomplete statistics, based on the collections of Herbarium of Lushan Botanical Garden, more than 500 new species of plants have been identified and published by researchers across the country. The inventory of plant resources in several provinces and regions, including the floristic investigation in Jiangxi province, has strongly contributed to promoting the conservation of diversity in Jiangxi province.

Benefit from the good collections from the Fungarium of Institute of Microbiology, Chinese mycologists have made significant progress on fungal systematics, which has enhanced the international influence of mycological systematics in China. For example, Umbilicariaceae (lichen) was separated from Leacnorales and proposed as a new order Umbilicariales, on the basis of morphological and molecular evidence. Similarly, Cocoonihabitaceae, a new family in Hypocreales (Ascomycota), was published to accommodate a single and distinct species, *Cocoonihabitus sinensis*.

Wood inhabiting fungus is an important group of higher fungi closely related to the terrestrial ecosystem. The taxonomy and systematics of wood-inhabiting fungi are the main fields of the Northeast Biological Herbaria. It is one of the two

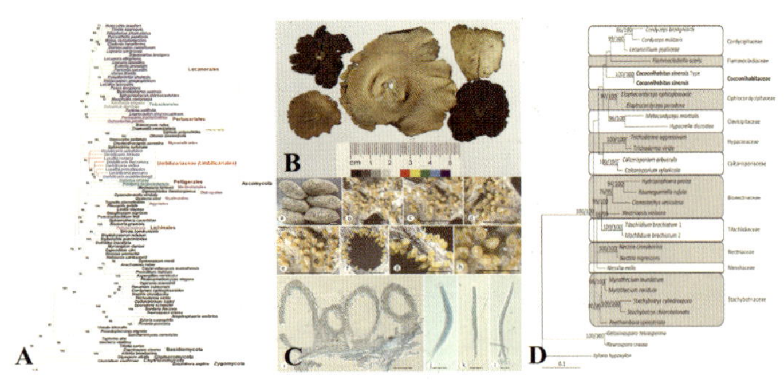

真菌新目及新科：石耳目（A、B）和茧壳菌科（C、D）。
New order and family: Umbilicariales (A, B) and Cocoonihabitaceae (C, D)

朽病原菌，具有小薄孔菌属特有的菌丝嗜蓝特征，但其子实体质地有所不同。科研人员在对东北生物标本馆馆藏标本进行形态学研究的基础上，基于多基因序列分析表明，与传统上属于多孔菌目的认知不同，*Antrodiella gypsea* 等类群属于与多孔菌目系统发育关系较远的锈革菌目，因此建立新科 Neoantridiellaceae。该研究颠覆了锈革菌目子实体形态学特征的界定范围，为进一步研究该目物种演化途径提供了重要的参考。

物种分类修订

随着对各个类群研究的不断深入和扩展，以及更加精确的研究分析技术的出现，旧的分类系统就必须做出调整和修订。这项工作更是离不开对大量馆藏标本的研究。

溪泥甲科（Elmidae）隶属于昆虫纲鞘翅目，是水生无脊椎动物的一个重要类群。该科昆虫为世界性分布，包含两个亚科，全球范围内报道溪泥甲科约 130 属，1400 种。但由于近缘属的属级鉴别特征适用范围有限，存在一些疑难物种的归属问题。东北生物标本馆对中国溪泥甲科的 17 个属共 70 个种类的 2 个核基因片段及 4 个线粒体基因片段进行了扩增，并与 GenBank 获得的部分同源序列一起，初步构建了我国溪泥甲科的系统发育树。序列分析表明 *Paramacronychus* 属和 *Urumaelmis* 属应并入 *Zaitzevia* 属中，短刻溪泥甲属 *Ordobrevia* 应并入狭溪泥甲属 *Stenelmis* 中。分子生物学手段的应用对于水生鞘翅目属级分类特征的界定和传统疑难物种的鉴定具有重要意义。

通过对成都生物研究所两栖爬行动物标本馆馆藏标本的研究，研究人员评估了异纹蛇属 *Pararhabdophis* 的系统地位，将其归并为东亚腹链蛇属 *Hebius* 的次定同物异名，并报道沙坝腹链蛇 *Hebius chapaensis* 在我国的新分布记录；确认了横纹后棱蛇 *Opisthotropis balteata* 的系统地位，恢复其所代表的环游蛇属 *Trimerodytes* 的有效性，并将华游蛇属 *Sinonatrix* 归并为其次定同物异名。

此外，标本馆科研人员还对中国爬行纲动物的分类体系和物种进行了系统的评估及修订，于 2015 年发表《中国爬行纲动物分类厘定》。该文章规范了中文和拉丁学名，整理出了中国爬行纲分类厘定名录 3 目 30 科 132 属 462 种。该文章被评为 2018 年度 F5000 论文（F5000, http://f5000.istic.ac.cn/）。以此为基础，并结合在 2017 年的被引频次而评选，该篇论文也被列入《生物多样性》2018 年度高影响力论文。

东北生物标本馆通过对馆藏标本的研究，提出苔藓植物新分类系统。

藓类植物新系统 在苔藓植物遗传、孢粉和系统发育研究中，标本馆高谦研究员发现了藓类植物传孢（粉）器官在长期生态环境过程中形成的主要形式和相应结构，将藓类的传孢类型分为腐媒传孢型、风媒传孢型、汽 - 风媒传孢型、水媒传孢型和虫媒传孢型等 5 种，反映了苔藓植物起源于绿藻及其内在演化过程，修正了近一个世纪以来广泛使用的 Brotherus (1924-1925) 系统，并提出了一个以科为单位的新系统表和系统图，该系统共 2 亚纲，17 目，86 科。

苔类系统 通过对馆藏标本的雌苞结构进行解剖和文献研究，标本馆研究人员提出苔纲（Hepaticae）起源于藓纲，并且苔纲的演化应以雌苞为主要特征，它贯穿苔纲各目、科之间，同时参考配子体的组织分化程度来判断它们的演化程度。按雌苞成分组成划分为：蒴帽型雌苞苔类、蒴萼型雌苞苔类、鞘萼型雌苞苔类、茎鞘型雌苞、蒴囊型雌苞苔类、总苞型雌苞苔类、托生总苞型雌苞苔类，并以此为依据重新划分了

important units in China for the study of fungal diversity and systematics of wood inhabiting fungi. In recent years, the number of new species described and the submitted molecular sequences is also at the forefront of the relevant institutions of wood-inhabiting fungal taxonomy in the world. The mycologists of the herbaria established 1 new family and 17 new genera, described 200 new species, proposed 59 new combinations, and found 386 new records. It has increased the species of wood-inhabiting fungi in China from 800 to more than 1,400, accounting for 60% of the world's total species and ranking first in the world. Besides, a total of 3 new genera, 115 new species, 3 families and 28 genera of bryophytes were newly found in China by the researchers.

Antrodiella was comprised by Ryvarden and Johansen in 1980, the type speciens is *Polyporus semisupinus*. At present, there are about 50 species in *Antrodiella*. *Antrodiella gypsea* and *A. thujae* are dry rot pathogens that grow on conifers. They have the unique characteristics of mycelial bluephilia, but the texture of their fruiting bodies is different. Based on the morphological study of the specimens collected in the Northeast Biological Herbaria, and based on the multigene sequence analysis, the researchers showed that, different from the traditional recognition of belonging to the order of Polypores, *Antrodiella gypsea* and other groups belong to the order Hymenochaetales, which is far away from the phylogeny of Polypores. Therefore, a new family Neoantridiellaceae was established. This study overturned the definition of the morphological characteristics of the fruiting bodies of Hymenochaetales, and provided an important reference for the further study of the evolutionary pathway of the order.

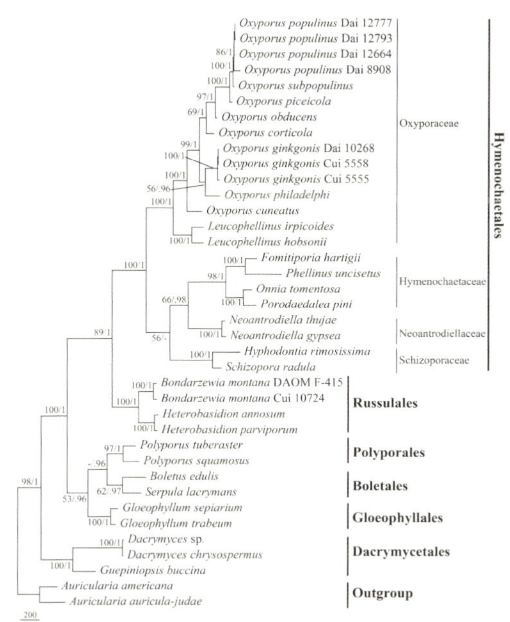

基于 *ITS* + *nLSU* + *RPB2* + *mtSSU* 基因序列的最大简约树显示 Neoantridiellaceae 科的系统发育地位

Maximum parsimony tree illustrating the phylogeny of Neoantridiellaceae and related species in Agaricomycetes based on *ITS* + *nLSU* + *RPB2* + *mtSSU* sequence data

Taxonomic revision

With the deepening and expansion of the research on each group, and the emergence of more accurate research and analysis technology, the old classification system must be adjusted and revised. This work is inseparable from a large number of collection specimens.

The Family Elmidae comprises about 130 genera, 1,400 species worldwide, which belongs to Coleoptera, Insecta. It is one of the most important groups in aquatic invertebrates. The families usually split into two subfamilies. Due to the limited identification characteristics between the related genera, some species were hard to assign to the genus level. In the study of Northeast Biological Herbaria, two nuclear and four mitochondrial genes were amplified for 70 species in 17 genera. Together with the sequences downloaded from the GenBank, a phylogenetic tree of Elmidae was constructed. The results showed that *Paramacronychus* and *Urumaelmis* should be merged into *Zaitzevia*, and *Ordobrevia* should be incorporated into *Stenelmis*. The molecular technique is of great significance for the limitation of the genus and identification of closely related species.

Based on the research of specimens collected in the Herpetological Museum of Chengdu Institute of Biology, the taxonomic status of *Pararhabdophis* is re-evaluated, synonymize the genus *Pararhabdophis* with *Hebius* and report *Hebius chapaensis* from China. The taxonomic status of *Opisthotropis balteata* is confirmed, the nomen *Trimerodytes* is resurrected from *Opisthotropis* and synonymize the junior generic nomen *Sinonatrix* with *Trimerodytes*.

The research staff of the museum also presented a comprehensive reassessment of the classification systems to the Chinese reptiles, updated corresponding Chinese scientific names and concluded the *A revised taxonomy for Chinese reptiles* in 2015. This analysis suggested that in China there are a total of 3 orders, 30 families, 132 genera, and 462 reptile species. This paper was rated as F5000 paper in 2018 (F5000, http://f5000.istic.ac.cn/). Combined with the cited frequency in 2017, this paper was also listed in the 2018 high impact paper of *Biodiversity Science*.

Based on the study of the collected specimens in the Northeast Biological Herbaria, a new classification system of bryophytes is proposed.

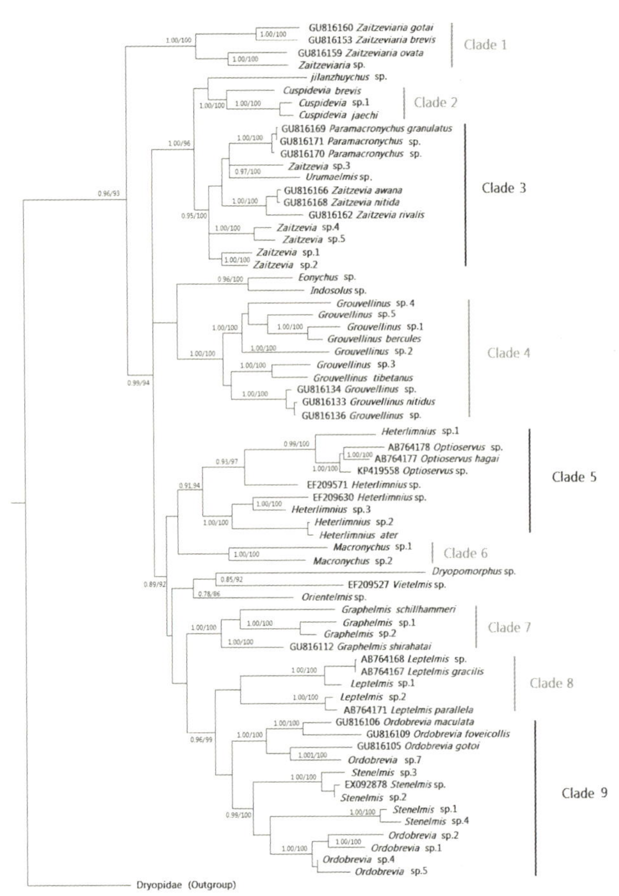

基于 *28S*、*18S*、*COI*、*16S*、*COXI*、*Cytb* 六个基因片段构建的 ML 系统发育树

ML tree based on the combined data matrix of *28S*, *18S*, *COI*, *16S*, *COXI* and *Cytb* genes

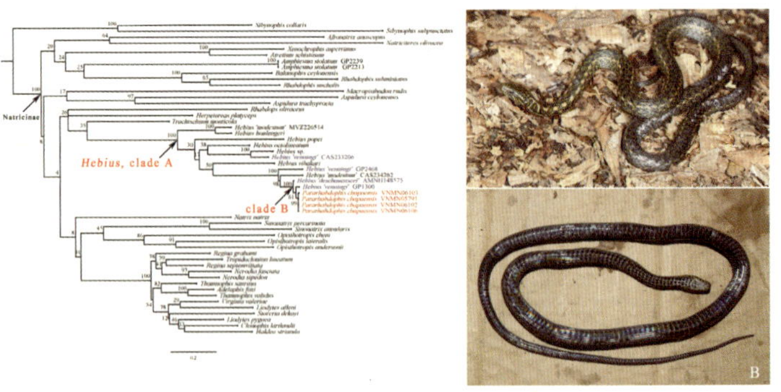

水游蛇科系统进化树，示异纹蛇属 *Pararhabdophis* 分子系统地位（左）、我国云南产沙坝腹链蛇（右）

Phylogenetic tree of Natricidae, showing molecular systematics status of *Pararhabdophis* (left); and *Hebius chapaensis* from Yunnan, China (right)

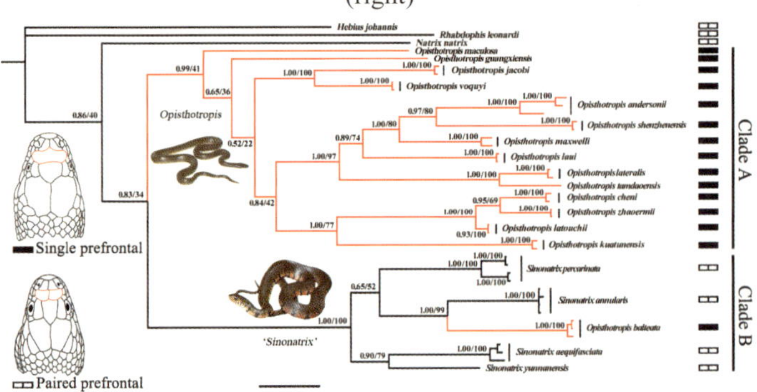

后棱蛇属 *Opisthotropis* 与华游蛇属 *Sinonatrix* 分子系统发育树

Phylogenetic tree depicting phylogenetic relationships of two genera, *Opisthotropis* and *Sinonatrix*

苔类系统。

依托菌物标本馆标本，研究人员针对世界范围的接合菌、子囊菌、担子菌和地衣等重要类群开展系统发育研究。发表根霉属、小克银汉霉属、毛壳科和亚隔孢壳科的世界性系统学专著。运用多基因分析，明确了担子菌各主要高阶类群的亲缘关系与分化时间，修订担子菌门分类系统为 4 亚门、18 纲、68 目、241 科、1926 属的分类体系。

菌物标本馆还构建了中国菌物名录数据库，已录入记录超过 28 万条，涉及 2.8 万余个不重复的菌物名称。编写出版了《中国生物物种名录》（印刷版）菌物卷 4 卷分册。建立国际菌物名称注册信息库 Fungal Names，与 Index Fungorum（英国）和 MycoBank（荷兰）菌物信息库成为国际真菌学协会正式认可的在菌物名称发表前必须注册的 3 个国际网站之一，表明我国菌物分类学已处于世界前沿，为我国生物分类学者增添了新的荣誉。

依托青藏高原生物标本馆，西北高原生物研究所对部分植物类群进行了分类修订。例如，对龙胆科中最大的主要分布于我国的龙胆族进行了世界范围的系统发育、起源中心、分布扩散、物种形成机制和专著性研究。英文专著 *A Worldwide Monograph of Gentiana* 不仅是我国第一部有关超过 300 种大属植物的专著性修订，也是世界上第一部对一个属发表种名超过 2000 种的类群进行的分类修订。“中国龙胆科植物研究”获 2004 年国家自然科学奖二等奖。英文专著 *A Worldwide Monograph of Swertia and Its Allies* 出版，研究了

后棱蛇属 *Opisthotropis* 与环游蛇属 *Trimerodytes* 物种形态特征对比

Morphological comparisons of species of *Trimerodytes* and *Opisthotropis*

生物多样性 2015, **23** (3): 365–382
Biodiversity Science

doi: 10.17520/biods.2015037
http: //www.biodiversity-science.net

中国爬行纲动物分类厘定

蔡 波 王跃招[*] 陈跃英 李家堂

(中国科学院成都生物研究所, 成都 610041)

摘要: 本文对中国爬行纲动物的分类体系和物种进行了系统的评估, 规范了中文学名, 给出了《中国爬行纲校正名录》, 结果表明: 中国现存爬行纲动物3目30科132属462种, 其中鳄形目(Crocodylia)1科1属1种, 龟鳖目(Testudines)6科18属33种, 有鳞目(Squamata)蜥蜴亚目(Lacertilia)10科41属189种, 有鳞目蛇亚目(Serpentes)13科72属239种。与《中国动物志 爬行纲 第一卷(总论、龟鳖目、鳄形目)》、《中国动物志 爬行纲 第二卷(有鳞目: 蜥蜴亚目)》和《中国蛇类》相比, 目和亚目无变化; 科级水平新增5科, 变更2科; 属级水平新增23属, 合并15属, 变更6属; 种级水平新增81种, 变动2种; 未收录同物异名12种、杂交6种、中国无分布7种。形态和分子系统发育研究结果在爬行动物不同分类阶元均有一定差异, 文章对这些争议进行了讨论, 并对名录的选择做了说明。

关键词: 爬行动物, 名录, 中国, 生物多样性

A revised taxonomy for Chinese reptiles

Bo Cai, Yuezhao Wang[*], Yueying Chen, Jiatang Li

Chengdu Institute of Biology, Chinese Academy of Sciences, Chengdu 610041

Abstract: Based on taxonomic and phylogenetic studies, we presented a comprehensive reassessment of the classification systems, updated corresponding Chinese scientific names and concluded the *Checklist of Chinese Reptilia*. Our analysis suggested that in China there are a total of 3 orders, 30 families, 132 genera, and 462 reptile species. The order Crocodylia includes one family, one genus, and one species. The order, Testudines includes 6 families, 18 genera, and 33 species. The order Squamata includes the suborder Lacertilia and Serpentes. Lacertilia includes 10 families, 41 genera, and 189 species and Serpentes includes 13 families, 72 genera, and 239 species. Compared to *Fauna Sinica (Reptilia 1): General Accounts of Reptilia, Testudoformes and Crocodiliformes, Fauna Sinica (Reptilia 2): Squamata (Lacertilia)* and *Snakes of China*, we added 81 species, revised 2 species and excluded 12 synonymous, 6 hybrid, and 7 undistributed reptile species. At the genus level, there are 23 additional, 6 revised, 15 synonymous genera. At the family level, there are 5 additional and 2 revised families. There is no change at the order or suborder levels. Studies of morphology and molecular phylogeny on taxonomic categories of reptiles reflects several divergences. We discusssed these controversies and explained choices of the *Checklist of Chinese Reptilia* in this article.

Key words: Reptilia, checklist, China, biodiversity

《中国爬行纲动物分类厘定》论文首页

The title page of the article entitled "A revised taxonomy for Chinese reptiles"

Moss: In the research of inheritance, sporopollenin and phylogeny of bryophyte, Prof. GAO Chien, the famous bryologist in the herbarium, found that the types of spore dispersal have an important significance in the evolution of mosses. They revised the widely used system of Brotherus since 1924-1925 and proposed a new evolution system on the family level (including two subclasses, 17 orders, and 86 families), based on the studies of the sporophyte, peristome teeth and the individual development of mosses. In this system, 5 types of spore dispersal were inducted: decay dispersal, wind dispersal, vapour-wind dispersal, water dispersal and insect dispersal. This new system reflects the evolutionary process within mosses and the origin of moss from green algae.

Liverworts: Based on the anatomical structure of liverworts perichaetium in the herbarium and literatures, researchers of the Herbaria proposed that Hepaticae is originaged from Musci, and the perichaetium component elements

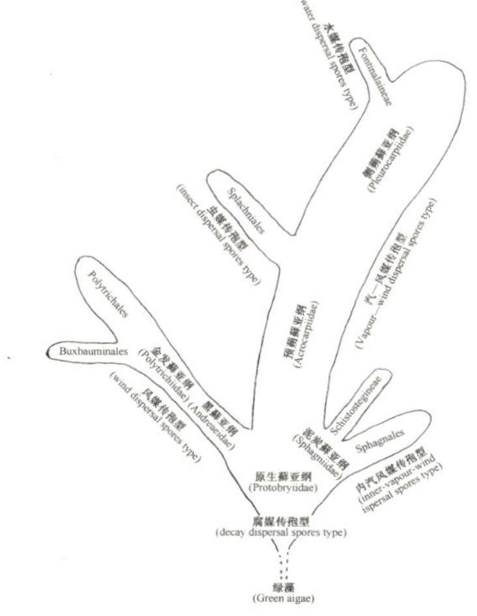

基于传孢类型的藓类植物新系统树状图

Newly proposed evolution system of families of mosses

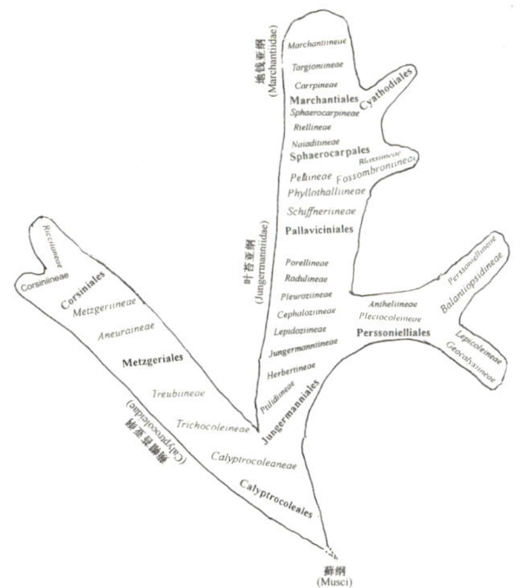

基于雌苞类型的苔类植物新系统

Newly proposed liverwort system based on perichaetium type

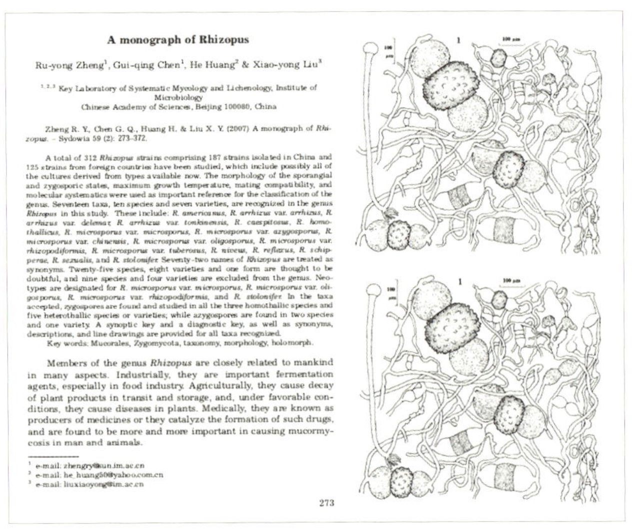

A monograph of Rhizopus

Ru-yong Zheng[1], Gui-qing Chen[1], He Huang[2] & Xiao-yong Liu[3]

[1,2,3] Key Laboratory of Systematic Mycology and Lichenology, Institute of Microbiology
Chinese Academy of Sciences, Beijing 100080, China

Zheng R. Y., Chen G. Q., Huang H. & Liu X. Y. (2007) A monograph of *Rhizopus*. – Sydowia 59 (2): 273-372.

A total of 312 *Rhizopus* strains comprising 187 strains isolated in China and 125 strains from foreign countries have been studied, which include possibly all of the cultures derived from types available now. The morphology of the sporangial and zygosporic states, maximum growth temperature, mating compatibility, and molecular systematics were used as important reference for the classification of the genus. Seventeen taxa, ten species and seven varieties, are recognized in the genus *Rhizopus* in this study. These include: *R. americanus*, *R. arrhizus* var. *arrhizus*, *R. arrhizus* var. *delemar*, *R. arrhizus* var. *tonkinensis*, *R. caespitosus*, *R. homothallicus*, *R. microsporus* var. *microsporus*, *R. microsporus* var. *azygosporus*, *R. microsporus* var. *chinensis*, *R. microsporus* var. *oligosporus*, *R. microsporus* var. *rhizopodiformis*, *R. microsporus* var. *tuberosus*, *R. niveus*, *R. reflexus*, *R. schipperae*, *R. sexualis*, and *R. stolonifer*. Seventy-two names of *Rhizopus* are treated as synonyms. Twenty-five species, eight varieties and one form are thought to be doubtful, and nine species and four varieties are excluded from the genus. Neotypes are designated for *R. microsporus* var. *microsporus*, *R. microsporus* var. *oligosporus*, *R. microsporus* var. *rhizopodiformis*, and *R. stolonifer*. In the taxa accepted, zygospores are found and studied in all the three homothallic species and five heterothallic species or varieties; while azygospores are found in two species and one variety. A synoptic key and a diagnostic key, as well as synonyms, descriptions, and line drawings are provided for all taxa recognized.

Key words: Mucorales, Zygomycota, taxonomy, morphology, holomorph.

Members of the genus *Rhizopus* are closely related to mankind in many aspects. Industrially, they are important fermentation agents, especially in food industry. Agriculturally, they cause decay of plant products in transit and storage, and, under favorable conditions, they cause diseases in plants. Medically, they are known as producers of medicines or they catalyze the formation of such drugs, and are found to be more and more important in causing mucormycosis in man and animals.

[1] e-mail: zhengry@sun.im.ac.cn
[2] e-mail: he_huang50@yahoo.com.cn
[3] e-mail: liuxiaoyong@im.ac.cn

273

根霉属分类学专著

The monograph on taxonomy of *Rhizopus*

基于多基因系统演化关系修订的担子菌门新分类系统

The revised taxonomy of Basidiomycota base on multi-locus phylogeny

中国菌物名录数据库、《中国生物物种名录——菌物卷》和菌物名称注册网站

Database of Chinese fungi, *Species Catalogue of China: Volume 3 Fungi*, and Fungus name registration website

世界上主要标本馆所藏该类群的上万份标本，整理了该类群涉及的上万篇分类学文献，将发表在獐牙菜属下的上千个物种名确定为198个物种，并将这些物种处理为5个属，其中两个属为发表的新属，是对一个发表物种名超过上千的大属进行世界性分类处理的研究，是我国分类学上在《中国植物志》研究基础上，在世界范围内进行大属分类学研究的典型范例，必将促进我国植物学工作者从世界范围内系统性完成一些大属的分类学工作。完成《昆仑植物志》四卷的全部出版工作，这部近420万字的专著，从筹划编撰到出版历时12年，由中国科学院西北高原生物研究所等单位40位植物分类学者，依据10多次野外考察成果及标本馆60余年标本积累著成，是目前对该地区最详尽的植物分类与区系研究成果。此外，《青海植物检索表》出版，共收录青海省境内野生、重要露天栽培维管植物129科723属3489种。

依托新疆生态与地理研究所标本馆馆藏的沙拐枣属标本，科研人员完成了4个沙拐枣属科研项目，对中亚地区沙拐枣属植物进行分类学修订，诠释了环境变化对该属植物的地理分布、遗传变异以及进化历史的影响，理论上填补了沙拐枣属植物亲缘地理学研究的空白，阐明了沙拐枣属植物在干旱区的生态分布的重要意义。另一项在研的国家自然科学基金项目“新疆灌木区系的系统发育研究”是通过构建新疆灌木植物区系宏系统发育树，来了解现代新疆灌木植物区系生物多样性格局和维持机制，从而采取更有效的保护措施。

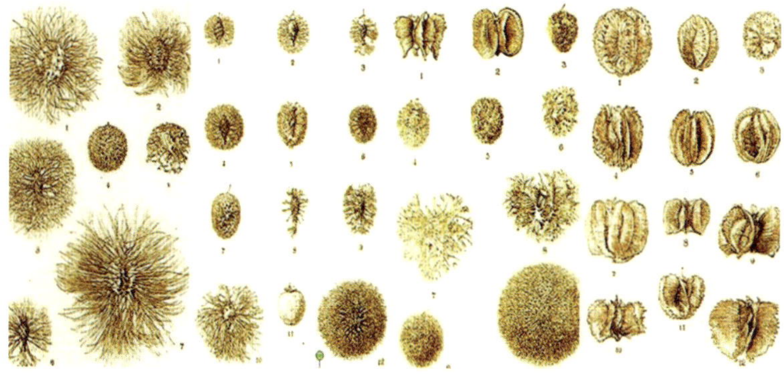

沙拐枣属植物分类学修订
The taxonomic revision of *Calligonum*

should be used as the main feature within evolution of liverwort, and differentiation degree of gametophyte as supplement. They reclassified the liverwort system into three subclass, eight orders, with seven types of perichaetium, including: including perichaetium of calyptra type, perichaetium of perianth type, perichaetium of coelocaule and perianth, perichaetium of coelocaule type, perichaetium of perigynium type, perichaetium of involucre type and perichaetium of carpocephalum involucre type.

Based on the specimens from the Fungarium, the researchers carried out phylogenetic studies on the important groups of zygomycetes, ascomycetes, basidiomycetes and lichens in the world. World monographs on economically important fungal groups *Rhizopus*, *Cunninghamella*, Chaetomiaceae and Didymellaceae were published by Fungarium. Phylogenic relationships and divergence times of high-leveled taxa in Basidiomycota were estimated based on multi-locus analyses. A new taxonomic system of Basidiomycota was newly provided, accepting four subphyla, 18 classes, 68 orders, 241 families and 1,926 genera.

The database of *Chinese fungi* was created, with more than 280,000 records referring to 28,000 non-repetitive species names included. Meanwhile, four volumes of "Species Catalogue of China. Fungi" were compiled and published. The website of "Fungal Names" was built up to fulfill the requirement of fungal name registration when describing a new fungal species. This database, together with the other two web databases, "Index Fungorum" (Britain) and "MycoBank" (Netherland), are the International Mycological Association officially accepted registration websites for valid publication of new fungal names, which demonstrated the great impact of the Chinese mycological community.

Based on the Qinghai-Tibet Plateau Museum of Biology, the researchers of Northwest Institute of Plateau Biology have revised some plant groups. *A Worldwide Monograph of Gentiana* was the first monograph in China that had taxonomic revision for over 300 genera, and it was also the first monograph that revised a huge genus within which more than 2000 species were reported. "Studies on Gentianaceae in China" was award the Second Class Prize of National Natural Science Award, 2004. Such a large-scale monographic study had not been published before in the alpine plant study area, and it was also very unusual even in studies on low altitude plants. *A Worldwide Monograph of Swertia and Its Allies* is the research achievement that Ho Ting-nong and Liu Shang-wu have done research in the western area under harsh conditions in China for 50 years. The thousands of species which was reported under the *Swertia* were assessed as 198 species, and they classified those species into five genera including *Swertia, Sinoswertia, Lomatogoniopsis, Veratrilla* and *Lomatogonium*. The *Sinoswertia* and *Lomatogoniopsis* are the new genera. This taxonomic treatment study of *Swertia*, which has thousands of published spices, is rarely seen in the world. It is the typical example of the large genus taxonomic treatment study in the scope of the world on the foundation of *Flora Reipublicae Popularis Sinicae*; it will definitely encourage Chinese botanists to finish the large genus taxonomic work systematically worldwide. Four volumes of *Flora Kunlunica* were published with 4,200,000 words. More than 40 scientists spent 12 years on it. More than 10 times of field surveys were carried out. And they were the most exhaustive research output about plant classification and flora of this area. Claves of Plantarum Qinghaiensis was published with a record of 129 Class, 723 Genus and 3,489 Species.

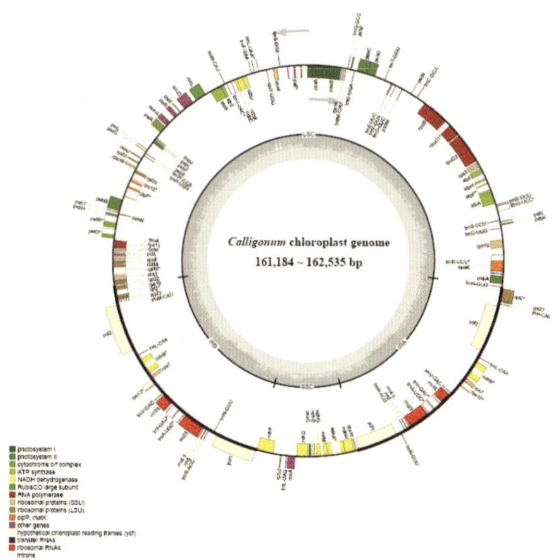

新疆灌木植物区系宏系统发育树
Macro-phylogenetic tree of shrub flora in Xinjiang

Relying on the specimens of *Calligonum* collected in XJBI, researchers have completed four scientific research projects of *Calligonum*, revised the taxonomy of *Calligonum* in Central Asia, interpreted the impact of environmental changes on the geographical

2）进化生物学领域研究进展

生物标本是开展生分类研究不可或缺的研究载体，而物种的分类研究是生物区系研究、物种系统演化、生态学、细胞遗传学、分子生物学等的研究基础。生物标本为进化生物学提供了大量研究素材。

近年来，植物研究所标本馆的分类学家充分利用馆藏标本，在植物物种进化、系统发育等方面得了重要进展，其中代表性工作如下。

（1）揭示中国被子植物进化的博物馆和摇篮 PE 科研人员及其合作者采用包括中国 92% 被子植物属的时间树和 140 余万条详尽的空间分布数据，揭示了中国被子植物的多样性格局和的进化历史。该研究结果发现，中国约 66% 的被子植物属是在早中新世之后出现。中国东部的植物区系分化古老，系统发育结构离散，具有较高的系统发育多样性；西部的区系相对年轻，系统发育结构聚集，具有较低的系统发育多样性。进一步的分析发现：对于草本植物，中国东部是博物馆，西部是进化的摇篮；而对于木本植物，中国东部既是博物馆也是摇篮。最后，通过识别中国被子植物系统发育多样性热点地区，提出了优先保护策略。

（2）揭示种子植物深层分支的进化关系 裸子植物四大支系间的进化关系以及裸子植物与被子植物的关系一直备受争议。PE 科研人员利用 1308 个直系同源基因重建了种子植物五大支系间的进化关系，结果强烈支持倪藤类是松科植物的姐妹群。同时，倪藤类与被子植物具有相似的分子进化速率，可能经历了相同的选择压。研究还发现，倪藤类与被子植物间发生了分子趋同或同塑进化，导致了以往研究中倪藤类系统位置的巨大争议。该研究首次确立了种子植物的深层进化关系，为未来种子植物的形态和分子研究提供了进化框架。

（3）维管植物重要类群的系统发生关系研究 PE 科研人员在维管植物重要类群的系统发生研究中，取得了重要进展：张宪春团队与合作者提出了蕨类植物新的分类系统，发表蕨类植物 2 个新科；汪小全团队重建了裸子植物主要类群系统发生关系；张树仁团队、金效华团队等重建了被子植物兰科、菊科、毛茛科、防己科、桔梗科等类群系统发育关系。

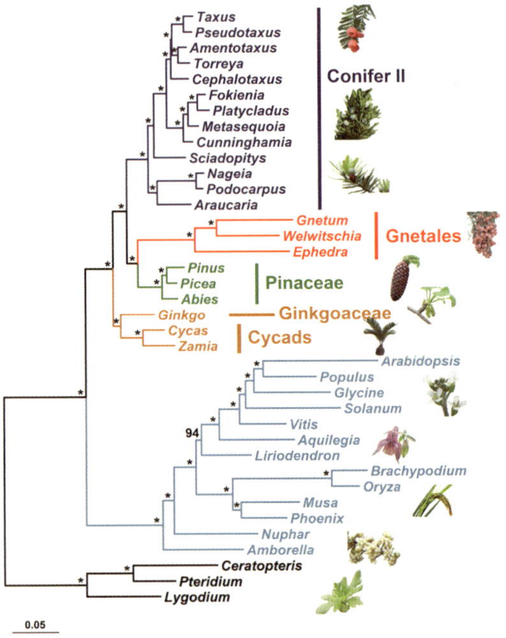

依托成都生物研究所植物标本馆，近十年来加强了植物进化及物种形成、系统发育方面的研究。利用植物形态特征和基因位点及基因组数据，重建了蕨类植物卷柏科、三叉蕨科、岩蕨科、瓶尔小草科、凤尾蕨科的凤尾蕨亚科、铁角蕨科的铁角蕨属、鳞毛蕨科的耳蕨属和复叶耳蕨属、水龙骨科的石韦属和瓦韦属，种子植物中豆科的山蚂蝗属、长柄山蚂蝗属、胡枝子属、杭子梢属、木蓝属等，蔷薇科蔷薇属、花楸属和悬钩子属，菊科的黄鹌菜属和假还阳参属，百合科的广义百合属等一系列重要科属的系统发育关系。在物种进化和物种形成方面，利用形态和 DNA 数据，在三叉蕨属网状进化、胡枝子属基因渐渗和不完全谱系发育、胡枝子属 ITS 进化不一致及杂交、蔷薇属植物的天然杂交、生物进化、谱系发育及物种形成等研究方面取得了代表性成果。相关文章发表在 *Cladistics*、*Taxon*、*Molecular Phylogenetics and Evolution*、*Scientific Reports* 等领域内高水平杂志。

西双版纳热带植物园标本馆依托馆藏及野外资源，与中国科学院植物研究所合作，构建了中国兰科植物系统树；基于覆盖所有分布区的 177 种冬青属植物的两个核基因，重建了冬青属植物的系统发育关系。

高分辨率和支持率的种子植物系统发育树揭示了倪藤类与被子植物的关系

A consistent and well-resolved phylogeny of seed plants shows the relationship of Gnetales and angiosperms

distribution, genetic variation and evolutionary history of the *Calligonum*, and theoretically filled the gap in the study of the phylogeography of *Calligonum*. And the significance of the ecological distribution of *Calligonum* in the arid area was clarified. Another NSFC project, "Study on the Phylogeny of Shrub Flora in Xinjiang", is to understand the biodiversity pattern and maintenance mechanism of modern shrub flora in Xinjiang by constructing a macro phylogenetic tree of shrub flora in Xinjiang, so as to take more effective protection measures.

2) Research progress in evolutionary biology

Biological specimens are indispensable for taxonomy research. The taxonomic study forms the basis of research in such disciplines as fauna study, species system evolution, ecology, cytogenetics, and molecular biology. Biological specimens provide a lot of research materials for evolutionary biology.

In recent years, taxonomists of PE Herbarium take great advantage of specimens of PE, and have made great contributions to systematics, evolutionary biology and biogeography, of which some outstanding studies are list as follows.

(1) Revealing the evolutionary museums and cradles for the Chinese angiosperms. With a dated phylogeny of 92% of the angiosperm genera from China and detailed spatial distribution data of ca. 1.4 million occurrence records, PE taxonomists and colleagues uncovered the biodiversity patterns and evolutionary history of the Chinese angiosperm flora. PE taxonomists found that 66% of the angiosperm genera in China did not originate until the early Miocene. The flora of eastern China bears a signature of older divergence, phylogenetic over dispersion and higher phylogenetic diversity. In western China, the flora shows more recent divergence, pronounced phylogenetic clustering and lower phylogenetic diversity. Their analyses indicate that eastern China represents a floristic museum, and western China an evolutionary cradle, for herbaceous genera; eastern China has served as both a museum and a cradle for woody genera. They also identify areas of high phylogenetic diversity and provide strategies for conservation priorities in China.

(2) Phylogenomics resolves the deep phylogeny of seed plants and indicates partial convergent or homoplastic evolution between Gnetales and angiosperms. The phylogenetic placement of order Gnetales remains one of the most controversial issues in seed plant evolution. PE taxonomists have studied the phylogeny, speciation and phylogeographical pattern of gymnosperms systematically. Recently, to resolve the deep phylogeny of seed plants and to address the sources of phylogenetic conflict, taxonomists conducted a phylotranscriptomic study with a sampling of all 13 families of gymnosperms and main lineages of angiosperms. Multiple datasets containing up to 1,308 loci were analyzed, using concatenation and coalescence approaches. The study generated a consistent and well-resolved phylogeny of seed plants, which places Gnetales as sister to Pinaceae and thus supports the Gnepine hypothesis. Cycads plus Ginkgo is sister to the remaining gymnosperms. The herbarium also found that Gnetales and angiosperms have similar molecular evolutionary rates, which are much higher than those of other gymnosperms. This implies that Gnetales and angiosperms might have experienced similar selective pressures in evolutionary histories. The study provides a robustly reconstructed backbone phylogeny that is important for future molecular and morphological studies of seed plants, in particular gymnosperms, in the light of evolution.

(3) Reconstruction of the phylogeny of many lineages of vascular plants. In the last decade, PE staff have made a great contribution in the reconstruction of the phylogeny of vascular plants. ZHANG Xianchun's group and his collaborators proposed a new classification system of ferns and their relatives, and published two new families. WANG Xiaoquan's group reconstructed the phylogeny of major lineages of gymnosperms. ZHANG Shuren's group and JIN Xiaohua's group and other teams reconstructed the phylogeny of Orchidaceae, Asteraceae, Ranunculaceae, Tetrandriaceae Campanulaceae.

Relying on the Herbarium of Chengdu Institute of Biology, researchers have strengthened their research on plant evolution, speciation and phylogeny in the past decade. Phylogenetic relationships of fern plant of Selaginellaceae, Tectariaceae, Woodsiaceae, Ophioglossaceae, Pteridoideae in Pteridaceae, *Asplenium* in Aspleniaceae, *Polystichum* and *Arachniodes* in Dryopteridaceae, *Pyrrosia* and *Lepisorus* in Polypodiaceae, and flowering plant of *Desmodium*, *Hylodesmum*, *Lespedeza*, *Campylotropis* and *Indigofera* in Leguminosae, *Rosa*, *Sorbus* and *Rubus* in Rosaceae, *Youngia* and *Crepidiastrum* in Asteraceae, *Lilum* (in the broad sense) in Liliaceae were reconstructed based on morphological characteristics, gene loci and genomic data. A series of representative research results on the evolution and speciation, such as reticulate evolution in fern genus *Tectaria*, introgression, incomplete lineage sorting, and ITS non-concerted evolution in the legume genus *Lespedeza*, and natural hybridization, biological evolution, speciation of the genus *Rosa* have been made by using morphological and DNA data. Relevant articles are published in a series of high-level magazines such as *Cladistics*, *Taxon*, *Molecular Phylogenetics and Evolution*, *Scientific Reports* etc.

Based on orchid specimens from both herbaria and the field, HITBC has cooperated with the Institute of Botany,

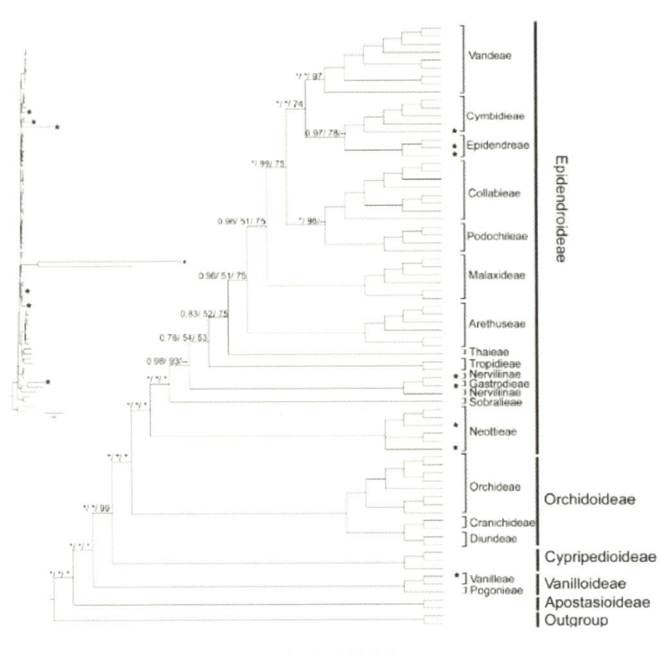

兰科系统树

Phylogenetic tree of Orchidaceae

依托华南植物园标本馆馆藏资源，主要进行了如下研究。

（1）喀斯特特殊生境地理隔离与物种形成 ①采用多基因片段重建了苦苣苔科报春苣苔属近 190 个物种的系统发育关系和生物地理历史，首次量化地揭示东亚季风气候对植物物种形成的影响，也为理解东亚地区生物多样性的起源、形成及维持机制奠定了理论基础。②对河池唇柱苣苔复合群开展物种界定研究发现，发现基于多物种溯祖统计分析（包括 BPP 及 iBPP）在对具有强烈遗传分化的类群进行物种界定分析时存在过多划分物种的风险，需进一步考虑其他数据支撑。

（2）代表性植物类群的分子进化与多样化研究 利用世界分布的珍珠菜族的数据研究发现最早定居时间和净分化速率与东亚物种多样性没有显著相关关系，但东亚类群却具有较早的定居时间和高的净分化速率，显示两者共同导致了该类群在东亚具有极高物种多样性。在微观进化方面，对 57 个荠菜个体的比较转录组学分析发现，亚基因组的表达漂变可能是重复基因进化的主要驱动力。

（3）“二型花柱的进化、维持与破损”的研究 以被子植物中二型花柱最为集中的茜草科、报春花科为主要研究材料，揭示了耳草属、玉叶金花属的二型花柱与隐性自交不亲和系统及其维持机制，玉叶金花属二型花柱向雌雄异株进化的细胞机制、向同型花柱进化适应的繁殖机制，海岛二型花柱植物海岸桐在特殊生境的繁育系统维持机制；完成了对玉叶金花属植物繁育系统进化研究，驳斥了雌雄异株是“进化死胡同”的假说；探讨了玉叶金花属不同繁育系统类型与花粉胚珠比的关系、玉叶金花属植物同域分布物种的生殖隔离机制、玉叶金花属植物与叶际真菌的共存模式；阐明迎阳报春性系统进化及遗传机制。

近 60 年来，昆明动物研究所及全球学者依托昆明动物博物馆馆藏标本在物种进化、系统发育等方向，结合传统形态分类和分子水平来研究生物多样性的演化及机制，阐明了哺乳类、鸟类、两栖爬行类、鱼类等动物类群系统与演化中的一些重要问题，如高黎贡白眉长臂猿的研究、中国大鲵保护策略、不同动物的分子系统学研究、白鹡鸰的谱系地理学研究、棕颈钩嘴鹛的分子系统学研究、蛙类的物种形成、分化、适应性进化和生物地理研究。

喀斯特生境中的报春苣苔属植物

Plants of the genus *Primulina* in Karst

基于青藏高原生物标本馆馆藏标本对虎耳草属、绣线菊亚科等高原物种的生物地理学研究表明，高原隆升及冰期气候波动是两个物种种间和种内居群间分化的主要驱动因子。裂腹鱼类系统进化研究、特有小哺乳动物的谱系地理研究，揭示了不同类群动物适应青藏高原气候与环境特点及种群历史动态。印象初院士提出了蝗虫类在高原上的适应

CAS, in the construction of the phylogenetic tree of Orchidaceae. Researchers constructed a phylogenetic tree of the hollies (*Ilex* L., Aquifoliaceae) based on two nuclear genes, representing 177 species spread across the geographical range.

Relying on the collection resources of the Herbarium of South China Botanical Garden, the following researches are mainly carried out:

(1) The geographical isolation and speciation of plants in Karst special habitats. 1) the phylogenetic relationships and biogeographical history of nearly 190 species of *Primulina* (Gesneriaceae) were reconstructed by using multi-gene fragments, which quantitatively revealed the impact of East Asian monsoon climate on plant speciation at the first time, and lays a theoretical foundation for understanding the origin, formation and maintenance mechanism of biodiversity in East Asia; 2) A study was conducted to delineate species-level divergence in the *Primulina hochiensis* complex, which suggested that there was a risk of over presence of species by using the multispecies coalescent methods (including BPP and iBPP) in the analysis of species delimitation with strong genetic differentiation, and thus further analysis with additional data was needed.

(2) The molecular evolution and diversification of representative plants. By using the plant tribe Lysimachieae which is world-wide distributed, the herbarium found that neither time nor diversification rates alone explain richness patterns among regions in Lysimachieae. Instead, a new index that combines both factors explains global richness patterns in the group and their high East Asian biodiversity. On the other hand of microevolution, the comparative transcriptome analysis of 57 individuals of *Capsella bursa-pastoris* revealed that the expression drift of subgenome might be the main driving force of repeated gene evolution.

(3) The evolution, maintenance and breakdown of distyly. The herbarium used Rubiaceae and Primulaceae, the two families distyly frequently occurred, as the main research materials. Firstly, researchers investigated the distyly and revealed the maintenance of cryptic self-incompatible systems in *Hedyotis* and *Mussaenda*, the cellular mechanism of how dioecy evolved from distyly polymorphism, the reproductive mechanism that distyly breakdown to homostyly in *Mussaenda*, and how the floral polymorphism and mating systems of *Guettarda speciosa* were preserved in the special island environment. Secondly, by constructing the phylogenetic relationship of *Mussaenda* and investigating the breeding system of multiple species, researchers rejected the hypothesis that dioecy is an "evolutionary dead end". Thirdly, researchers investigated the relationship between mating patterns and pollen/ovule ratios, the reproductive isolation mechanism of sympatric species, and co-occurrence patterns between *Mussaenda* plants and phyllosphere fungi. At last, this study clarified the evolution and genetic mechanism from distyly to homostyly in *Primula oreodoxa*.

In the past 60 years, relying on specimens collected by Kunming Natural History Museum of Zoology, KIZ and global scholars have studied the evolution and mechanisms of biodiversity in species evolution, phylogeny and other disciplines, combining traditional morphological classification and molecular analysis, and having clarified the problems concerning the systematics and evolution of mammals, birds, amphibians, reptiles, and fish. This includes studies on the Gaoligong white-browed gibbon, protection strategies of Chinese giant salamander, molecular systematics on various animals, genealogy of the white salamander, molecular systematics of the brown-necked hook-billed salamander, and frog species formation, differentiation, adaptive evolution and biogeography.

The biogeographical research based on the specimen of Genus *Saxifraga*, Spiraeoideae and other alpine species from Qinghai-Tibet Plateau Museum of Biology indicated that uplift of Qinghai-Tibet Plateau and climatic vacillation in ice age were the driving factors of differentiation among species and populations. The studies on the phyletic evolution of fishes of Schizothoracinae and endemic small mammals indicated characteristic adaptions to the climate and environment of Qinghai-Tibet Plateau in different animals and historical population dynamics. Academician. Yin Xiangchu proposed some new ideas about the adaptability and evolution ways of locusts in the highlands and the type of plateau zorotypus. The strong wind in the plateau is not suitable for the flight of locusts what leads to the degradation of the locust's wing. Locust's wing is one of its vocal organs. The degradation of the locust's wing leads to the vocal organs' degradation what causes the degradation or disappearance of the locust's hearing organs. It is showed that the species which are lack of wings, vocal organs and hearing organs living in the high attitude areas are the most evolutional ones and the unique species in the Qinghai-Tibet Plateau.

迎阳报春的不同花型结构

Flower Structure of *Primula oreodoxa*

性、演化途径和高原缺翅型等新见解；阐明了高原上风大不适于蝗虫飞行导致翅的退化，翅是蝗虫的发音器官构造之一，翅的退化导致发音器的退化，发音器的退化和消失又导致听觉器官的退化和消失。在高海拔地区生存的缺翅、缺发音器、缺听器的种类是最进化的种类，也是青藏高原的特有种类。

国家动物博物馆标本馆近年来注重动物组织样品的收集和整理，为采用分子生物学等新技术、新方法开展研究提供了丰富的材料，取得了突出的成绩。例如：科研人员基于对馆藏标本的形态研究和基因序列测定与分析，进行了属（姬鼠属、鼠兔属）、亚属（耗兔亚属、鼠兔亚属）、种（羚牛、塔里木兔、黑腹绒鼠、白腹鼠）、物种复合体（白腹鼠复合体）等的不同阶元的进化及其历史、分类与分化问题的探讨；开展了一些物种种群遗传结构、种群数量历史动态研究（如间颅鼠兔），以及物种间的基因渗透研究（如草兔与塔里木兔）；对一些分类单元（如鼹形鼠科、三趾跳鼠属）开展了系统发育研究；评估了山雀科的系统发育、海拔和物种相互作用对其形态特征进化的贡献；确认了斑嘴鹈鹕（*Pelecanus philippensis*）在中国的历史分布，认为这些标本可为将来成功地将该物种重新引入中国和以前地理范围内的其他地方提供信息。此外还有对气候、海拔梯度、栖息地、地理等信息在物种的演化过程所发挥作用的研究与探讨。

科研人员充分利用成都生物研究所两栖爬行动物标本馆馆藏标本数据，在物种演化历史和极端环境适应方面做了大量工作。

（1）解析欧亚大陆和北美大陆之间的生物交换 结果表明了共 61 次从欧亚大陆向北美大陆和 31 次北美大陆向欧亚大陆的扩散事件，揭示了欧亚大陆和北美大陆之间存在长时间不对称量级的生物扩散。研究进一步揭示了欧亚大陆和北美大陆之间扩散事件随时间变化的动态模式。多变的气候环境和相关类群的系统发育多样性造成了欧亚大陆到北美大陆的物种扩散事件比北美大陆到欧亚大陆的物种扩散事件普遍较多。扩散事件的不对称性主要体现在晚渐新世升温事件（LOWE），以及始新世末期（TEE）降温事件、中新世气候平台期（MMCO）之后的降温事件时期，扩散的动态模式与气候变化动态紧密相关。这一研究表明白令陆桥在生物扩散过程中起到了重要作用，并为古地理和古气候变动研究提供了新的视角。

（2）温泉蛇基因组解析及其高海拔适应的分子机制 温泉蛇（*Thermophis baileyi*）是世界上分布海拔最高的蛇类之一。研究通过对温泉蛇进行全基因测序，在温泉蛇属 3 个物种中发现了 27 个共有氨基酸替换。其中与 DNA 修复相关的 *FEN1* 基因的突变型相对于野生型具有更强的稳定性，推测突变有助于温泉蛇在高海拔环境中对紫外线的抵抗。温泉蛇 *EPAS1* 基因的突变减弱了其调节下游基因 *EPO* 表达的能力，进而导致血红蛋白浓度处于较低水平，是温泉蛇适应高海拔低氧条件的重要原因。

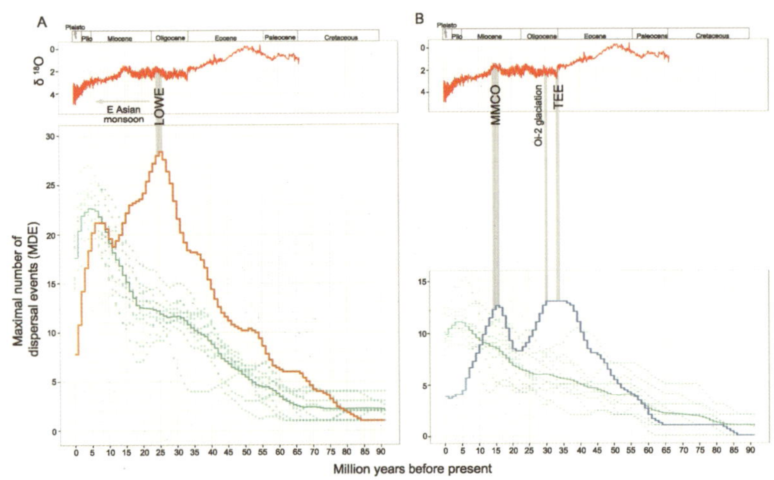

白令陆桥两侧大陆生物交换速率，（A）代表由欧亚大陆向北美大陆的扩散事件；（B）代表由北美大陆向欧亚大陆的扩散事件
Rates of biotic exchange on both sides of Bering Land Bridge (A) from Eurasia to North America and (B) from North America to Eurasia

（3）海蛇基因组解析及其海洋适应的分子机制 海蛇是由陆生蛇类在大约 1000 万年前从陆地重新返回海洋生境中的，是研究适应性演化的理想模型。研究以平颏海蛇（*Hydrophis curtus*）为对象，基于高质量的基因组数据，揭示出平颏海蛇基因家族扩张时间大约为 2000 万年前，可能与海洋环境的适应有密切关系。选择压分析与分子趋同演化分析鉴别出了部分快速演化的蛋白质编码基因，且可能与其他海洋动物共享特有氨基酸残基替换位点。此外，保守非编码元件（CNE）分析发现平颏海蛇 108 个 CNE 发生快速演化，这些

In recent years, the collection of the National Zoological Museum of China has paid attention to the collection and arrangement of animal tissue samples, which has provided rich materials and made outstanding achievements for the research using new technologies and methods such as molecular biology. For example, based on the morphological study and gene sequencing and analysis of collected specimens, researchers have carried out the evolution of different stages of genera (*Apodemus*, *Ochotona*), subgenera (*Pika*, *Ochotona*), species (*Budorcas taxicolor*, *Lepus yarkandensis*, *Eothenomys melanogaster*, *Niviventer coxingi*), species complex (*Niviventer coxingi* complex), and their history, classification and differentiation In addition, the genetic structure, population number history and dynamics of some species (e.g. *Ochotona cansus*) as well as gene penetration among species (e.g. grass hare and Tarim hare) were studied, and the phylogeny of some taxa (e.g. family Spalacidae, genus *Dipus*) was studied; The contribution of phylogeny, altitude and species interaction to the evolution of morphological characteristics of Paridae was evaluated; The historical distribution of *Pelecanus philippensis* in China was confirmed and it is believed that these specimens can provide information for the successful reintroduction of the species into China and other places in the past. In addition, the role of climate, altitude gradient, habitat, geography and other information in the evolution of species is also studied.

Researchers have made full use of the specimen data collected by the Herpetological Museum, Chengdu Institute of Biology and have done a lot of work in the aspects of species evolutionary history and extreme environmental adaptation as follows.

(1) Biotic interchange between Eurasian and North America. Researchers have compiled 92 dispersal events between Eurasia and North America and revealed a long-lasting phase of asymmetric biotic interchange across the Bering land bridge between Eurasia and North America. The overall rate of biotic exchange was higher for dispersal from Eurasia to North America (61 events) compared to dispersal in the reverse direction (31 events). Furthermore, researchers revealed dispersal events between Eurasia and North America had a dynamic pattern changing with time. Shifting environmental conditions in North America and Eurasia, combined with associated phylogenetic diversification of taxa in these areas, thus potentially account for most of the shifting asymmetries of biotic interchange between North America and Eurasia. Although rates of exchange from Eurasia to North America have generally been higher than the reverse throughout the Cenozoic, that asymmetry was most evident during the late Oligocene warming event and was least evident during cooler periods at the end of the Eocene and in the Miocene. These results showed that Bering Land Bridge played an important role in the biological diffusion between Eurasia and North America and provided a new perspective from studies in paleogeography and paleoclimate change.

(2) Genome analysis and molecular mechanism of high-altitude adaptation of *Thermophis baileyi*. *Thermophis baileyi* is one of the world's highest-altitude snakes. The museum revealed the genetic mechanism of ectothermic animals' adaption to extreme high-altitude environments on the basis of genomic data of hot-spring snakes. Studies founded that there were 27 substitutions of common amino acids in 27 different proteins in the three species of *Thermophis* and molecular function experiments confirmed that the mutation of *FEN1* (petal endonuclease-1) gene related to DNA repair

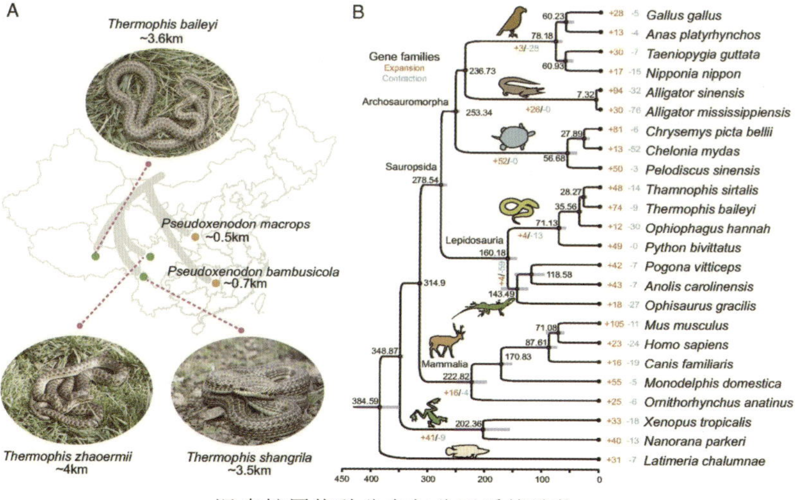

温泉蛇属物种分布与分子系统地位

Geographical distribution of genus *Thermophis* and phylogenetic position of *Thermophis baileyi*

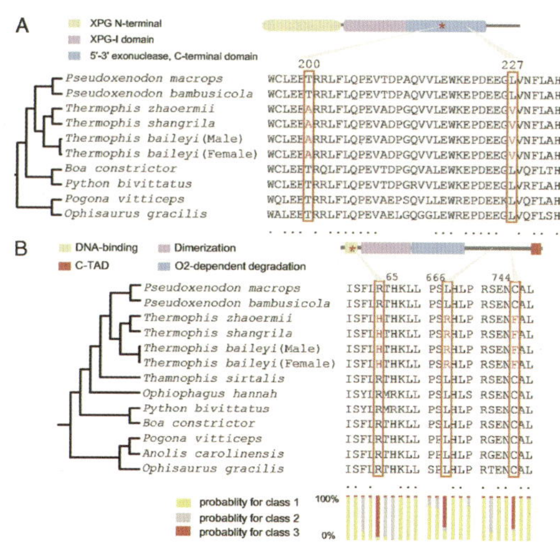

温泉蛇属 3 个物种中共有的氨基酸替换

Shared amino acid replacements in genus *Thermophis* genomes

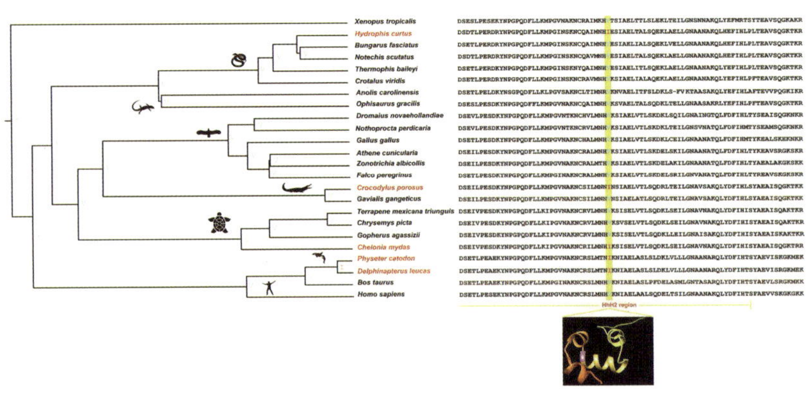

海蛇与其他海洋动物共有的特有氨基酸残基替换位点
The convergent replacement of amino acid of seasnakes and other marine organisms

可能与表型变化有关。基因组与转录组联合分析表明海蛇与陆生蛇类毒液成分不同，海蛇毒腺中具表达量最高的 *PLA2* 毒素转录本。研究结果有助于加深对于由陆地次生性进入海洋爬行动物环境适应性的理解，同时对研究海蛇毒液功能提供了重要启示。

水生生物方面，自 20 世纪 50 年代起，水生生物研究所陈宜瑜院士和曹文宣院士曾带队多次进入青藏高原，开展鱼类资源调查，收集了大量高原鱼类标本保存于水生生物博物馆，开创性地发现裂腹鱼类体表鳞片的有无、须的数量等性状与其分布海拔有关，据此可大体将其分为 3 个类群，其演化过程的 3 个阶段与青藏高原隆起的 3 个主要阶段存在着对应关系。水生生物研究所几代学者对这些高原鱼类标本继续开展研究，采用形态学与多组学相结合的手段揭示了青藏高原地区鱼类高海拔适应性进化的分子机制和遗传基础等。

依托水生生物博物馆馆藏标本，研究人员所做的工作有骨鳔鱼类适应辐射的分子机制、东亚地区淡水鱼类多样性历史的重建过程、鲤科鱼类系统发育新格局等。此外，依托长江鱼类标本，建立长江鱼类 DNA 条形码数据库和分子鉴定体系，截止到 2019 年底共收集长江鱼类 2830 条条形码序列，涉及 16 目、40 科、135 属、238 种。通过构建 DNA 条形码参考数据库，对长江流域鱼类进行了全面的分子评估，涉及的物种约占目前长江已知物种的 64.2%，几乎包含大家所知的鱼类。还利用矛尾鱼标本生物信息学方案在基因组中鉴定了 472 个“反转座拷贝”，通过研究它们的进化年龄、选择压力、表达模式以及相关的基因功能，探讨了它们对脊椎动物“登陆”事件以及矛尾鱼进化的意义。

依托南海海洋生物标本馆馆藏标本，科研人员针对长期以来鲽亚目地位存在的争议，通过对 50 个种类的分析和比较，结合形态结果和分子生物学研究结果最终确定鲽亚目和鲽亚目有更近的系统发育关系，揭示鲽形目是单系群；发现日本须鳎（*Paraplagusia japonica*）拥有已报道以外的其他色型，并查明了具有不同花色的个体体长范围及成熟水平；明确了日本须鳎眼侧的多种花色特征，为须鳎属物种厘定奠定了基础。以海洋生物进化和物种保护研究的旗舰类群——海龙科鱼类为研究对象，基于多组学研究，在国际上率先解密了海马腹鳍丢失之谜，阐明海马竖直游泳的适应进化机制，阐释珊瑚礁物种（海马）对岛礁环境的适应性。通过比较线粒体基因组学对牡蛎、扇贝等重要的经济贝类种类进行系统学解析，阐释了种间基因顺序和基因组变异产生与演化关系；深入揭示了双壳类基因组多样的结构及其丰富的变异。

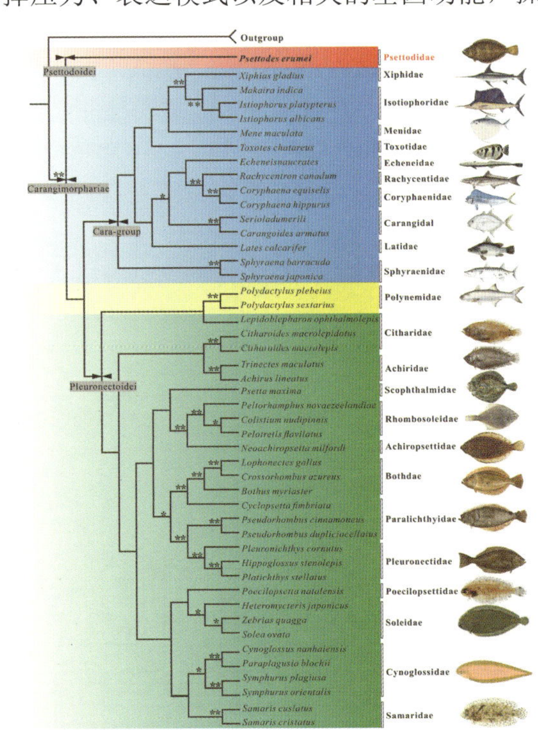

鲽形目鱼类系统发育关系
The phylogenetic relationship of Pleuronectiformes

原尾虫、跳虫、双尾虫和昆虫的系统关系对于阐明六足动物的早期演化以及昆虫的起源等问题具有重要的意义，但

was more stable under ultraviolet irradiation than wild type (low-altitude species). As a species of the plateau, mutation of the *EPAS1* (endothelium PAS domain containing protein-1) gene of the hot-spring snake attenuates its ability to regulate the expression of downstream gene erythropoietin, leading to a lower hemoglobin concentration. That is an important reason for the hot-spring snake's adaptation to high-altitude and low-oxygen conditions.

(3) Genome analysis of sea snake and its molecular mechanism of marine adaptation. The transition of terrestrial snakes to marine life approximately 10 million years ago (Ma) is ideal for exploring adaptive evolution. The museum reported the first genome of *Hydrophis curtus* and use it to investigate sea snake secondary marine adaptation. Based on the high-quality genome, gene family analyses date a pulsed coding-gene expansion to about 20 Ma, and these genes associate strongly with adaptations to marine environments. Analyses of selection pressure and convergent evolution discover the rapid evolution of protein-coding genes, and some convergent features. Additionally, 108 conserved non-coding elements (CNE) appear to have evolved quickly, and these may underpin the phenotypic changes. The integration of genomic and transcriptomic analyses indicates independent origins and different components in sea snake and terrestrial snake venom; the venom gland of the sea snake harbours the highest *PLA2* expression in selected elapids, and these genes may organize tandemly in the genome. These analyses provide insights into the genetic mechanisms that underlay the secondary adaptation to marine and venom production of this sea snake.

In the 1950s, academicians CHEN Yiyu and CAO Wenxuan of the institute of Hydrobiology led teams to explore the Qinghai-Tibet Plateau many times. Fish resources investigation was carried out, and a large number of plateau fish specimens were collected. It was found that the number of scales on the body surface and the number of barbels were related to the height of the fish distribution. Accordingly, fish in the Qinghai-Tibet Plateau could be roughly divided into three groups, and the three stages of their evolution are corresponding to the three main stages of the Qinghai-Tibet Plateau uplift. By combing methods of morphology and multi-omics, several generations of scholars of IHB, CAS continued to study these plateau fish specimens and revealed molecular mechanism and genetic basis of fish adaptive evolution at the high altitude in the Qinghai-Tibet Plateau.

Many scientific problems were solved based on the collection of MHBS: the molecular mechanism of the adaptive radiation of Ostariophysan fishes was revealed based on the specimens collected in MHBS; the reconstruction process of freshwater fish diversity in East Asia based on East Asia fish specimens; a new phylogenetic pattern of cyprinid fishes based on the cyprinid specimens; and the DNA barcode database and molecular identification system for the Yangtze River fishes were established based on Yangtze River fish specimens. By the end of 2019, 2,830 barcode sequences of Yangtze River fish had been collected, including 238 species belonging to 135 genera, 40 families, and 16 orders. A comprehensive molecular assessment of Yangtze fish was conducted by using a DNA barcode reference database. The species involved were about 64.2% of the known species in the Yangtze River, including nearly every common fish. Based on the coelacanth specimen in MHBS, 472 copies of inverted loci were identified in the genome through bioinformatics protocols. Through studies of their evolutionary age, selection pressure, expression patterns, and related gene functions, the significance of coelacanths' evolution and "landing" events of vertebrates was revealed.

Combining the morphological and molecular results, the researchers propose that the initial diversification time of Psettodoidei was very close to the primary diversification time of Pleuronectoidei, and confirm the monophyly of Pleuronectiformes which supports the explanation for the difficulty in determining the phylogenetic position of psettodes. It was found that there was new pattern of the *Paraplagusia japonica* never been reported before, and the body length range and maturity level of individuals with different patterns were identified; the characteristics of various patterns on the eye side of the species were identified. Based on multi-omics research, the researchers decrypt the mystery of hippocampal ventral fin loss, clarify the adaptive evolution mechanism of hippocampal vertical swimming, and explain the adaptability of coral reef species

日本须鳎新色型
New color pattern of *Paraplagusia japonica*

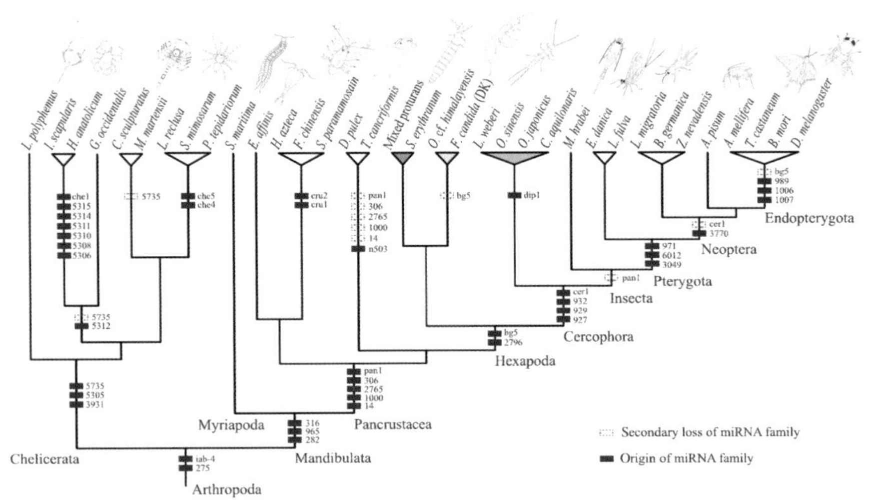

节肢动物的演化伴随 microRNA（miRNA）的起源与丢失
microRNA (miRNA) origins and losses mapped onto the phylogenetic tree of Arthropoda

至今悬而未决。依托上海昆虫博物馆多年来采集和饲养的馆藏标本，研究人员测定了原尾虫、跳虫和双尾虫代表类群的转录组及 miRNA 数据，构建进化树，首次发现两个 MIR-2796 和 MIR-bg5 为六足动物特有，还有四个 miRNA 家族为双尾虫和昆虫特有，支持六足动物单系，双尾虫与昆虫近缘，并证明了使用 miRNA 数据解决困难演化问题的有效性。

（1）原尾虫　研究人员测定了首条原尾虫红华蚖的线粒体基因组，发现其 *trnL2* 的排布位置不同于甲壳动物和六足动物的祖先模式，而与多足动物和螯肢动物类似，暗示原尾虫演化历史久远。

（2）跳虫　长期饲养 12 种跳虫，阐明其发育历程，以跳虫为不变态类群的模式生物，开拓了跳虫发育 - 演化（Evo-Devo）研究方向，首次从分子层面探讨跳虫眼的发育与演化，提出跳虫眼是单起源的，生活在土壤中的跳虫的无眼特征是适应性演化。

（3）双尾虫　双尾虫包括康虮、原铗虮和铗虮三大类。它们的系统关系和双尾虫的单系性一直争论不休。原铗虮的分类地位非常关键，但因罕见而鲜有研究。我们在广东采集到原铗虮代表类群—— 中国八孔虮，通过线粒体基因组分析支持双尾纲的单系，原铗虮与康虮亲缘关系更近，而若去掉原铗虮，即使改进数据分析方法，增加康虮和铗虮类群，也无法重建双尾虫单系，说明关键类群对构建系统演化关系至关重要。

古脊椎动物与古人类研究所标本中心收藏有大量从全国各地收集来的化石标本，涵盖了从鱼到人的各个类群。基于这些标本，研究人员在脊椎动物包括人类起源演化方面得到了一系列具有重要影响力的突破性的成果。其中部分具有代表性的成果如下。

（1）浙江曙鱼与颌的起源　颌的出现，标志着被动捕食向主动捕食的转化，是脊椎动物演化中一个非常重要的事件。浙江曙鱼的鼻垂体系统已经发生分裂，口鼻腔两侧生长了一对鼻囊，且与垂体管不相连。这一现象与分子和发育生物学长期以来所推测的，在颌的起源之前所发生的最为关键的一次演化事件极为相似，从而为颌的起源提供了条件。

（2）潇湘动物群与硬骨鱼的起源　硬骨鱼是脊椎动物的一个重要类群，可细分为辐鳍鱼和肉鳍鱼，这两个类群间存在巨大的形态学差异，并且硬骨鱼类与其他有颌类也存在形态学差异，长期以来，一直缺乏过渡类型。潇湘动物群产出的一系列鱼类化石，以梦幻鬼鱼、初始全颌鱼、钝齿宏颌鱼等为代表，拥有原始有颌类脊椎动物特征组合，很大程度上填补了鸿沟。它的发现将有颌类进化的一系列重大分歧事件从泥盆纪推回到志留纪，同时提示早期脊椎动物进化的这段历史仍疑云密布。

（3）三叠纪海生爬行动物　贵州西南部及邻近云南部分地区的海相三叠系产出丰富的海生爬行动物化石，代表类群包括鱼龙类、始鳍龙类、楯齿龙类、海龙类、原龙类、主龙类和龟鳖类等，其中以半甲齿龟为代表的龟类化石记录了龟类演化早期的一系列原始特征，如牙齿及独立存在的腹甲等，在龟起源的研究中起到了重要的作用。

(seahorses) to the reef environment. Systematic analysis by comparing mitochondrial genomics explains the inter-species gene sequence and the generation and evolution of genomic mutations of important economic shellfish species such as oysters and scallops, and reveal the diverse structures of bivalve genomes and their rich variations.

The phylogenetic interrelationships among four hexapod lineages (Protura, Collembola, Diplura and Insecta) are pivotal to understanding the origin of insects and the early diversification of Hexapoda, but they have been difficult to clarify. Based on the collections in the Shanghai Entomological Museum, researchers sequenced the transcriptomes and miRNA data of hexapod representatives and constructed the phylogenetic tree. For the first time, researchers identified two miRNA families (MIR-2796 and MIR-bg5) unique to Hexapoda, and four miRNA families that exist exclusively in Diplura and Insecta, suggesting a close relationship between Diplura and Insecta as well as the monophyly of Hexapoda. This study demonstrates the effectiveness of miRNA in resolving deep phylogenetic problems.

(1) Protura. The mitochondrial genome of *Sinentomon erythranum* was sequenced, as the first Proturan species to be reported. It underwent highly divergent evolution, showing many different features from other hexapod and arthropod mitochondrial genomes. The most remarkable finding suggests a long evolutionary history for the Protura.

(2) Collembola. Researchers culture 12 collembolan species and study their development and evolution. This study provides the first molecular evidence for the monophyletic origin of collembolan eyes and indicate the eye degeneration of collembolans is caused by adaptive evolution.

(3) Diplura. Diplura are traditionally classified into three major groups: Campodeoidea, Projapygoidea, and Japygoidea. The interrelationships of these three groups and the monophyly of Diplura have been much debated. Few previous studies included Projapygoidea due to their rare distribution. Researchers collected *Octostigma sinensis* (Octostigmatidae, Projapygoidea) from Guangdong, and studied the complete mitochondrial genomes from the Campodeidae, Parajapygidae, and Japygidae. The phylogenetic analyses retrieve significant support for a monophyletic Diplura, with Projapygoidea more closely related to Campodeoidea than to Japygoidea. Another key finding is that monophyly of Diplura cannot be recovered unless Projapygoidea is included in the phylogenetic analyses; this explains the dipluran polyphyly found by past mitogenomic studies. This finding provides an example of how proper sampling is significant for phylogenetic inference.

Collection Center collected tons of fossil specimens from nearly all parts of China, covering all groups of vertebrates from fish to humans. Based on these materials, researchers contributed a series of influenced breaking accomplishments in the origin and evolution of vertebrates, including humans. Here are some representative work and their related specimens:

(1) *Shuyu zhejiangensis* and origin of the jaw. The appearance of the jaw marks the transition from passive predator to active predator and is an extremely important event in the evolution of vertebrates. *Shuyu zhejiangensis* have paired nasal sacs in the braincase, and the hypophyseal duct opens anteriorly towards the oral cavity. These structures were thus already independent of each other. Therefore, have the condition that current molecular and developmental models regard as prerequisites for the development of jaws.

(2) Xiaoxiang Fauna and origin of Osteophyte. Osteophyte is an important group of vertebrates, which can be subdivided into ray-finned fish and lobe-finned fish. There are huge morphological differences between these two taxa, and they are different from other gnathostomes as well. There has been a lack of transition types for a long time among those groups. Represented by *Guiyu oneiros*, *Entelognathus primordialis*, *Megamastax amblyodus*, Xiaoxiang Fauna yield a series of fish fossils, which possessed many mosaic characters of primitive gnathostomes, filled the majority of the gap among those groups. By pushing a whole series of branching points in gnathostome evolution out of Devonian and into Silurian, the discovery of Xiaoxiang Fauna also implies that a significant part of early vertebrate evolution is unknown.

(3) Triassic marine reptiles. The marine Triassic of southwestern Guizhou and adjacent area of Yunnan yield amount of marine reptile fossils, including ichthyosaurs, eosauropterygians, placodonts, thalattosaurs, protorosaurs, archosaurs and turtles, in which the turtle fossils represented by *Odontochelys semitestacea* recorded a series of primitive such as teeth and independent abdominal beetle, which play a very significant role in the origin of turtle by the existence of teeth and plastron only.

(4) Feathered dinosaur and origin of bird. Birds are different from other vertebrate animals by feather and powerful flight, while its origin is always absent of detailed fossil materials until the discovery of Jehol Biota, which yield amount of feathered dinosaurs and birds to provide unprecedented materials for the origin of bird. A huge amount of fossil materials, such as *Sinosauropteryx*, *Microraptor*, *Epidendrosaurus*, *Epidexipteryx*, *Yi qi*, *Ambopteryx*, *Anchiornis*, *Jeholornis*, *Confuciusornis*, *Sapeornis*, *Protopteryx*, *Yanornis*, provide a detailed evolution process, and features in the early evolution of the turtle have reached the scientific consensus that birds are a group of theropod dinosaurs that originated during the Mesozoic Era.

（4）带羽毛恐龙和鸟类起源 鸟类因具有羽毛及强大的飞行能力而区别于其他脊椎动物，但其起源一直缺乏足够的化石材料。热河生物群的发现，产出了大量带羽毛恐龙和早期鸟类化石，为解决这个问题提供了前所未有的化石材料。大量的化石材料，如中华龙鸟、小盗龙、树栖龙、耀龙、奇翼龙、混元龙、近鸟龙、热河鸟、孔子鸟、会鸟、原羽鸟、燕鸟等，提供了详细的演化过程，使得鸟类的兽脚类恐龙中生代起源说得到认可。

（5）早期哺乳动物的演化及多样性 近几十年，热河生物群产出了大量早期哺乳动物，对哺乳动物的演化及多样性提供了重要的支撑。例如，五尖张和兽展示了早期哺乳动物娇小的形态，远古翔兽将哺乳动物的飞行提前 7000 万年，胡氏辽尖齿兽、盖氏热河俊兽为哺乳动物中耳的演化提供了证据，李氏源掠兽展现了听觉和咀嚼结构的分离。

（6）走出西藏与冰期巨动物起源 冰期动物具有体型巨大、身覆长毛等特征，如猛犸象、披毛犀，其起源与灭绝一直广受关注。产自西藏札达盆地哺乳动物化石的大量发现改变了人们对冰期动物的起源地的认识。西藏披毛犀、札达三趾马、喜马拉雅原羊、布氏豹、雪山豹鬣狗、邱氏狐等均为各个类群的早期祖先类型，从而为冰期动物的西藏起源提供了充足的依据。

（7）东亚现代人起源演化新认识 现代人在东亚的起源与演化一直是古人类学的研究热点。近来在我国南方发现许多新的化石，如距今 10 万年左右的道县人、黄龙洞人、崇左人等。基于现有材料提出，早期现代人和完全现代类型的人类至少 10 万年前在华南地区已经出现，对深入研究现代人在欧亚地区的出现及扩散具有重要意义。同时，揭示现代人在东亚大陆起源与演化过程中具有非常复杂的多样性和区域间差别。

华南植物园标本馆还在协同进化领域有所突破：①发现了两类全新的植物与传粉昆虫共生系统：五味子科植物与瘿蚊共生传粉系统和叶下珠科植物与小蛾类专性共生传粉系统。研究成果成为英国本科生和研究生重要教材 *The Ecology of Tropical East Asia* 的经典案例。②榕与榕小蜂协同物种形成及协同进化的研究发现，在整个东南亚区域宿主榕树为同一种，传粉榕小蜂有 9 种，其中 8 个是姐妹种，由地理隔离产生，1 个是宿主转移种，这些物种主要是异域或邻域分布。这是目前为止在同一种宿主榕树上发现最多的传粉榕小蜂物种，有助于更好地理解榕与榕小蜂的物种形成及进化机制。

3）生物多样性保护领域研究进展

生物标本是生物多样性重要的具体体现和凭据，是认识生物多样性、保护生物多样性研究的重要材料。各馆积极参与国家有关生物多样性的重大项目，为生物多样性保护领域积累了大量研究材料。例如，动物研究所、植物研究所、昆明动物研究所、昆明植物研究所、成都生物研究所、微生物研究所等标本馆服务先导专项 A“地球大数据科学工程”和科技部“科技基础性工作专项”（近 10 项），参与并支撑物种多样性信息平台建设，包括动物、植物和微生物物种名录，基于标本的物种地理分布数据，基于标本的物种形态特征获取，基于标本的物种图像与 CT 图像获取等。还参与建设 DNA 条形码凭证标本库和 DNA 材料库，支撑 DNA 条形码数据平台建设，涉及 2.5 万余种、DNA 条形码凭证标本 7.4 万余号 / 份，重要生物 DNA 材料 22.5 万余号 / 份，其中“中国植物 DNA 库”保存维管植物 DNA 材料 19.5 万份，覆盖 424 科 6504 属 48 253 种（85%）。

成都生物研究所两栖爬行动物标本馆参与了“中国脊椎动物红色名录·两栖纲和爬行纲”评估工作。《爬行动物红色名录》和《两栖动物红色名录》同其他脊椎动物红色名录一起，于 2015 年 5 月 23 日以中华人民共和国环境保护部、中国科学院 2015 年第 32 号公告形式发布。在《爬行动物红色名录》评估的 461 种中，有 2 种区域灭绝（RE）、34 种极危（CR）、37 种濒危（EN）、66 种易危（VU）、78 种近危（NT）、175 种无危（LC）、69 种数据缺乏（DD）。在《两栖动物红色名录》评估的 408 种中，有 1 种灭绝、1 种区域灭绝、13 种极危、46 种濒危、117 种易危、76 种近危、102 种无危、52 种数据缺乏。两栖爬行类生物多样性和受

(5) Evolution and diversity of early mammal. Lots of mammalian fossils yield from Jehol Biota during the past few decades provides details in evolution and diversity of early mammals. *Zhangheotherium quinquecuspidens* indicates the overall small size of early mammals, *Volaticotherium antiquum* pushes flight ability in mammals 70 million years earlier than before, *Liaoconodon hui* and *Jeholbaatar kielanae* provide new evidence for evolution of middle ear, *Origolestes lii* shows the separation moment of hearing and chewing modules in mammalian evolution.

(6) Out of Tibet and origin of Ice Age megafauna. Ice Age megafauna are chartered by large body size, covered by long hair, like woolly mammoths and woolly rhinos, and their origin and extinction are always being concerned. Amount of mammal fossils from Zanda Basin, Tibet, have changed the traditional view of their original place. As their ancestral types separately, *Coelodonta thibetana*, *Hipparion zandaense*, *Protovis himalayensis*, *Panthera blytheae*, *Chasmaporthetes gangsriensis*, *Vulpes qiuzhudingi*, gathered together to provide enough evidence for the Tibet origin of Ice Age megafauna.

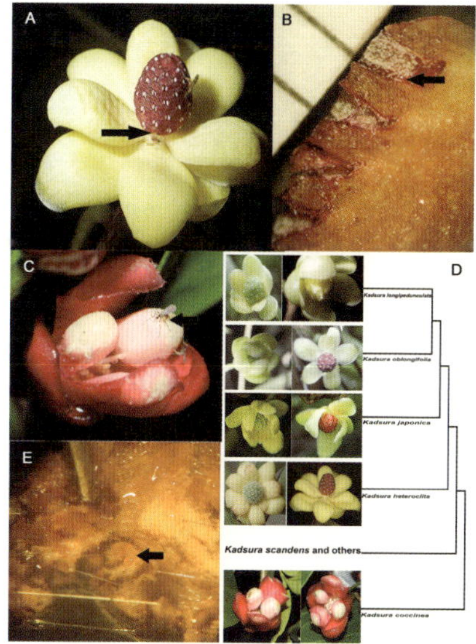

五味子科植物的花及其传粉瘿蚊
Flowers of Schisandraceae and their pollinator gall midges

(7) Modern humans' origin and evolution in East Asia. Origin and evolution of modern humans in East Asia have always been one of the main researches in paleoanthropology. It is proposed early and fully modern humans should have appeared in southern China at least 100,000 years ago, which is significant in further in-depth study of the appearance and dispersal of modern humans in Eurasia, based on many new fossils found in southern China recently, such as Daoxian, Huanglongdong, Chongzuo. At the same time, it reveals that the origin and evolution of modern humans in East Asian have very complex diversity and regional differences.

The Herbarium of South China Botanical Garden made a breakthrough in the field of coevolution: 1) Two new symbiotic systems of plants and pollinators have been found. One is the symbiotic pollination system of Schisandraceae and gall midges, and the other is the specialized symbiotic pollination system of Phyllanthaceae and moths. Both the results become classic cases cited in the textbook *The Ecology of Tropical East Asia*, which is for British undergraduate and graduate students. 2) The study on the co-speciation and coevolution of *Ficus* and *Ficus* wasps has found that one same host of *Ficus* in the whole Southeast Asia has co-evolved with nine different pollinating *Ficus* wasps, of which eight are sister species produced by geographical isolation and the other one is from host transfer species. These *Ficus* wasps species are mainly distributed in allopatry or neighborhood regions. It is the first time to find that so many pollinating *Ficus* wasps co-evolved in the same host *Ficus*, which will be helpful to better understand the speciation and evolution mechanism of *Ficus* and Ficus wasps.

3) Research progress in biodiversity conservation

Biological specimens are the important embodiment and evidence of biodiversity, and important material for understanding and protecting biodiversity. The collections have actively participated in major national biodiversity projects and accumulated a large number of research materials for biodiversity protection. For example, the collections of Institute of Zoology, Institute of Botany, Kunming Institute of Zoology, Kunming Institute of Botany, Chengdu Institute of Biology, Institute of Microbiology and other collections service the "Earth Big Data Science Project" of CAS and nearly 10 "Science and Technology Basic Work Project" of Ministry of Science and Technology, participate in and support the construction of species diversity information platform, including animal, plant and microbial species lists, specimen-based species geographic distribution data, morphological characteristics of species based on specimens, species images and CT images based on specimens, etc. The collections also participated in the construction of DNA barcode voucher specimen bank and DNA material bank, and supported the construction of DNA barcode data platform, involving more than 25,000 species and more than 74,000 DNA barcode voucher specimens, and more than 225,000 important biological DNA materials, including 195,000 vascular plant DNA materials preserved by "China Plant DNA Bank", covering 48,253 species of 6,504 genera, 424 families (85%).

Herpetological Museum participated in the assessment of "Red List of Vertebrates in China, Amphibia and Reptilia". The red list of reptiles and amphibians, together with the red list of other vertebrates, was released on May 23, 2015, in the form of Announcement No. 32, 2015 of the Ministry of Environmental Protection of People's Republic of China and

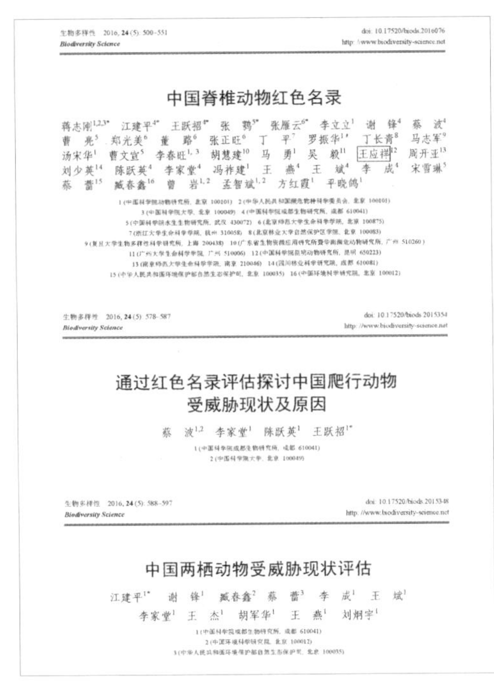

生物多样性 2016, 24 (5): 500–551 doi: 10.17520/biods.2016076
Biodiversity Science http://www.biodiversity-science.net

中国脊椎动物红色名录

蒋志刚[1,2,3*] 江建平[4*] 王跃招[4*] 张鹗[5*] 张雁云[6*] 李立立[1] 谢锋[4] 蔡波[4] 曹亮[5] 郑光美[6] 董路[6] 张正旺[6] 丁平[7] 罗振华[1*] 丁长青[8] 马志军[9] 汤宋华[1] 曹文宣[5] 李春旺[1,3] 胡慧建[10] 马勇[1] 吴毅[11] 王应祥[12] 周开亚[13] 刘少英[14] 陈跃英[4] 李家堂[4] 冯祚建[1] 王燕[4] 王斌[4] 李成[4] 宋雪琳[5] 蔡蕾[15] 臧春鑫[16] 曾岩[1,2] 孟智斌[1,2] 方红霞[1] 平晓鸽[1]

1 (中国科学院动物研究所, 北京 100101) 2 (中华人民共和国濒危物种科学委员会, 北京 100101)
3 (中国科学院大学, 北京 100049) 4 (中国科学院成都生物研究所, 成都 610041)
5 (中国科学院水生生物研究所, 武汉 430072) 6 (北京师范大学生命科学学院, 北京 100875)
7 (浙江大学生命科学学院, 杭州 310058) 8 (北京林业大学自然保护区学院, 北京 100083)
9 (复旦大学生物多样性科学研究所, 上海 200438) 10 (广东省生物资源应用研究所暨华南濒危动物研究所, 广州 510260)
11 (广州大学生命科学学院, 广州 510006) 12 (中国科学院昆明动物研究所, 昆明 650223)
13 (南京师范大学生命科学学院, 南京 210046) 14 (四川林业科学研究院, 成都 610081)
15 (中华人民共和国环境保护部自然生态保护司, 北京 100035) 16 (中国环境科学研究院, 北京 100012)

生物多样性 2016, 24 (5): 578–587 doi: 10.17520/biods.2015354
Biodiversity Science http://www.biodiversity-science.net

通过红色名录评估探讨中国爬行动物受威胁现状及原因

蔡波[1,2] 李家堂[1] 陈跃英[1] 王跃招[1*]

1 (中国科学院成都生物研究所, 成都 610041)
2 (中国科学院大学, 北京 100049)

生物多样性 2016, 24 (5): 588–597 doi: 10.17520/biods.2015348
Biodiversity Science http://www.biodiversity-science.net

中国两栖动物受威胁现状评估

江建平[1*] 谢锋[1] 臧春鑫[2] 蔡蕾[3] 李成[1] 王斌[1] 李家堂[1] 王杰[1] 胡军华[1] 王燕[1] 刘炯宇[1]

1 (中国科学院成都生物研究所, 成都 610041)
2 (中国环境科学研究院, 北京 100012)
3 (中华人民共和国环境保护部自然生态保护司, 北京 100035)

红色名录及两栖爬行动物受胁评估等方面的论文

Papers related to red list and assessment of threatened amphibians and reptiles

威胁状况的研究，是迄今为止研究对象最广、信息最全、参与专家人数最多的一次评估，对中国两栖爬行动物研究和资源保护与利用具有重要意义。

水生生物博物馆近年来对长江上游特有鱼类现状评估及其保护措施开展研究，通过野外调查，收集大量鱼类标本，利用种群生存力分析、单位补充量模型等方法对中华鲟、厚颌鲂、黑尾近红鲌等珍稀特有鱼类的保护生物学展开研究，并对三峡水库蓄水后长江上游鱼类资源的变化进行分析，为制定和实施相应的保护措施提供科学依据；金沙江上游水生生态及水生生物多样性调查与评价，从生态学角度探讨分析了长江上游水利工程实施后将对长江上游珍稀鱼类、特有鱼类、经济鱼类以及相关栖息生态环境产生的综合影响，提出了相应的保护对策，为金沙江上游水电规划提供理论参考；开展长江下游干流鱼类资源调查，洞庭湖、鄱阳湖湖区鱼类资源调查，收集鱼类标本，摸清这些区域生态环境及鱼类资源现状，为资源保护和可持续发展提供理论依据；持续在青藏高原开展野外调查和资源收集，包括藏北高原湖泊裂腹鱼类资源多态性与生态物种形成研究，青藏高原溪流鱼类群落生态学研究，藏东南动物资源综合考察与重要类群资源评估等，分析研究青藏高原优势种、特有种和入侵种，环境适应机制，物种保护策略等；此外，在三峡水库干支流、杨浦水库、四湖（洞庭湖、龙感湖、洪湖、鄱阳湖）、南水北调干渠等地开展调查，采集藻类标本 1 万余号。

从 20 世纪 50 年代开始，水生生物研究所对全国进行水生生物基础调查，开创了鱼类学、藻类学、原生动物学等基本学科，发现并命名了一批新物种。目前水生生物博物馆保存有 310 多种鱼类模式标本，并还在不断发现新物种。利用最近 10 多年来采集到的长江鱼类标本，首次对长江流域鱼类进行了全面的多样性评估，并建立了长江鱼类条形码数据库，让长江鱼也有了“身份证”。

基于这些工作以及对长江鱼类长期不间断的监测，发现生态环境遭到严重破坏，鱼类资源过度开发，推动建成长江上游珍稀特有鱼类国家级自然保护区，曹文宣院士从 2006 年开始呼吁长江“十年禁渔”。2017 年中央一号文件提出“率先在长江流域水生生物保护区实现全面禁捕”。2018 年，国务院办公厅要求到 2020 年，长江流域重点水域实现常年禁捕。2019 年，农业农村部、财政部与人力资源和社会保障部三部委联合公布《长江流域重点水域禁捕和建立补偿制度实施方案》。至此，长江“十年禁渔”成为现实。目前，曹文宣院士带领的团队仍致力于长江生态修复理论、技术与法规等研究。

国家动物博物馆自建馆以来，一直将发现新物种、为生物多样性保护提供本底材料和数据作为一项重要的工作。在鱼类资源与多样性保护研究方面，科研人员基于对馆藏标本的研究，发现数十个新物种，推进了我国的鱼类学研究以及对鱼类资源和多样性等数据的掌握。近年来，科研人员通过野外调查结果与标本馆藏记录和相关研究材料的对比分析，发现相关调查水系及流域的鱼类物种多样性正在急剧下降，其物种组成成分与馆藏记录和文献书籍记录相比有所减少，有针对性地提出了保护和恢复原有土著鱼类资源的建议，如进一步做好鱼类资源本底调查和物种多样性追踪、进一步加强渔政管理、筹建水生生物保护区、建立种质资源库、订立地方性水生野生动物保护名录、加强对水生野生动物保护的宣传教育、严控引入鱼种等，为我国鱼类资源的保护、可持续发展和利用献计献策。

在昆虫资源方面，保藏有中国首次发现的缺翅目和蛩蠊目标本，以及其他国家一、二级保护的昆虫标

CAS. In 461 species of "the red list of reptiles", 2 species were assessed as Regionally Extinct (RE), 34 species as Critically Endangered (CR), 37 species as Endangered (EN), 66 species as Vulnerable (VU), 78 species as Near Threatened (NT), 175 species as Least Concern (LC), 69 species as Data Deficient (DD). In 408 species of "the red list of amphibians", 1 species was assessed as Extinct (EX), 1 species as Regionally Extinct (RE), 13 species as Critically Endangered (CR), 46 species as Endangered (EN), 117 species as Vulnerable (VU), 76 species as Near Threatened (NT), 102 species as Least Concern (LC), 52 species as Data Deficient (DD). The study on the biodiversity and threat status of amphibians and reptiles is the most comprehensive assessment with the widest research object, the complete information and the most experts. It is of great significance to the study on amphibians and reptiles and the conservation and utilization of resources in China.

In order to evaluate the current status of endemic fishes upstream of Yangtze River and protect them effectively, MHBS collected a large number of fish specimens by carrying out field investigations, studied the conservation biology of rare and endemic fish, including *Acipenser sinensis*, *Megalobrama pellegrini*, and *Ancherythroculter nigrocauda* by using the population viability analysis and the yield per recruit model, and analyzed changes in the fish resources upstream of the Yangtze River after water impoundment of Three Gorges Reservoir. MHBS also implemented surveys in the upstream of Jinsha River and evaluated the aquatic ecology and biodiversity of rare fishes, endemic fishes, and economic fishes affected by reservoirs, providing theoretical strategy for fish protection and practical suggestion for hydropower planning. Moreover, MHBS conducted the investigation of fish resources in the lower reaches of the Yangtze River, Dongting Lake, and Poyang Lake, collected fish samples, and studied the ecological and fish resources in these areas, providing a theoretical basis for resource protection and sustainable development. Besides, MHBS are continuing to implement field surveys and resource collection in the Qinghai-Tibet Plateau, including studies on polymorphism and ecological speciation of fish resources in lakes in northern Tibet, ecological studies of stream fish communities in the Qinghai-Tibet Plateau, comprehensive investigation of animal resources and evaluation of important group resources in southeast Tibet and resource assessment of important groups, etc. studying dominant species, endemic species, and invasive species, analyzing the mechanism of environmental adaptation, and providing the strategies of species conservation. In addition, MHBS has undertaken investigations in the main and tributaries of the Three Gorges Reservoir, the Yangpu Reservoir, Dongting Lake, Longgan Lake, Honghu Lake, Poyang Lake, and the main canals for the South-North Water Transfer Project and collect 10,000 algae specimens.

Started from the 1950s, basic surveys of aquatic organisms were carried out by the Institute of Hydrobiology throughout the country, and basic disciplines such as ichthyology, phycology, and protozoology were initiated. Many new species were found at that time. At present, there are more than 310 type specimens of fish in MHBS, and new species are still being discovered. Based on the Yangtze River fish samples collected in recent 10 years, the first comprehensive diversity assessment of the Yangtze River fish was conducted, and the barcode reference database of the Yangtze River fish was established. According to the barcode reference database, Yangtze River fish have identification cards similar to those of human beings.

Based on the long-term uninterrupted monitoring of the Yangtze fish, scientists found that the ecological environment has deteriorated, and the fish resources are overexploited. They promoted the establishment of a national nature reserve for rare and unique fish in the upper reaches of the Yangtze River. For instance, Academician Cao Wenxuan has appealed for a "ten-year fishing ban" in the Yangtze River since 2006. In 2017, the No.1 Document of the Central Government proposed to "take the lead in implementing a comprehensive ban on fishing in aquatic organisms in Yangtze River Basin". In 2018, the General Office of the State Council of the People's Republic of China (PRC) called for a year-round fishing ban in important water areas along the Yangtze River. In 2019, the Ministry of Agriculture and Rural Affairs, the Ministry of Finance and the Ministry of Human Resources and Social Security jointly announced *Implementation Plan for Banning Fishing and Establishing Acompensation*. At this point, "ten-year fishing ban" in the Yangtze River has become a reality. At present, Academician Cao Wenxuan's team is still committed to the study of theory, technology, and regulation of ecological restoration of the Yangtze River.

"曹文宣院士与长江十年禁渔"展区

The Special Exhibition of "Academician Cao Wenxuan and ten-year fishing ban of the Yangtze River Basin" in MHBS

Since the establishment of the National Zoological Museum of China, it has been an important work to discover new species and provide background materials and data for biodiversity conservation. In the research of fish resources and

阿波罗绢蝶
Parnassius apollo

硕步甲
Carabus (Apotomopterus) davidis

本，如金斑喙凤蝶（*Teinopalpus aureus*）、双尾目伟铗虮（*Atlasjapyx atlas*）、彩臂金龟（*Cheirotonus* spp.）、叉犀金龟（*Allomyrina davidis*）、双尾褐凤蝶（*Bhutanitis mansfieldi*）、三尾褐凤蝶（*Bhutanitis thaidina dongchuanensis*）、中华虎凤蝶（*Luehdorfia chinensis huashanensis*）、阿波罗绢蝶（*Parnassius apollo*）、拉步甲 [*Carabus* (*Coptolabrus*) *lafossei*]、硕步甲 [*Carabus* (*Apotomopterus*) *davidis*] 等。这些珍贵的标本不但为展示中国昆虫多样性提供了实物凭证，而且为科研人员开展资源研究、政府部门制定保护规划等提供了重要的材料和依据。

在鸟类濒危物种保护方面，标本馆收藏有 1957 年采自陕西洋县的朱鹮标本，这为考察队在 20 世纪 70-80 年代开展朱鹮野外调查提供了重要的线索，并最终在 1981 年在陕西洋县发现 7 只当时全球唯一的野生朱鹮种群，成为朱鹮能够得以保护并延续的转折点。

标本馆长期收集的兽类标本为《中国濒危动物红皮书——兽类》（1998 年出版）、《中国生物多样性红色名录——脊椎动物卷》编制提供了大量翔实的资料，并在我国开展青藏铁路规划设计和施工建设过程中，为减少包括藏羚羊迁徙在内的对野生动物的影响提供了重要的依据。

海洋生物标本馆依托科技部基础工作专项“我国近海海洋生物 DNA 条形码数据库构建”，建立了国内规模最大的海洋生物 DNA 条形码凭证标本库，保藏了中国近海 20 000 余号海洋生物 DNA 凭证标本。建设的“中国海洋生物 DNA 条形码数据共享平台”上线运行。平台内有 2607 个物种的 15 944 条标准数据的 DNA 条形码数据，成为我国近海常见海洋生物物种和 DNA 条形码的标准数据共享平台，为我国近海生态系统现状和演变提供了快速、便捷的生物物种查询、鉴定平台。

依托成都生物研究所植物标本馆，研究人员承担了国家重大工程建设的环评调查任务，如雅鲁藏布水电站建设、川藏铁路建设等，并提出了工程实施区域的重点保护野生植物和狭域特有植物的迁地保育建议规划，为国家重大工程建设和植物多样性的保护做出了应有的贡献。并长期致力于生物多样性的保护工作及其相关政策的制定讨论，多次为国际山地综合发展中心周边国家的学员授课，提高相关国家的人员对生物多样性保护的认知。还参与了《四川省重点保护野生植物》和《国家重点保护野生植物》的审定，并参与了《四川省极小种群调查规则》、《四川省林业种质资源》等有关生物多样性保护规则的讨论。

diversity protection, researchers found dozens of new species based on the research of collected specimens, which promoted the research of Ichthyology in China and the mastery of data on fish resources and diversity. In recent years, through the comparative analysis of field survey results, specimen collection records and relevant research materials, researchers found that the diversity of fish species in the relevant investigation water system and drainage basin is declining sharply, and the species composition is less than that in the collection records and literature books records, and put forward suggestions for the protection and restoration of the original indigenous fish resources. Such as investigation of fish resources and the tracking of species diversity, further strengthen the management of the fishery, prepare for the establishment of aquatic biological reserves, establish a germplasm resource bank, establish a local list of aquatic wildlife protection, strengthen the publicity and education of aquatic wildlife protection, and strictly control the introduction of fish species, so as to provide suggestions for the protection, sustainable development and utilization of fish resources in China.

In terms of insect resources, the collection has preserved the specimens of Zoraptera and Blattella first found in China, as well as the first and second class protected insect specimens, such as *Teinopalpus aureus*, *Atlasjapyx atlas*, *Cheirotonus* spp., *Allomyrina davidis*, *Bhutanitis mansfieldi*, *Bhutanitis thaidina dongchuanensis*, *Luehdorfia chinensis huashanensis*, *Parnassius apollo*, *Carabus* (*Coptolabrus*) *lafossei*, *Carabus* (*Apotomopterus*) *davidis*, etc. These precious specimens not only provide physical evidence for the display of insect diversity in China, but also provide important materials and basis for researchers to carry out resource research and government departments to formulate conservation plans.

In terms of bird endangered species protection, the collection have Crested Ibis specimens collected from Yangxian County, Shaanxi Province in 1957, provided an important clue for the investigation team to carry out the field investigation of Crested Ibis in the 1970s-1980s, and finally found seven wild Crested Ibis populations in Yangxian County, Shaanxi Province in 1981, which became the turning point for Crested Ibis to be protected and continued.

The long-term collection of animal specimens in the collection provides a large number of detailed materials for the compilation of the *Red Book of Endangered Animals in China - Mammals* (published in 1998) and the *Red List of Biodiversity in China - Vertebrate*. And in the process of planning, design and construction of the Qinghai Tibet railway, it provides an important basis for reducing the impact on wild animals, including Tibetan antelope migration.

Relying on the special project for basic work of the Ministry of Science and Technology"Construction of DNA Barcode Database of Offshore Marine Organisms" of the Ministry of Science and Technology, the Marine Biological Museum has established the largest sample bank of DNA barcode of marine organisms in China and preserved more than 20,000 samples of DNA of offshore marine organisms in China. The "DNA Barcode Data Sharing Platform for China's Marine Organisms" has been built and operated online. There are 15,944 DNA barcode data of 2,607 species in the platform, which has become a standard data-sharing platform for common marine species and DNA barcode in China's coastal areas, providing a fast and convenient platform for biological species query and identification for the status and evolution of China's coastal ecosystem.

Relying on the Herbarium of Chengdu Institute of Biology, some environmental-impact assessments (EIAs) for national key project construction were undertaken, such as the hydropower station at the Brahmaputra river, Sichuan-Tibet railway, etc., and put forward the ex-situ conservation plan for the national key preserved wild plants and narrow endemic plants in these project implementation regions. In general, the herbarium has made contributions for national major project constructions and the protection of plant diversity. The herbarium has long been dedicated to biodiversity conservation and formulation of the relevant policies, and trained students from neighboring countries of the International Centre for Integrated Mountain Development (ICIMOD) to improve their understanding of biodiversity conservation. The herbarium also participated in the review and approval of "Key protected wild plants in Sichuan Province" and "Key protected wild plants in China", and participated in the discussion of biodiversity protection rules such as "Methods of investigation on plant species with extremely small populations (PSESP) in Sichuan Province", "Forest germplasm resources in Sichuan Province", and so on.

In the 1980s and 1990s, based on HITBC staff's suggestions, Naban River Watershed National Nature Reserve, Bulong State Nature Reserve, and Yiwu State Nature Reserve were established in Xishuangbanna. HITBC took part in assessments of the conservation status of vascular plants in China and re-introduction works supported by the Chinese Union of Botanical Gardens. In 2014, researchers launched the general investigation of national key protected species in order to find out the resource situations of these plants. HITBC took part in this project, and investigated 37 key protected species in South Yunnan, including Xishuangbanna and Pu'er. In 2014, HITBC began to take part in the biodiversity expeditions in Southeast Asia led by the Southeast Asia Biodiversity Research Institute, CAS (CAS-SEABRI). So far, HITBC has completed nine large Scale China-Myanmar joint field biodiversity investigations. Since 2017, HITBC has started the biodiversity investigations in karst areas of Yunnan Province. So far, HITBC has updated the plant checklist of these areas. Since 2017, HITBC staff have investigated germplasm resources of plant species with extremely small

物种资源和多样性调查
Species resources and diversity survey

极小种群物种苗木繁育
Seedling breeding of plant species with extremely small populations (PSESP)

20 世纪 80-90 年代，在西双版纳热带植物园标本馆的建议下，在西双版纳傣族自治州新建立了纳板河流域国家级自然保护区、布龙州级自然保护区和易武州级自然保护区。在植物园联盟的支持下，标本馆参加全国维管植物的濒危状况评估和再引种工作。2014 年启动全国重点保护植物资源调查，全面摸清重点保护物种的资源状况。标本馆负责滇南的西双版纳傣族自治州、普洱市，对分布于这两个地州的 37 种重点保护植物进行了全面的调查。2014 年起，在东南亚中心的领导下，标本馆参与了东南亚地区生物多样性调查工作，目前已经进行了 9 次大规模的中缅联合生物多样性野外科考工作。2017 年起，开展云南省喀斯特地区生物多样性保护研究工作，标本馆目前已经系统地整理出一份石灰岩地区的种子植物名录。2017 年启动极小种群物种滇西南地区野生种质资源的调查，标本馆负责其中的 6 个物种的调查任务，目前已经收集到 3 个物种的种质资源及 2 个物种的无性繁殖材料，已经通过人工繁育获得一批珍贵的苗木，计划进行野外回归工作。2019 年开始启动全国兰科植物资源普查，标本馆负责滇南 4 个地州，通过样方、样线调查，全面摸清兰科植物的生存状况，对受威胁严重的物种建立永久监测样方。2019 年启动西双版纳生物多样性热点地区的植物资源调查与评估，开展关键地区的生物多样性调查与编目工作，标本馆负责景洪市的植物调查。2019 年启动西双版纳傣族自治州国家级自然保护区第二次本底资源综合科学考察，标本馆负责勐养和尚勇两个子保护区的植物调查任务。2010 年以来，开展热带代表性类群的调查、分类和系统学研究，发表新科 1 科，新属 7 属，新种 142 种。

以昆明植物研究所标本馆人员为主，完成了《中科院专家关于我国防范外来物种入侵面临形势、问题的分析及对策建议》、《关于中亚 - 中国生物多样性研究中心的建设思路》、《商业采挖等致高山冰缘带植物陷生存危机》等多份专题咨询报告，为国家和地方政府决策提供咨询建议及立法支撑。

上海昆虫博物馆关注盐沼湿地的研究。盐沼湿地具有非常重要的生态系统功能，在营养再生、初级生产、野生动物栖息生境保障和海岸线保护方面起着重要作用。昆虫是盐沼生态系统生物多样性的主要组成部分，然而在国内却一直乏人研究。研究团队自 2009 年来，一直关注长江口九段沙湿地国家级自然保护区昆虫多样性和保护研究。在研究人员的努力下，截至 2017 年，九段沙湿地记录的昆虫物种数已由 99 种增加至 364 种。由于盐沼湿地是非常容易受生物入侵影响的脆弱生态系统，我们对外来植物互花米草入侵可能对盐沼湿地昆虫造成的进化和生态影响也进行了高度关注。我们的研究表明，互花米草入侵已经降低了九段沙湿地昆虫的多样性，并显著改变了湿地昆虫功能群的组成和动态。与植食昆虫相比，天敌昆虫更容易受到互花米

populations (PSESP) in Southwest Yunnan. HITBC is in charge of the investigations of six species of the total target. Recently HITBC has collected germplasm resources of three species and vegetative propagation materials of two. Based on artificial propagation techniques, researchers effectively enlarged the population size of these species and reintroduced them to the wild. HITBC also took part in a general investigation of Orchidaceae plant resources, which was launched in 2019, and is in charge of four autonomous prefectures in Southern Yunnan. Researchers can find out the conservation status of the plants, and construct plots for seriously threatened species, key protected species, and valuable economic species. These moves will establish the foundations for species surveys and biodiversity monitoring. The project of investigation and assessment of the plant resources from the biodiversity hot-spot regions in Xishuangbanna was launched in 2019. HITBC is taking part in these biodiversity investigations in the key regions and making a checklist of the plant species. In 2019, the second general scientific investigation of background biodiversity resources in Xishuangbanna National Nature Reserve was launched. HITBC is in charge of the plant surveys in Mengyang subordinate Nature Reserve, and Shangyong subordinate Nature Reserve. Since 2010, HITBC has been doing field investigation, taxonomy, and phylogeny researches on tropical representative taxa. One new family, seven new genera, and 142 new species have been published.

With staffs of KUN, several advisory reports, including "Chinese Academy of Sciences Experts' Analysis on the Situation and Problems of China's Prevention of Invasion of Alien Species and Suggestions for Countermeasures", "About Central Asia-China Biodiversity Research Center Construction Thoughts" and "The survival crisis of alpine periglacial plants caused by commercial mining" were submitted to governments, and supported for national and local government decision-making.

Shanghai Entomological Museum also focuses on the research of salt marsh wetlands. Salt marshes have great ecological value for the ecosystem, namely in nutrient regeneration, primary production, habitat for wildlife species, and as shoreline stabilizers. Insects represent a dominant component of biodiversity in salt marsh ecosystems, yet they have largely been neglected in studies in China. Since 2009, the museum has focused the insect diversity and conservation in Jiuduansha Wetland National Nature Reserve in Yangtze Estuary, East China. Through their efforts, the number of insect species recorded in Jiuduansha has increased from 99 to 364 in 2017. Since the salt marsh systems are highly susceptible to biological invasion, the museum also pays attention to the potential ecological and evolutional effects of *Spartina alterniflora* invasion on insect community in wetlands of Yangtze Estuary. These studies showed that *S. alterniflora* invasion has reduced the insect biodiversity and significantly changed the structural characteristics and seasonal dynamics of insect functional groups in Jiuduansha wetland. In comparison with insect herbivores, natural enemy insects are more susceptible to the *S. alterniflora* invasion. However, some generalist herbivores have developed adaption mechanisms to *S. alterniflora*. Especially, both larval survival experiment and feeding performance observation in field and lab suggested that the enemy-free space provided by the *S. alterniflora* is important in driving the host expanding of a native generalist *Laelia coenosa*. These results are important for assessing the invasion consequences of *S. alterniflora* on salt marsh systems in China. The museum also has put great concern on the maintenance of insect diversity in salt marsh systems. The studies revealed that in common reed dominated salt marshes parasitoids have strong top-down effects on population dynamics of insect herbivores, and the strength of their effects is positively correlated with the host density. Comparatively, effects of plant quality and spider predation are more varied among herbivores with different seasonality. Generally, these results highlight the importance of top-down effects in shaping the populations of insect herbivores in salt marsh systems.

Northeast China is one of the regions with the richest biodiversity at the same latitude in the world. It is also an important area in the "Two barriers and three belts" of China's ecological security strategic pattern. The northeast region includes four in eight vegetation regions in China. The cold temperate coniferous forest region and the temperate coniferous and broad-leaved mixed forest region are unique in China. Especially in the temperate coniferous and broad-leaved mixed forest region, there are alpine tundras in the alpine zone of Changbai Mts., which is one of the only two alpine tundras in China. The temperate grassland area is the easternmost part of China's grassland, and the warm temperate deciduous broad-leaved forest area is the northernmost part of that in China. Therefore, the biological species in Northeast China are very rich and complex, which occupies a very important position in the country.

Over the past 60 years since the Northeast Biological Herbaria establishment, 368 new vascular plant species have been discovered, and more than 40 monographs have been published, including *Herbaceous Flora of Northeast China*, *Flora of Liaoning Province*, *Key List of Plants of Northeast China*, etc. In recent years, the flora of wild vascular plants in the whole northeast region has been systematically sorted. The herbaria compiled the book *Atlas Northeast Plant Distribution*, which has a total of 3.4 million words, and determined more than 200 county-level production areas of all vascular plants in the northeast region; in the book *The Color Atlas of Northeast Forest Plants*, 1,000 plant species aaccounting for 1/3 of the wild plants in Northeast China are introduced in detail; in the book of *Handbook of Identification of Grassland Plants in Northeast China*, 200 species of grassland plants in Northeast China are introduced. The series of works provide a scientific basis for

草入侵的影响。不过，一些广食性昆虫已产生了对互花米草的适应机制。其中，通过野外幼虫存活率试验和室内幼虫取食发育观察实验，表明互花米草入侵造成的“天敌释放空间”是导致土著广食性昆虫素毒蛾寄主范围扩展的重要原因。上述结果对于评价互花米草入侵对中国盐沼生态系统的影响具有重要参考价值。在盐沼昆虫多样性的稳定维持机制方面，我们也开展了一些研究工作。其中，我们的研究表明，在芦苇湿地寄生天敌对植食昆虫种群动态具有非常强的下行控制作用，并且其作用强度与寄主昆虫的密度正相关。相比之下，植物质量和蜘蛛捕食作用更易受植食昆虫物候特征差异的影响。总体上，该研究结果强调了下行控制作用对芦苇湿地的植食昆虫种群具有非常重要的调控作用。

中国东北地区是全球同纬度物种多样性最丰富的区域之一，也是我国生态安全战略格局“两屏三带”中一个重要的地带。东北地区包括全国八个植被区域中的四个植被区域，寒温带针叶林区域和温带针阔叶混交林区域为国内特有，尤其在温带针阔叶混交林区域内，在长白山高山带分布有高山冻原，为我国仅有的两处高山冻原之一。温带草原区域为我国草原的最东端，暖温带落叶阔叶林区域为我国暖温带落叶阔叶林的最北端。因此，中国东北生物物种十分丰富、复杂，在全国占有十分重要的位置。

东北生物标本馆自建馆 60 多年来，在维管植物研究方面，共发现新种 368 个，出版专著 40 余部，包括《东北草本植物志》、《辽宁植物志》、《东北植物检索表》等。近年来，植物组对野生维管植物在整个东北地区的分布进行系统整理，编撰了迄今为止国内最翔实、最完整的植物分布著作《东北植物分布图集》，全书共 340 万字，确定了东北所有维管植物在区域内的 200 多个县级产地；此外，《东北森林植物原色图谱》详细介绍了东北地区森林植物 1000 种，占东北野生植物的 1/3；《东北草地野生植物识别手册》详细介绍了东北草地植物 200 种。系列著作为东北地区生物多样性研究和保护、合理开发利用东北植物资源，为合理进行林业、草业和农业区划提供科学依据。

水生鞘翅目昆虫是东北生物标本馆昆虫研究的特色类群，从 1993 年开始，持续开展水生鞘翅目的分类及系统发育工作，保藏的水生鞘翅目昆虫标本 2 万余号，模式标本 142 种 567 号，在我国同类标本馆中处于领先地位。与奥地利维也纳自然历史博物馆于 1995-2003 年合作出版 *Water Beetles of China* 3 卷，10 年间报道了 1 个新科、8 个新属、200 余个新种。近 10 年发表水生鞘翅目新记录属 2 属，新种 34 种，我国大量水生甲虫的报告，使人们对中国的水生鞘翅目昆虫有了全新的认识，为水生生物作为环境指示物种来评价

2019 年新出版专著
Three new published books in 2019

沙鲁里山区大型真菌生物多样性调查与评估
Biodiversity survey and assessment of macrofungi survey in the Shaluli Mountains

2018 年 5 月 22 日,《中国生物多样性红色名录——大型真菌卷》发布
Redlist of China's Biodiversity: Macrofungi was published on 22 May 2018

the research and protection of biodiversity in Northeast China, the development and utilization of plant resources and the rational division of forestry, grass and agriculture.

Water beetle is a unique collection of the Northeast Biological Herbaria. Taxonomic research of water beetles began in 1993, and about 20,000 specimens, 142 type species, 567 type specimens were collected, with a leading position among similar herbaria in China. Three volumes of *Water beetles of China* were published from 1995 to 2003, one new family, eight new genera and about 200 new species were described during the lateral cooperation with the Natural History Museum of Vienna, Austria. Two new record genera, thirty-four new species were reported in the past ten years. It improves the understanding of aquatic beetles and lays a solid foundation for the evaluation of ecosystem health using aquatic insects as bio-indicator.

The Fungarium actively took part in the national project "The Biodiversity Survey and Assessment", founded by the Ministry of Ecology and Environment, China. Investigations were focused on macrofungi and carried out in the Taihang Mountains (in Beijing and Hebei Province) and the southwest of Sichuan Province. A great number of fungal collections were added into the Fungarium, and useful information about the local macrofungal diversity of macrofungal was supplemented into the database.

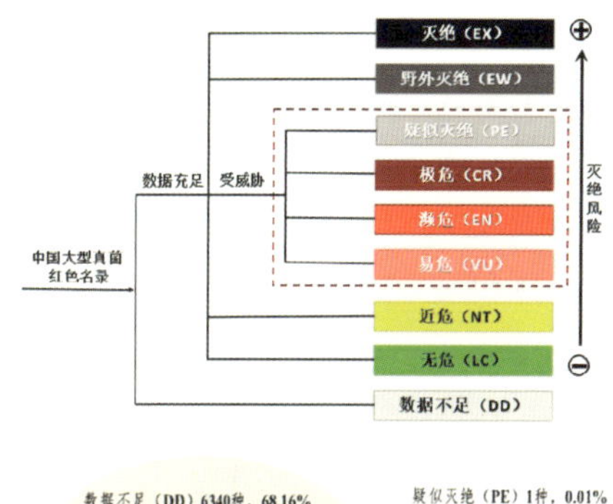

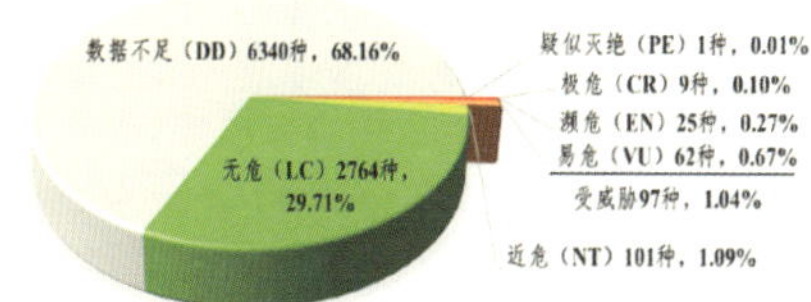

中国大型真菌红色名录评估等级及比例
Grade and proportion of the assessments of *Redlist of China's Biodiversity: Macrofungi*

藜科植物系统学研究
Systematical studies of Chenopodiaceae

水生生态系统的健康打下了坚实的基础。

菌物标本馆参与国家生态环境部资助的“生物多样性调查与评估”项目，针对京冀太行山区和四川西南部山区进行大型真菌野外调查，收集大量标本及菌物多样性相关信息和数据，阐明了当地大型真菌物种多样性，积累了重要的标本材料和信息数据。2018 年 5 月 22 日，由菌物标本馆牵头完成编写《中国生物多样性红色名录——大型真菌卷》，由生态环境部和中国科学院联合正式发布。该名录由 140 多名国内菌物学者共同编制，是我国第一部官方真菌红色名录，涵盖了我国 9302 种大型真菌和地衣，分属于 2 门 14 纲 62 目 227 科 1298 属，其中 97 种被评估为受威胁物种。

新疆生态与地理研究所标本馆开展了中国藜科植物系统学研究和新疆猪毛菜属植物多样性及其地理分布特征研究，并与中亚各国合作开展新疆及周边国家特殊植物种质资源调查。干旱区植物资源迁地保育及其生态建设应用研究获得了 2015 年新疆科学技术进步奖一等奖。

4.3 服务国家需求

近年来，中国科学院生物标本馆加强服务意识，服务国家需求，支撑国家经济建设，面向国家重大战略、重大研究计划和国门生物安全，在生物资源的可持续利用、生物多样性监测、有害生物防控、入侵生物预警等方面取得了丰硕成果。

除了在服务国家战略方面，各馆积极组织资源考察和采集，为国家积累了大量战略生物资源外，在服务国家重大研究计划方面，各馆积极参与并执行，为科学研究和国家生物资源库的建设提供有力支撑；在服务国门生物安全方面，标本馆为保障国门生物安全提供了强有力的技术支持，对于科研工作发挥了巨大的支撑作用。“十三五”以来各馆主持或者支撑的院内外项目共计 491 项，经费达 5.7 亿元。

1）生物资源可持续利用

生物资源是人类生存和发展的基本物质基础，是食品、医药、工业等各个行业原材料的重要来源。我国拥有占全世界 10% 的生物资源，是生物资源最丰富的国家之一，对生物资源利用的历史也十分悠久。但随着生态环境问题日益严峻，对生物资源利用的可持续性也越来越引起人们的重视。各馆也依靠自身优势，积极投身于生物资源可持续利用的研究和开发工作中。

水生生物博物馆长期服务于长江水生生物多样性的保护与利用，主要有以下几项工作。

（1）为国家大型水利工程建设提供决策咨询服务 通过持续监测长江三峡等大型水利工程建设及蓄水后长江水文因子的改变，研究其对中华鲟等长江珍稀特有水生生物、长江上游特有鱼类的影响；对四大家鱼产卵场的影响；对鱼类早期资源的影响；对生境连续性的影响；对鱼类种群和生物群落的影响等。进而探讨是否需要修建过鱼设施，如何开展生态调度、增殖放流、生态修复等问题，为长江三峡等大型水利工程建设提供决策咨询服务，从而协调水电开发和物种保护之间的矛盾，实现水电可持续发展和物种永续繁衍。

（2）生物多样性监测及生物资源可持续利用 连续三十余年对长江流域水生生物进行监测，对水生生物资源的时空大尺度变化进行分析，为长江水生生物保护提供理论依据。在水生生物研究所专家的呼吁下，中华鲟、长江江豚保护取得重大进展，长江十年禁渔成为现实。

（3）政府特别许可咨询服务 2016 年以来，参与由农业农村部渔业局委托的，对全国新建的近百家水族馆进行现场考察和综合评估，为国家水生野生动物保护主管部门特别许可提供决策咨询服务。

广东植物资源种类丰富，但资源研究与利用少。华南植物园标本馆牵头组织 6 家单位开展了“广东省特色植物资源利用与产业化关键技术研究与应用”项目，查清了广东植物资源现状，发现新种 26 个，以新

Redlist of China's Biodiversity: Macrofungi was officially and jointly announced by the Ministry of Ecology and Environment of China and CAS on 22 May 2018. This is the first official fungal red list of China, compiled with joint efforts of more than 140 Chinese mycologists organized by the Fungarium. In this work, 9,302 species of macrofungi and lichens, belonging to two phyla, 14 classes, 62 orders, 227 families and 1,298 genera, were accessed, and 97 of them were listed as threatened.

The Specimen Museum of Xinjiang Institute of Ecology and Geography has carried out systematical studies of Chenopodiaceae in China and studies on the diversity and geographical distribution characteristics of *Salsola* in Xinjiang and cooperated with Central Asian countries to carry out the investigation of special plant germplasm resources in Xinjiang and its surrounding countries. In 2015, the first prize of Xinjiang Science and Technology Progress was awarded for the research on the ex situ conservation of plant resources and the application of ecological construction in arid areas.

新疆维吾尔自治区科学技术进步奖

奖励证书

为表彰在科学技术进步中做出突出贡献的单位或集体，特颁发自治区科学技术进步奖，以资鼓励。

获奖成果：干旱荒漠区植物资源迁地保育研究及其生态建设应用
主要完成单位：中国科学院新疆生态与地理研究所
奖励等级：一等奖
奖励年度：2015年度

新疆维吾尔自治区人民政府
2016年3月25日

证书编号：J20150024

获得2015年新疆科技进步奖一等奖

Won the first prize of Xinjiang Science and Technology Progress in 2015

4.3 Serving national needs

In recent years, the Biological Collections of CAS has strengthened its service awareness, supported national economic construction by serving national needs, and made fruitful achievements in sustainable utilization of biological resources, biodiversity monitoring, pest prevention and control, and early warning of invasive organisms in the face of national major strategies, major research plans and national biosafety.

In addition to serving the national strategy, all collections actively organize resource investigations, accumulating a large number of strategic biological resources for the country, actively participate in and implement major national research plans, providing strong support for scientific research and the construction of the national biological resource bank, and in serving the national biological security, the collections provide support for safeguarding the national biological security strong technical support has played a huge supporting role in scientific research. Since the 13th Five-Years Plan, 491 projects have been presided over or supported by each collection, with a total expenditure of 570 million yuan.

1) Sustainable utilization of biological resources

Biological resources are the basic material basis for human survival and development, and an important source of raw materials for food, medicine, industry and other industries. China has 10% of the world's biological resources, is one of the most abundant countries in biological resources and has a long history in the utilization of biological resources. But with the increasingly serious problems of the ecological environment, more and more attention has been paid to the sustainable utilization of biological resources. Depending on their own advantages, the collections are also actively engaged in the research and development of sustainable utilization of biological resources.

The Museum of Hydrobiological Sciences has been serving the protection and utilization of aquatic biodiversity in the Yangtze River for a long time, mainly including the following works:

(1) MHBS has been providing decision-making advisory services for the construction of large-scale water conservancy projects in China: MHBS is continuing to monitor the ecological impact of large-scale water conservancy projects, including the Three Gorges Project. After water impoundment, MHBS studies the hydrological influences on rare and endemic aquatic organisms, such as Chinese Sturgeon and endemic fishes in the upper reaches of the Yangtze River. The influences on spawning grounds of "four major fish" (Asian carps), early fish resource, habitat continuity, fish populations and biological communities, etc., are also studied. Then to discusses whether we the need to build fish facilities and how to conduct ecological scheduling, stock enhancement, ecological restoration, etc. MHBS also provides decision-making consulting services for the construction of large-scale water conservancy projects, including the Three Gorges Project so as to coordinate the contradiction between hydropower development and species protection and achieve sustainable development of hydropower and species reproduction.

(2) Biodiversity monitoring and sustainable utilization of biological resources: MHBS has been monitoring aquatic organisms in the Yangtze River basin for more than 30 years and analyzing large-scale spatiotemporal variability of aquatic biological resources which has provided a theoretical basis for the protection of aquatic organisms in the Yangtze River. With the appeal of experts from the Institute of Hydrobiology, the conservation of Chinese Sturgeon and the Yangtze finless porpoise have made significant progress, and the Yangtze River fishing ban has become a reality for a

获得 2018 年度广东省科技进步奖一等奖
First Prize of Scientific and Technological Progress of Guangdong

种为材料培育出新品种 10 个；对蔷薇科等进行了专科研究，挖掘出 62 种新优植物用于园林绿化，重点对樱花属和檀香等进行了栽培繁殖等技术研究与应用；开展了特殊生境的适生植物的评价，筛选出用于生态修复植物 42 种，攻克了岩质和矿山边坡的生态修复及水污染净化等关键技术难题。成果在全国 20 多个省份进行推广应用，累计新增销售额 104 亿元，产生了良好的经济和社会效益。该研究获得 2018 年度广东省科技进步奖一等奖。

成都生物研究所植物标本馆致力于服务国家需求，如有害生物防治、生物多样性监测、生物资源的可持续利用、国家重大工程建设环评等。例如，地奥心血康胶囊是中国科学院第一个从植物薯蓣中提取的天然化合物制成的药物，并成立了地奥制药公司，现在已经发展成为集医药和化妆品生产的地奥集团。植物标本馆长期为地奥集团和其他相关企业提供植物及粉碎样品鉴定工作。此外还完成成兰铁路、川藏铁路和多项水电项目的植物多样性调查与影响评估，为国家重大工程的生物多样性保护决策和措施实施提供科学依据。

西双版纳热带植物园标本馆自建立以来，一直为国家经济建设提供支撑。1959 年中国著名植物学家蔡希陶教授在西双版纳勐腊县勐仑镇建立版纳植物园，是为了研究中国热带植物可持续利用资源，当时的“植物资源组”即现在的标本馆，对云南乃至全国的橡胶产业、烟草产业等做出了巨大贡献。

昆明动物博物馆也服务云南烟草产业，完成云南四大产区烟草天敌种类和天敌数量的动态调查、天敌标本制作、天敌鉴定、天敌生物学习性等研究工作；馆藏烟草昆虫标本 2 万余号，也开始着手烟草害虫 DNA 条码测序工作，为鉴定和绿色防控云南烟草天敌提供支撑服务。

植物研究所标本馆科研人员历经多年积淀，以标本馆馆藏标本材料为基础，建立了目前世界上最大的植物 DNA 库——“中国植物 DNA 库”。目前，该 DNA 库保存了维管植物 DNA 材料 195 253 份，覆盖 424 科、6504 属、48 253 种 (不含亚种和变种)。其中，中国产 DNA 材料 144 775 份，包含 345 科、3903 属、26 480 种，覆盖中国已知维管植物物种的近 85%。在中国植物 DNA 库的基础上，以分子鉴定技术为基础，为社会提供服务，鉴定材料主要涉及药用饮片原植物物种鉴定、植物物种定名、辨别用料真伪、毒品鉴定等方面。近两年来，植物分子鉴定平台为北京市公安局等 60 余家单位提供了鉴定服务，鉴定了查获的毒品原植物材料等近 200 份样品。此外，标本馆还对医药产业有所贡献。

中国植物 **DNA** 库中国产 **DNA** 材料类群分布情况

DNA materials in China of Plant DNA Bank of China

	石松 Lycopodiales+ 蕨类 Ferns	裸子植物 Gymnosperm	被子植物 Angiosperm	合计 Total
科 Family	47	12	286	345
属 Genus	396	81	3 426	3 903
种 Species	2 895	436	23 149	26 480
份 Duplicates	11 236	8 071	125 468	144 775

中国科学院微生物研究所菌物标本馆所在的微生物研究所真菌室于 1958 年率先实现灵芝人工栽培并推广。如今，灵芝产业年产值已达约 200 亿元。支持开展针对蘑菇目、牛肝菌目、多孔菌目等重要经济真菌类群的分类及相关研究，相关研究成果获得国家教育部科技进步一等奖 1 项和西藏自治区科学技术奖一等

decade.

(3) Special Government Examine and Approve Advisory Service: Since 2016, commissioned by the Fisheries Bureau of the Ministry of Agriculture and Rural, MHBS has participated in the on-site inspection and comprehensive evaluation of nearly 100 newly built aquariums in China and provided decision-making consulting services for the special examining and approving of the State Aquatic Wildlife Protection Administration.

Guangdong is rich in plant resources, but the research and utilization of plant resources are few. Herbarium of South China Botanical Garden took the lead in organizing 6 units to carry out the "Research and Application of Key Technologies for Utilization and Industrialization of Characteristic Plant resources in Guangdong Province" program, investigate the current situation of plant resources in Guangdong, found 26 new species, and cultivated 10 new varieties with new species as materials; carried out specialized research on Rosaceae, excavated 62 new excellent plants for landscaping, focused on the cultivation and propagation of the genus *Cerasus* and *Santalum album*, and carried out the evaluation of suitable plants in special habitats, 42 species were selected for ecological restoration, and the key technical problems such as ecological restoration of rock and mine slope and water pollution purification were overcome. The achievements have been popularized and applied in more than 20 provinces in China, with a total new sales volume of 10.4 billion yuan, producing good economic and social benefits. This research work won the first prize of Scientific and Technological Progress of Guangdong Province in 2018.

The Herbarium of Chengdu Institute of Biology has long been dedicated to meeting national needs, including control of harmful species, biodiversity monitoring, sustainable utilization of biological resources, and environmental-impact assessment for key national projects. For example, the Di'aoxinxuekang capsule is the first drug made from the natural compounds extracted from *Dioscorea panthaica* by CAS. Therefore, Di-Ao Pharmaceutical Company was established, and now it has developed into Di-Ao Group for the production of medicines and cosmetics. CDBI has been providing the identification of plant and pulverized samples for Di-Ao Group and other related enterprises. The herbarium also has accomplished the plant diversity investigations and impact assessments of Chengdu-Lanzhou railway, Sichuan-Tibet railway and a number of hydropower projects, which provided the scientific basis for the decision and implementation of key national projects on biodiversity conservation.

Since its establishment, HITBC has been devoted to supporting national economic construction. In 1959, the famous botanist Prof. TSAI Hse-Tao organized to build XTBG in Mengla County, Xishuangbanna Town in order to research the sustainable uses of the tropical plants in China. The Plant Resources Group at that time was the predecessor of the present HITBC. This group made great contributions to the rubber and tobacco industries, and other fields.

KNHMZ has served the tobacco industry and has completed researches on natural tobacco pests in such four major production areas in Yunnan as pest specimen preparation, identification, ecology and biology. With 20,000 tobacco insect specimens collected, KNHMZ begins the DNA barcode sequencing of tobacco pests, which has laid a solid foundation for future identification and the green prevention and control of Yunnan tobacco pests.

"Plant DNA Bank of China" in PE has been established after hard work based on the specimen materials collected in the PE. It is the largest plant DNA bank in the world, including 195,253 DNA materials of vascular plants, covering 424 families, 6,504 genera, and 48,253 species (excluding subspecies and variants). Among them, 144,775 DNA materials (Table 43) were produced in China, including 345 families, 3,903 genera, and 26,480 species, covering nearly 85% of known vascular plant species in China. Based on Plant DNA Bank of China and molecular technology, PE provides rapid

中国植物 DNA 库
Plant DNA Bank of China

中亚地区药用植物收集和保育合作研究
Cooperative research of collection and conservation of Medicinal Plant in Central Asia

奖1项。

基于菌物标本馆标本、实地调查和文献考证，研究人员阐明了冬虫夏草的分布范围。采用物种分布模型，分析冬虫夏草对气候变化的反应，对其未来分布区变化进行了预测，也为其他大型真菌红色名录的评估和保护提供了参考。

西北高原生物研究所青藏高原生物标本馆评估青海省各类保护生物及特有动植物的生存现状，并对未来的保护策略等工作提出了建议。其中，青藏高原有5100种野生植物，超过50%的是药用植物。通过对20余种药用植物的抚育工作，提出应建立完善的资源保护、引种和繁育办法及合理的可持续资源利用体系。

昆明植物研究所标本馆支撑从事植物资源开发的企业、个人及国家执法部门植物标本/样品鉴定。主要包括：开展各类植物的标本、活体及器官组织、分子遗传材料的物种鉴定业务，如植物类药材、食品原料的鉴定；花卉种类及品种鉴定等，并提供植物的濒危等级、保护级别及相关国家法律法规咨询事宜。

新疆生态与地理研究所标本馆对中亚地区药用植物阿魏种质资源进行收集和保育合作研究，为地方政府部门出具植物鉴定报告，建立了生物多样性监测数据，并构建了78种维药基源植物数据库。

2）生物多样性监测与生态环境治理

近年来，国家十分重视生态环境建设，提出了“生态文明建设”和“绿色发展”的理念。在这一思想的指导下，逐步发展基于生物资源的绿色低碳循环经济体系，生物多样性的监测与生态环境的保护工作将越来越重要。而馆藏标本除了能够直接展示物种的丰富程度外，还能够从侧面反映生物多样性的变化，并为国家对生态环境的治理提供建设性意见。因此，部分标本馆积极参与国家或地方的生物多样性监测和环保项目，如昆明动物研究所、昆明植物研究所、西北高原生物研究所等近10个标本馆参加了国家重大科研计划——“第二次青藏高原综合科学考察”项目，负责生物资源的调查；植物研究所标本馆还主持了“美丽中国”生态文明科技工程A类战略性先导专项中一个项目。各标本馆凭借所保藏历史标本的资源优势和人才优势，未来将为“美丽中国”和“生态文明”建设提供大量服务，发挥巨大的推动作用。

依托馆藏标本和数据库，菌物标本馆参与国家生态环境部、科技部、中国科学院等部门组织的《中国生物多样性战略与行动计划（2011—2030年）》、《生物物种资源监测概论》、《生物物种监测技术指南——大型真菌》、《中国生物多样性国情研究报告》和《中国生物种质与实验材料资源发展报告2017—2018》中菌物多样性相关部分的编写工作。

华南植物园主编的《广东植物志》为全面了解广东省植物多样性奠定了重要基础，而华南植物园标本馆为其编研提供了大量研究标本。依托标本馆，华南植物园近年来承担了广东省第二次全国重点保护野生植物资源调查和植物生态恢复理论与实践研究。为了更好地对全省植物的濒危状况进行全面掌握，根

identification service to society and the scientific community. The identified materials include original plant species of medicinal decoction pieces, plant species for new resource foods, the authenticity of materials, and drug plant materials. In the past two years, PE has provided identification services to more than 60 units, including the Beijing Public Security Bureau, and has identified nearly 200 samples of seized drug plant materials. In addition, the herbarium also contributes to the pharmaceutical industry.

In 1958, the fungus room of the Institute of Microbiology where the Fungarium of Institute of Microbiology is located, took the load in realizing the artificial cultivation and promotion of "Lingzhi". Currently, the annual income of "Lingzhi" industry has reached 20 billion Yuan (RMB). Similarly, achievements on taxonomy and application of economically important *Agaricale*, *Boletales* and *Polyporales* and some other groups were gained by different researchers, which were awarded the First Class Prize for the Scientific and Technological Progress of the Ministry of Education, China, and the First Class Prize for Scientific and Technology of the Tibet Autonomous Region.

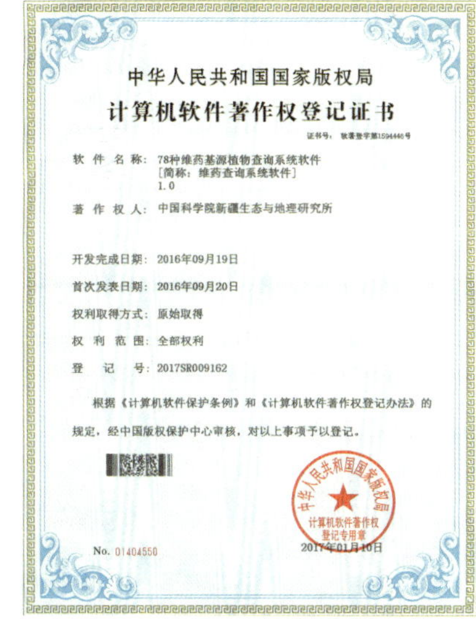
中华人民共和国国家版权局
计算机软件著作权登记证书
证书号：软著登字第1594446号
软件名称：78种维药基源植物查询系统软件
[简称：维药查询系统软件]
1.0
著作权人：中国科学院新疆生态与地理研究所
开发完成日期：2016年09月19日
首次发表日期：2016年09月20日
权利取得方式：原始取得
权利范围：全部权利
登记号：2017SR009162
根据《计算机软件保护条例》和《计算机软件著作权登记办法》的规定，经中国版权保护中心审核，对以上事项予以登记。
中华人民共和国国家版权局 计算机软件著作权登记专用章
No. 01404550
2017年01月10日

78 种维药植物数据库

Database of 78 species of Uygur medicinal plants

The distribution of *Ophiocordyceps sinensis* was clarified according to sampling data of massive collections in the Fungarium. The species distribution modeling method was used to analyze the responses of *O. sinensis* to global climate change, as well as to predict the future distribution of this rare medicinal fungus. These data provided essential references to red list assessment and conservation of other macrofungi.

The QTPMB stimated current living situation of kinds of protected living beings and endemic species and gave advice on protection strategy in future. More than 50% were medical plants in about 5,100 plant species in Qinghai-Tibet Plateau. Qinghai-Tibet Plateau Museum of Biology suggested building the system of conservation, introduction, breeding and sustainable resource utilization according to the experiences of breeding of more than 20 medical plants.

KUN supports enterprises and individuals engaged in the development of plant resources and also national law enforcement department in plant specimen and sample identification. It mainly includes: conducting species identification of specimens of various plants, living plants, organ tissue and plant molecular materials, such as the identification of botanical herbs, food materials, and flower species etc., and provide advice on the endangerment level, protection level and relevant national laws and regulations of plants.

The Specimen Museum of Xinjiang Institute of Ecology and Geography carried out the cooperative research of conservation on the germplasm resources of *Ferula officinalis* in Central Asia, issued the plant identification report for the local government departments, established the biodiversity monitoring data, and constructed the database of 78 species of Uygur medicinal plants.

2) Biodiversity monitoring and ecological environment management

In recent years, the state attaches great importance to the construction of the ecological environment and puts forward the concepts of "Ecological Civilization Construction" and "Green Development". Under the guidance of this idea, gradually develop the green low-carbon recycling economy system based on biological resources, biodiversity monitoring and ecological environment protection will become more and more important. In addition to displaying the richness of species directly, specimens can also reflect the changes of biodiversity from the side, and provide constructive suggestions for the national governance of the ecological environment. Therefore, some collections actively participated in national or local biodiversity monitoring and environmental protection projects, such as nearly 10 collections of Kunming Institute of Zoology, Kunming Institute of Botany, Northwest Institute of Plateau Biology, and other institutes participated in the national major scientific research plan- "The Second Comprehensive Scientific Investigation of the Qinghai-Tibet Plateau" project, responsible for the investigation of biological resources; the Herbarium of Institute of Botany also presided over one of the class A strategic pilot projects of "Beautiful China" ecological civilization science and technology project. Depending on the resource and talent advantages of the historical specimens, each collection will provide many services for the construction of "Beautiful China" and "Ecological Civilization" in the future and play a huge role in promoting.

Relying on the specimens and databases participated in several attempts to compile important nation-wide advices on biodiversity, e.g., *The Biodiversity Strategy and Action Plan of China (2011–2030)*, *The Introduction to Monitoring of*

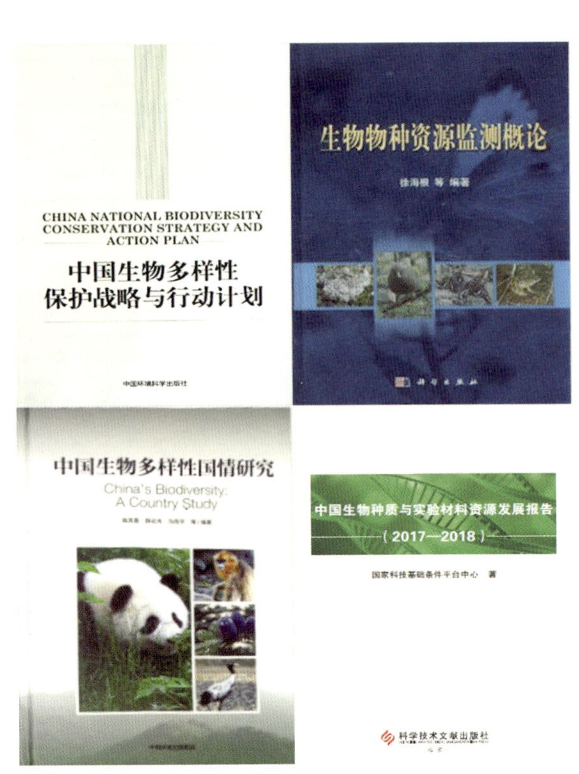

菌物多样性保护国家战略规划、研究报告和相关论著的编写

Contributions to the National Strategic Plan on biodiversity

据IUCN标准和要求，完成了对广东省6049种野生植物的全面濒危等级评估，制订了《广东省重点保护野生植物名录（第一批）》，并被省政府于2018年11月予以正式公布。在国际植物园保护联盟（Botanic Gardens Conservation International, BGCI）以及国家、广东省和各级地方政府的支持下，先后开展了国家一级重点保护野生植物伯乐树、报春苣苔、水松以及广东省重点保护野生植物观光木、走马胎、杜鹃红山茶、猪血木等30种珍稀濒危植物的保育和野外回归研究。“华南珍稀濒危植物的野外回归研究与应用”获得了2012年度广东省科学技术一等奖。

依托成都生物研究所植物标本馆馆藏和采集标本以及历史文献资料，研究人员建立了《四川植物多样性及其分布数据库》、《西藏植物多样性数据库》，以及《攀枝花市植物资源数据库》、《九顶山植物多样性数据库》、《都江堰植物多样性数据库》、《康定植物多样性数据库》、《画稿溪国家级自然保护区植物多样性数据库》、《鞍子河省级自然保护区植物多样性数据库》等一系列的植物多样性基础数据资料；为西藏环保厅编写了《西藏生物多样性保护行动计划》（植物部分）；在2008年地震后第一时间为四川省相关部门提供了《地震灾区植物多样性名录及其分布》的基础数据；为了推动植物多样性的保护，长期为四川省林业厅保护处提供相关生物多样性保护建议，退休老专家印开蒲研究员还提出了建立大熊猫保护生态走廊以及建立大熊猫国家森林公园的建议，其编写的《百年追寻》著作，用生动的图片展示了四川植物多样性景观的百年变迁。

2019年，庐山植物园在标本馆和森林生态研究组主导下，建设了25公顷庐山常绿落叶阔叶混交林的森林生态观测大样地，完善了中国森林生物多样性监测网络，填补了亚热带森林生物多样性监测大样地在江西地区的空白。

位于昆明的滇池，其湖面面积300余平方公里，是云南面积最大的高原湖泊。20世纪90年代，滇池水质逐渐恶化为劣V类，成为中国污染最严重的湖泊之一。昆明动物博物馆团队受中国科学院委托将“花—鱼—蚌”科学模型成果进行科普展示，获得了有关部门认可，促进了该科学方案在滇池保护治理和生态恢复中的推广应用。

上海昆虫博物馆多年来一直从事小节肢类土壤动物分类、鉴定和物种多样性的研究。近年来，由于社会和经济发展等原因，土壤生态系统退化日趋严重。为适应这一问题出现和发展的需求，博物馆开始将物种多样性研究与退化生态系统相结合，以期作为土壤生态系统主要功能群的小节肢类土壤动物在退化生态系统预防和修复中起到应有的作用。经过近些年的研究，在评价和监测指标方面有了较好的进展。例如，随着城市化的推进，城市绿地面积不断扩大，已成为不同于自然生态系统和农田生态系统的新生态系统类型。我们对上海城市绿化带不同环境土壤蜱螨目和其他小节肢类土壤动物的群落结构及物种多样性进行了调查。结果表明：城市绿化带土壤蜱螨目群落结构特征在很大程度上既不同于自然土壤也不同于农业土壤，反映了城市绿化带为一新型生态系统；土壤蜱螨目、甲螨亚目无翼类、低等类和中气门亚目是土壤质量评价的较好

Biological Resource, *Technological Guide to Monitoring of Species: Macrofungi*, *China's Biodiversity: A Country Study* and *Report of Progress in Resources of Biological Germ Plasm and Experimental Materials in China, 2017–2018*. In the compiling of these important white books, Fungarium is responsible for the fungal parts.

《百年追寻》

Tracing One Hundred Years of Change

The monograph *Flora of Guangdong* laid an important foundation for a comprehensive understanding of plant diversity in Guangdong Province. In recent years, the Herbarium of South China Botanical Garden has undertaken the second investigation of national key protected wild plant resources and the theory and practice research of ecological restoration in Guangdong. In order to better know the endangered situation of plants in the whole province, according to the IUCN standards and requirements, the comprehensive assessment of the endangered level of all 6,049 wild plants in Guangdong Province was completed, and the first list of key protected wild plants in Guangdong was officially announced by the Guangdong Provincial Government in November 2018. With the support of BGCI and the state, Guangdong Province and local governments at all levels, the herbarium has successively carried out the conservation and field return research on 30 rare and endangered plants, such as *Bretschneidera sinensis*, *Primulina tabacum*, *Glyptostrobus pensilis* , *Michelia odorant*, *Ardisia gigantilolia*, *Camellia azalea*, *Euryodendron excelsum*. In 2012, the Study and Application of Field Regression of Rare and Endangered Plants in South China won the first prize of Science and Technology of Guangdong Province.

A series of databases of the plant diversity in southwest China were established based on the specimen and literature in CDBI, including *Plant diversity database of Sichuan*, *Plant diversity database of Tibet*, *Plant diversity database of Panzhihua*, *Plant diversity database of Jiuding Mountain*, *Plant diversity database of Dujiangyan*, *Plant diversity database of Kangding*, *Plant diversity database of Huagaoxi National Natural Reserve*, *Plant diversity database of Anzihe Provincial Natural Reserve*. CDBI also compiled *Tibet biodiversity conservations strategy and action plan (Plants)* for Tibet Environmental Protection Department. After the 2008 Wenchuan earthquake, CDBI provided the list of plant diversity and distribution in the earthquake disaster area as the basic data to the related department of Sichuan Province for the first time. In order to promote the protection of plant diversity, CDBI has long offered proposals for biodiversity conservation to the Forestry Department of Sichuan Province. Additionally, YIN Kaipu, a retired ecological researcher in CIB, put forward the proposal on establishing Giant Panda National Forest Park. In his book *Tracing One Hundred Years of Change*, vivid photos were used to illustrate the environmental changes in Sichuan during one hundred years.

In 2019, 25 hectares of large-scale forest ecological observation plot of evergreen and deciduous broad-leaved mixed forest in Mt. Lushan was built under the leadership of LBG and forest ecology research group. This observation plat has significantly improved the Chinese Forest Biodiversity Monitoring Network (CForBio, www.cfbiodiv.cn) and filled the gap of subtropical Forest Biodiversity Monitoring area in Jiangxi province.

土壤取样以进行监测

Soil sampling for monitoring

Dianchi Lake, located in Kunming, covers an area of more than 300 km^2 and is the largest plateau lake in Yunnan. In the 1990s, the water quality of Dianchi Lake gradually deteriorated to inferior class V, becoming one of the most polluted lakes in China. According to the national ecological civilization construction ideas, the KNHMZ team was commissioned by CAS to display the scientific results of the “flower-fish-clam” scientific model, which was recognized by relevant departments and has actively promoted the popularization and application of this scientific program in Dianchi Lake protection and management and ecological restoration.

The Shanghai Entomological Museum has been engaged

指标。

西北高原生物研究所依托青藏高原生物标本馆所开展的三江源国家公园大中型有蹄类动物和猛禽等栖息地保护和优先保育区规划的研究，为三江源国家公园物种保护管理措施和优先保育区规划提供理论依据。中国科学院三江源国家公园研究院由中国科学院、青海省政府依托西北高原生物研究所共同建设。馆藏动植物标本为三江源国家公园野生动物本底调查提供数据支持。

青藏高原生物标本馆还为可可西里世界自然遗产地申请、国家公园建设提供基础资料。研究人员以标本馆丰富的馆藏为依托，查阅文献和标本，牵头编写了可可西里生物多样性资源报告。2017 年 7 月 7 日，中国青海省可可西里被列入世界自然遗产名录。

武汉植物园标本馆拥有最丰富的神农架地区和三峡库区的植物标本，近年来为三峡工程和南水北调中线工程建设对环境的影响研究与生态安全，以及宜昌大老岭国家森林公园的建设等项目提供了翔实的植被类型和植物区系资料。此外，还为华中地区，尤其是川东 - 鄂西一带的稀有、濒危和特有植物的保育遗传学研究做出了巨大贡献。

3）有害生物防治

各馆还向社会提供专业的鉴定、咨询与技术等服务。6 年来接待国内外社会各界专业咨询共计 1.8 万余人次，网站访问量超过 2 亿人次。特别是对农业、林业、工业有害昆虫的防治，为国家挽回了大量损失。

成都生物研究所两栖爬行动物标本馆协助开展新疆北部地区草原蛇伤防治工作。20 世纪 70 年代，新疆北部地区草原上的毒蛇数量较多，经常咬伤牛羊及牧民，对当地畜牧业产生了较大影响，也给牧民的生命安全带来一定危害。1976 年至 1977 年，赵尔宓院士等应邀赴天山和北疆开展蛇类调查，摸清了当地两种主要毒蛇草原蝰和中介蝮的种群数量、食性、生活习性和繁殖规律，提出一些防治措施，有效地减少了当地因蛇伤而带来的畜牧业损失，并研制抗蛇毒血清，解决了牧民的蛇伤问题。当时因研究需要而保存和解剖的草原蝰（*Vipera renardi*）及中介蝮（*Gloydius intermedius*）标本，均保藏于标本馆内。

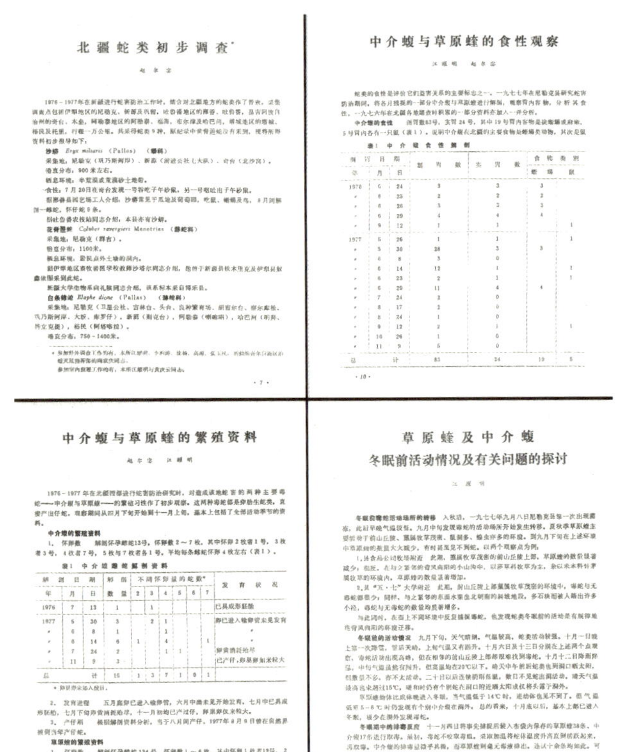

北疆蛇类初步调查

中介蝮与草原蝰的食性观察

中介蝮与草原蝰的繁殖资料

草原蝰及中介蝮
冬眠前活动情况及有关问题的探讨

发表于 1978 年的《两栖爬行动物研究资料》，关于草原蝰和中介蝮研究的部分论文

Papers of *Vipera renardi* and *Gloydius intermedius* on *Materials for Herpetological Research* published in 1978

成都生物研究所植物标本馆通过开展成都市外来入侵植物灾害化现状、外来入侵植物风险评价体系以及外来入侵植物的防控与资源化利用等方面的研究工作，构建了入侵植物防控及资源化利用技术体系，推进成都市外来入侵植物资源化利用试点的建设，为成都市相关管理部门提供了决策依据和为其他城市外来入侵植物的评估、防控与治理提供技术支撑和示范样板。

东北生物标本馆通过对馆藏标本分布数据的分析，对我国重要的经济真菌——桑黄的潜在分布范围进行模拟，提出了优先保护区域；对东北地区的多个林木干基腐朽病原真菌进行了潜在分布范围的模拟，为多个病原真菌的防治提供了科学依据。

甘肃省定西市是全国高品质特色中草药原产地和

in the study of the taxonomy, identification and species diversity of soil microarthropods for many years. In recent years, soil ecosystem degradation has been becoming more and more serious due to social and economic development or other reasons. In order to meet the needs for the emergence and development of these problems, the museum has started to relate the researches on species diversity to the degraded ecosystems, hoping that the soil microarthropods, as main functional groups of soil ecosystems, would play their due role in prevention and remediation of the degradation ecosystems. Large progress of the researches in evaluation and monitoring indicators for quality of soil ecosystems has been made. For example: City greenbelt area is strongly increasing with citifying on a large scale recently in China. The city greenbelt is becoming a new type of soil ecosystem that is different from the natural and agricultural ecosystems. In the study, the structure of soil mites and other microarthropods of greenbelts around Shanghai city were investigated to understand the role of the soil animals in the assessment of the soil quality and to improve the management practices for the greenbelts. The results indicate that soil mite community structure in city greenbelts was strongly different from that either in natural or in agricultural soil, suggesting that city greenbelt is a new ecological system; soil mite is a good bio-indicator group in the assessment of soil quality of city greenbelts.

Relying on the specimens in QTPMB, researches of Northwest Institute of Plateau Biology, CAS on the ungulate and raptors about the habitat and total priority conservation area in Sanjiangyuan National Park provide theoretical support for the selection of potential habitats for the animals in the national parks. The institute of Sanjiangyuan National Park was constructed jointly by CAS and the government of Qinghai Province. The specimen in QTPMB provided the fundamental information of the Catalogue of Life and distribution areas of plants and animals.

Qinghai-Tibet Plateau Museum of Biology provided the initial information for applying world natural heritage sites and constructions of National Park. The report on Hoh Xil's biodiversity resources was prepared relying on the specimen in QTPMB, literature and intensive expeditions. On July 7, 2017, Hoh Xil in China's western Qinghai Province had been successfully listed as a UNESCO world natural heritage site.

The Herbarium of Wuhan Botanical Garden has the most abundant specimens in the Shennongjia area and the Three Gorges Reservoir area. In recent years, it has provided detailed vegetation types and flora data for the research on the environmental impact and ecological security of the Three Gorges Project and the construction of the Middle Route Project of South to North Water Diversion, as well as the construction of the Dalaoling National Forest Park in Yichang. In addition, it has made great contributions to the conservation genetics of rare, endangered and endemic plants in Central China, especially in the East Sichuan to West Hubei region.

3) Pest control and management

The collections also provide professional identification, consultation and technical services to society. Over the past six years, it has received more than 18,000 professional consultations from all walks of life at home and abroad, and over 200 million visits to the website. In particular, the control of harmful insects in agriculture, forestry and industry has saved a lot of losses for the country.

Herpetological Museum of Chengdu Institute of Biololgy assist in the prevention and control of snake bites in the grasslands of northern Xinjiang. In the 1970s, there were a large number of venomous snakes on the grassland in northern Xinjiang. They often bit cattle, sheep and herdsmen, which had a great impact on the local animal husbandry and also brought some harm to the life safety of herdsmen. From 1976 to 1977, Prof. ZHAO Ermi and others were invited to Tianshan Mountains and northern Xinjiang to carry out snake surveys. They found out the population, feeding habits, living habits and reproduction rules of *Vipera renardi* and *Gloydius intermedius*, the two species of venomous snakes in the local, and put forward some control measures, which effectively reduced the local animal husbandry losses caused by snake bite, produced anti-snake venom serum, and solved the problem of herdsmen's snake bites. The specimens of *V. renardi* and *G. intermedius*, which were preserved and dissected for research purposes, were deposited in the museum.

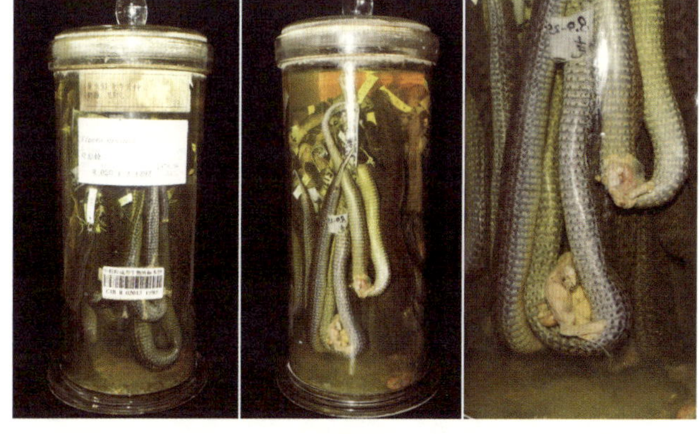

馆藏蛇伤防治工作期间采集的草原蝰（*Vipera renardi*）标本

Preserved *Vipera renardi* specimens collected during assist in the prevention and control of snake bit

Systems for intrusion prevention and resource

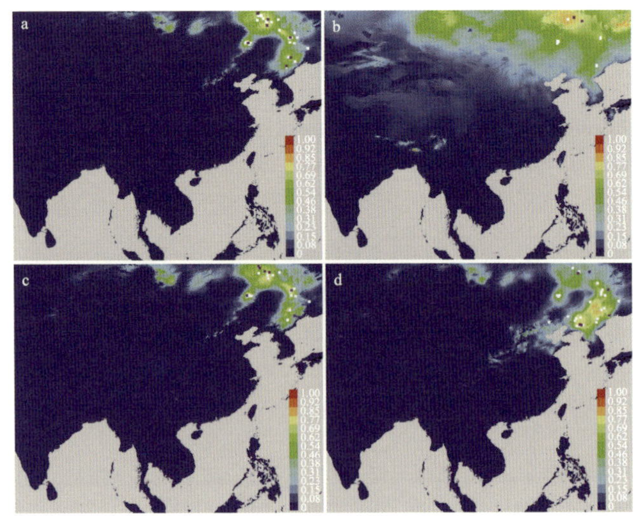
珍稀药用菌和干基腐朽病原真菌的潜在分布范围
The potential distribution ranges of rarely medicinal and butt-rot pathogenic fungi in China

重要商品生产地及交易集散地，但因连作种植普遍，虫害病害频发，已严重影响当地中草药的产量和品质，因此，对当地病虫害进行普查，并制定合理的病虫草防治策略是当地中草药生产的当务之急。应当地科研机构的邀请，国家动物博物馆标本馆组织专业人员前往甘肃陇南，深入田间详细调查中药材害虫及危害情况，与当地扶贫干部和村民座谈，了解有关情况，并提出监测、防控害虫以及发展壮大中药材产业的建议，同时达成持续帮扶当地防控中药材害虫的意向。

4）生物入侵与国门生物安全

生物入侵是导致生物多样性丧失的重要因素之一，与动植物栖息地丧失和全球气候变化一起被看作是影响全球环境的三大难题。生物入侵给我国造成了巨额经济损失，已经成为严重的生态与经济问题。调查表明，目前入侵我国的外来物种已超过 400 种，其中 39.6% 的物种是有意引进造成的，43.9% 为无意携带进入造成的，经自然扩散进入我国境内的仅占 3.1%。这一结果表明人为活动是导致生物入侵的重要因素。

中国科学院生物标本馆体系服务国门生物安全，共主持院重点部署项目 2 项——“面向我国出入境口岸动植物检疫工作的技术服务体系构建”和“国门入侵生物预防与控制技术”。“十三五”期间获取院内经费 1300 万元，院外争取经费 873 万元。构建了 5 个动植物检验检疫实体库和 3 个信息资源库；搭建了面向口岸对接的服务平台和国门生物安全动态响应平台；构建了重要检疫及入侵物种快速鉴定平台，汇集 DNA 条形码 30 573 条；提供 30 种有害象虫鉴定技术，新截获 10 种动物和 11 种植物的风险分析及评估报告；建立了 1 个国家标准、2 个植物检疫行标，申请国家标准 2 项；建立东北马铃薯甲虫阻截防控体系等。

中国科学院生物标本馆体系还为相关单位提供了鉴定、咨询、培训等服务。例如，积极参加检验检疫部门组织的“专家学者口岸行”等活动，到一线口岸实地调研存在的问题；针对一线工作人员的需求，充分挖掘实物标本和信息数据资源，整合潜在入侵物种和检验检疫物种数据，构建实物库和信息库，搭建物种快速鉴定平台，通过 DNA 条形码技术开展物种快速鉴定工作；组织专家力量，为各地检验检疫局鉴定截获动植物，并出具鉴定报告；借助国际合作途径，收集入侵物种和检验检疫物种标本及信息，充实平台资源库，并用于分析被截获物种的入侵风险。接待检验检疫工作人员到各馆检视、借阅标本，下载文献资料，并提供技术指导和培训。

基于以上服务成果，部分场馆还组织举办了国门生物安全展览。例如，昆明植物研究所标本馆联合昆明动物博物馆在昆明长水国际机场举办“铸就国门生物安全防线争当生态文明排头尖兵”大型国门安全科普展览，被云南省电视台云南新闻联播作为当日头条进行宣传报道，作为常设展在机场布展，服务出入境旅客约 150 万人次，并在展览期间接待国内外领导人。由于服务工作业绩突出且宣传成效显著，该成果入选“率先行动，砥砺奋进——‘十八大’以来中国科学院创新成果展”。

国家动物博物馆早期积极参与国家生物资源调查，为国家初步探明生物资源本底情况做出了重要的贡献。近年来，更是面向国门生物安全，一方面充分发挥馆藏检验检疫物种标本的价值，为科研人员发现、鉴定入侵有害昆虫提供比对凭证，另一方面承担科技部、中国科学院相关项目，开展入侵动物调查，构建

utilization technology were built by the Herpetological Museum as a result of the studies on the disaster situation, risk assessment, prevention and resource utilization of alien invasive plants in Chengdu. These works provided a decision-making basis for the related management departments in Chengdu as well as for the assessment, prevention and control of alien invasive plants in other cities.

Based on the analysis of the distribution data of the collected specimens, the potential distribution ranges of *Sanghuangporus sanghuang*, a group of important economic fungi in China, were simulated by Northeast Biological Herbaria, and the priority protection area was proposed; the potential distribution ranges of several butt-rot pathogens in Northeast China were simulated, which provided a scientific basis for the control of these pathogenic fungi.

赴甘肃调查中药材害虫
Investigation on pests of Traditional Chinese Medicine in Gansu Province

Dingxi City, Gansu Province, is the country of origin of high-quality Chinese herbal medicine and an important commodity production and distribution center. However, due to the widespread continuous cropping and the frequent occurrence of insect pests and diseases, it has seriously affected the yield and quality of local Chinese herbal medicine. Therefore, it is an urgent task for local Chinese herbal medicine production to conduct a general survey of local pests and diseases and formulate reasonable pest control strategies. At the invitation of local scientific research institutions, the herbarium of the National Zoological Museum of China organized experts to go to Longnan County, Gansu Province, to investigate in-depth the pests and hazards of traditional Chinese medicine in the field, discuss with local poverty relief cadres and villagers, understand the relevant situation, and put forward suggestions for monitoring and controlling pests and developing the large and medium-sized medicine industry. At the same time, it reached the goal of continuously supporting the local pest control of traditional Chinese medicine Intention.

4) Biological invasion and national biosafety

Biological invasion is one of the important factors leading to the loss of biodiversity, together with the loss of animal and plant habitats and global climate change, is regarded as the three major problems affecting the global environment. Biological invasion has caused huge economic losses to China and has become a serious ecological and economic problem. There are more than invasive plant species in China, of which 39.6% of caused by intentional introduction, 43.9% by unintentional introduction, and only 3.1% by natural dispersal. This result indicates that human activities are an important factor leading to biological invasion.

The biological collection system of CAS serves the national biosafety and presides over two key deployment projects of CAS: "Construction of Technical Service System for Animal and Plant Quarantine at Entry-exit Ports of China" and "National Invasive Biological Prevention and Control Technology". During the 13th Five Year Plan period, the system obtained 13 million yuan of CAS funds and 8.73 million yuan of other funds. Five animal and plant inspection and quarantine entity databases and three information resource databases have been constructed; a service platform for port docking and a dynamic response platform for national biosafety have been built; a rapid identification platform for important quarantine and invasive species has been built, with 30,573 DNA barcodes collected; 30 pest weevils identification technologies have been provided, and risk analysis and assessment reports of intercepted 10 animals and 11 plants have been provided; and one national standards were adopted, two plant quarantine standards, two national standards applied, and the prevention and control system of potato beetle in Northeast China was established.

到一线口岸实地调研
Investigation at first-line ports

It also provides identification, consultation, training and other services for relevant units. For example, actively participate in the "Experts and Scholars Visit the Port" and other activities organized by the inspection and quarantine department to investigate the existing problems at the front-line port; for the work needs of front-line staff, fully tap the physical samples and information data resources, integrate the data of potential invasive species and

“铸就国门生物安全防线争当生态文明排头尖兵”大型国门安全科普展览
National security science popularization exhibition: “Building a National Biosafety Defense Line and Striving to Be the Leader of Ecological Civilization”

“国门生物安全宣传教育基地”揭牌仪式
Opening ceremony of “National Biosafety Publicity and Education Base”

标本数据库和物种数据库，搭建入侵动物物种鉴定快速响应平台，为海关口岸一线提供物种鉴定服务。

国家动物博物馆还与原动植检司联合在博物馆内专门设立“国门生物安全展厅”，内容涵盖了与国门生物安全有关的法律法规、截获总体情况和相关案例等。该展厅已成为国家动物博物馆内一个令观众感觉既新奇又感慨的专题展厅。自开展以来，已接待观众超过 30 万人次，并配合原国家质检总局相关部门开展科普专题活动 5 次，举行新闻发布会 3 次。原国家质检总局还将国家动物博物馆指定为“国门生物安全宣传教育基地”，有关领导利用休息时间到展厅义务讲解，向社会大众宣传国门生物安全知识。

基于长期积累的丰富馆藏、良好的保藏条件和较高的管理水平，2019 年国家动物博物馆被科技部和财政部认定为“国家生物种质与实验材料资源库”之一的“国家动物标本资源库”，将继续为国家经济建设提供重要的动物标本资源服务和支撑。

植物研究所标本馆近年来先后与多个海关检验检疫部门合作，开展中国口岸外来入侵植物监测，并与江苏泰州有关部门建立联合实验室，对进口粮食、矿砂、原木、羊毛、入境旅客等载体所携带的外来入侵植物开展快速鉴定，对海关的检验检疫专门人员进行了培训。在与海关合作的过程中，标本馆科研人员发现许多新到的具有入侵趋势的植物物种，并构建了中国外来入侵植物 DNA 条形码数据库，为国门安全起到了良好的支撑作用。目前，在口岸外来入侵植物监测中，已截获并鉴定出 100 余种外来植物，其中 2/3 的种类为国外已入侵并大规模扩散的恶性入侵植物；为国家环境保护部发布的第三、四批外来入侵物种名单提供本底数据支持。该工作的开展为保障国门安全、国家生态文明建设提供了坚实的数据支持。

近年来，成都生物研究所两栖爬行动物标本馆协助云南濒科委司法鉴定中心、云南省森林公安局、成都市森林公安局等部门，对查获的两栖爬行动物进行快速鉴定，鉴定出缅甸蟒、圆鼻巨蜥等国家 I 级重点保护动物，虎纹蛙、大壁虎等国家 II 级重点保护动物，棘胸蛙、中华蟾蜍、黑眉锦蛇、乌梢蛇等国家“三有”保护动物，共计约 50 种；鉴定出辐射陆龟、孟加拉巨蜥等 CITES 附录 I 物种，马里王者蜥、中美洲刺尾鬣蜥、

inspection and quarantine species, build the physical specimen database and information database, build a rapid species identification platform, and pass the DNA barcode technology to carry out rapid identification of species; to organize experts to identify and intercept animals and plants for local inspection and quarantine bureaus and issue identification reports; to collect specimens and information of invasive species and inspection and quarantine species by means of international cooperation, enrich the platform resource base and analyze the invasion risk of the intercepted species. Receive inspection and quarantine staff to inspect and borrow specimens, download literature, and provide technical guidance and training.

Based on the above service achievements, some collections have also organized national biosafety exhibitions. For example, the Herbarium of Kunming Institute of Botany and the Kunming Natural History Museum of Zoology held a large-scale national security science popularization exhibition in Changshui International Airport, which is called "Building a National Biosafety Defense Line and Striving to Be the Leader of Ecological Civilization". As the Yunnan news broadcast of Yunnan provincial television station, it publicized and reported as the headline of the day. As a permanent exhibition, it was set up in the departure hall of the airport to serve domestic and foreign visitors. There are about 1.5 million foreign tourists and leaders at home and abroad during the exhibition. Due to the outstanding service performance and outstanding publicity effect, the achievement was selected as "Take the Lead in Action, Forge Ahead — Innovation Achievement Exhibition of CAS since the 18th National Congress".

At the early stage, the National Zoological Museum of China actively participated in the national investigation of biological resources and made an important contribution to the preliminary exploration of the background of biological resources. In recent years, it is more oriented to national biological safety. On the one hand, it gives full play to the value of the specimen collection of inspection and quarantine species, provides comparative evidence for the identification of invasive harmful insects by researchers. On the other hand, it undertakes the relevant projects of the Ministry of Science and Technology and CAS, carries out the investigation of invasive animals, builds the specimen database and species database, and builds a rapid response platform for the identification of invasive animal species, provide species identification services for the front line of customs port.

The National Zoological Museum of China and the Department of Animal and Plant Inspection jointly set up the "National Biosafety Exhibition hall" in the museum, which covers the laws and regulations related to national biosafety, the overall situation of interception and relevant cases. The exhibition hall has become a special exhibition hall in NZMC, which makes the audience feel both novel and emotional. Since its launch, it has received more than 300,000 visitors, cooperated with relevant departments of the General Administration of Quality Supervision, Inspection and Quarantine of the People's Republic of China (AQSIQ) to carry out five special activities on science popularization and held three press conferences. AQSIQ also designated the NZMC as the "National Biosafety Publicity and Education Base", and relevant leaders used the rest time to explain in the exhibition hall and publicize the national biosafety knowledge to the public.

Based on the long-term accumulation of rich collections, good preservation conditions and high management level, in 2019, the National Zoological Museum of China was recognized by the Ministry of Science and Technology and the Ministry of Finance as one of the "National Biological Germplasm and Experimental Material Resources Bank", "National Animal Collection Resource Center", which will continue to provide important animal specimen resources services and support for national economic construction.

In recent years, PE has cooperated with several customs inspection and quarantine departments to monitor of invasive plants at Chinese ports and rapid identification of invasive plants carried by imported food, mineral sand, logs, wool, inbound passengers and other carriers. PE has established a joint laboratory with the customs inspection and quarantine departments of Taizhou, Jiangsu Province, to collect and identify exotic species in imported goods and to train local staff in customs. Many new arrived species with a tendency to invade China have been identified in time in port. A database of DNA barcodes of invasive plants in China has been established and plays an important role in supporting national security. Currently, more than 100 alien plants have been intercepted and identified, 2/3 of which are foreign invasive plants that have invaded and spread on a large scale, which provided background data to support the third and fourth batches of the list of invasive species released by the Ministry of Environmental Protection. The development of this work provided solid data support for the security of the country and the construction of national ecological civilization.

In recent years, the Herpetological Museum of Chengdu Institute of Biology group has assisted the Yunnan Binkewei Center for Judicial Forensic, Yunnan Forestry Public Security Bureau, Chengdu Forestry Public Security Bureau and other departments in rapid identification of the amphibians and reptiles. The museum identified about 50 native species, such as the National First Class Protected Animal: *Python bivittatus*, *Varanus salvator*, etc.; the National Second Class Protected Animal: *Hoplobatrachus chinensis*, *Gekko gecko*, etc. The museum also identified about 30 trade species, such

在海关进行物种鉴定和指导
Species identification and teaching for customs

红尾蚺、球蟒等 CITES 附录 II 物种，以及拟鳄龟、大鳄龟等适应力强的入侵物种，共计约 30 种。准确、快速的鉴定为森林公安部门和司法鉴定部门提供了重要的参考依据，为我国及国际野生动物保护、生态环境保护等方面工作，以及国家生态文明建设提供了坚实的技术支持。

水生生物博物馆通过分析中国淡水鱼类分布格局在外来鱼类入侵前后的变动，探讨了入侵种对中国淡水鱼类区系格局的影响。揭示了不以国家界线来界定划分的外来种造成中国生物区系严重同质化的问题，有助于全面地理解生物入侵对中国生物多样性所造成的危害，并从宏观尺度上为中国水生生物的保护提供重要的科学依据。还为农业农村部主办的面向海关边检、工商管理、渔政管理人员培训班做了多场讲座，参加培训人员数百人。

华南植物园于 2007 年成立“华南植物鉴定中心”，以植物科学研究中心和华南植物园标本馆为支撑，为广东省、华南地区及国内其他地区的植物保护和执法工作提供快速、准确的物种鉴定和技术服务。针对近年来公安机关、海关边防、林业部门和社会团体等鉴定红木植物原树种的需求，鉴定中心还专设“红木 DNA 鉴定平台”，这是国内首个利用 DNA 技术准确鉴定海南黄花梨和越南黄花梨等贵重红木原树种的植物鉴定平台。鉴定中心自成立以来，每年为林业局、自然保护区、森林公安、法院、海关等单位提供 100 多批次物种鉴定服务。

昆明动物博物馆收集口岸截获动物标本，整理馆藏外来入侵实体标本，建立和完善了国门生物安全西南地区动物分馆。围绕西南地区外来入侵物种，昆明动物博物馆在物种鉴定、科普教育、展览等方面与司法鉴定、海关、边防、出入境检验检疫、野生动植物保护、生物多样性保护等组织机构合作开展科研、科普工作。2019 年全球重大害虫草地贪夜蛾第一次大范围入侵我国，昆明动物博物馆团队依托馆藏标本、分类鉴定、物种采集、科学实验、科普等优势，深入分析草地贪夜蛾对我国粮食安全的挑战并提

标本馆协助鉴定的马里王者蜥
Uromastyx dispar maliensis, identified with assistance of museum

2019 年 6 月在广西北海开展农业农村部渔业局培训讲座
Training Courses at the Fishery Bureau of Ministry of Agriculture and Rural Affairs, Beihai, Guangxi, June 2019

as the CITES I species: *Astrochelys radiata*, *Varanus bengalensis*, etc.; CITES II species: *Uromastyx dispar maliensis*, *Ctenosaura similis*, *Boa constrictor*, *Python regius*, etc.; and Invasive species *Chelydra serpentine* and *Macroclemys temminckii*, etc. Accurate and rapid identification provides important reference for Forest Public Security Department and judicial identification department and provides solid technical support national and international wildlife conservation, ecological environment conservation and national ecological civilization construction.

MHBS analyzed the changes of the distribution pattern of freshwater fishes in China before and after the invasion of exotic fishes and discussed the influences. Then MHBS revealed the problem of serious homogeneity of Chinese biota caused by alien species which are not confined by national boundaries, which helps to understand the damage of biological invasion to China's biodiversity and to provide an important scientific basis for the protection of aquatic organisms in China from a macro-scale view. At the request of the Ministry of Agriculture and Rural Affairs, several training courses for the staff members of customs border inspection, business administration, and fisheries management have been delivered in different areas in recent years. Hundreds of personnel participated in the training.

South China Plant Identification Center (SCPIC) was found in 2007 and supported by the plant science research center and the Herbarium of IBSC. It provides rapid and accurate species identification and technical services to the plant conservation and law enforcement agencies in Guangdong Province, the South China region and other regions in China. In recent years, in response to the needs of public security, customs, forestry departments and social organizations to identify the original species of mahogany, *"Hongmu" DNA Identification Platform* was set up, which is the first identification platform in China to accurately identify the redwood species such as *Delbergia odorifera* and *D. tonkinensis* by using DNA technology. Since its establishment, the identification center has provided more than 100 batches of services every year.

KNHMZ has collected animal specimens intercepted at the ports, has organized the collection of specimens of invading entities from the collection, and has established and improved the branch animal museum of the Southwest Region of National Biosafety. Around invasive alien species in the Southwest, KNHMZ cooperated with organizations such as judicial identification, customs, border control, entry-exit inspection and quarantine, wild animal and plant protection and biodiversity protection in such areas as species identification, science education and exhibitions. For the first time, a large-scale invasion of the world's major pest *Spodoptera frugiperda* appeared in China in 2019. The KNHMZ team took advantage of the preserved specimens, classification and identification, species collection, scientific experiments and science popularization, analyzing in-depth about the *Spodoptera frugiperda*'s *challenge* to China's food security and made recommendations, all of which were adopted and approved by the national and local departments. The Chinese first live science popularization project, *Spodoptera frugiperda* damaging corn, was held lively, and nearly 100,000 visitors were well-received during the year.

Taking full advantage of the collection of specimen resources and professionals, Marine Biodiversity Collections of South China Sea have established an intelligent identification system for zooplankton in the South China Sea using zooplankton sample scanning technology, so as to build a fast and accurate zooplankton identification platform for marine ecological monitoring and environmental assessment; provide species identification services for smuggling rare

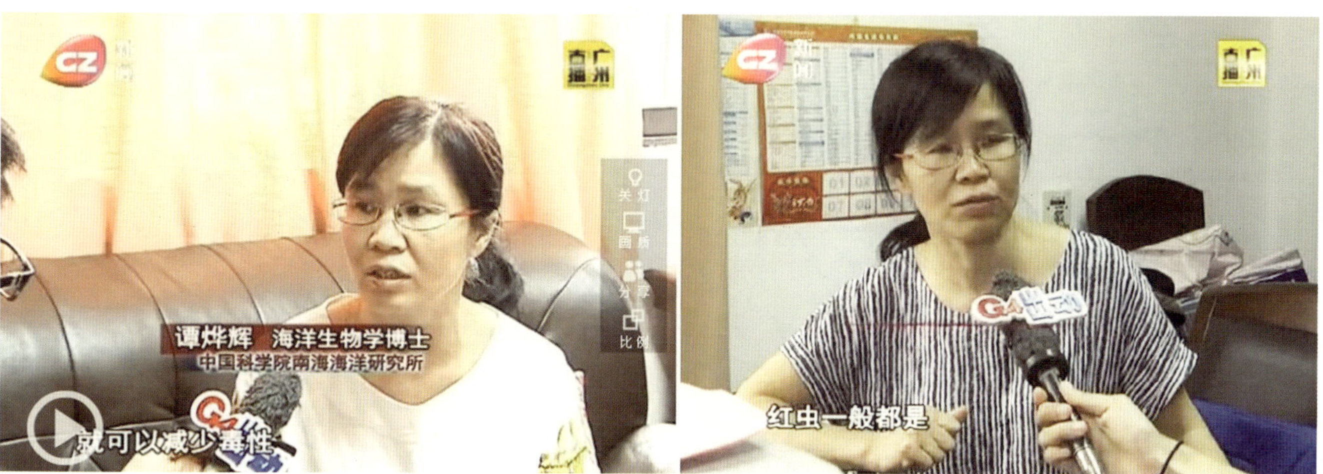

接受广州电视台采访
Interviewed by Guangzhou TV

物种鉴定意见书

受海南省公安边防总队海警第二支队的委托【琼公边（海二）鉴聘字（2017）01 号 】对“8.30”非法猎捕、收购、运输珍贵濒危野生动物案中查获的疑似波纹唇鱼类进行鉴定。对照该支队制作的琼琼海渔 01056 船鱼类统计清单（附后），经现场清点，通过观察其外部形态特征，鉴别结果如下：

物种鉴定意见：

这批 319 尾鱼类全部属于鲈形目、隆头鱼科、唇鱼属的波纹唇鱼 *Cheilinus undulates*；别名：苏眉。根据鱼的头上没有隆肉的形态学特征鉴定这批鱼类均为幼体。

保护级别：

依据《濒危野生动植物种国际贸易公约》（CITES）附录 II，本种属于二类保护动物。

样品的典型形态特征：

鱼类长度 23-55 厘米，身体长而侧扁，头背轮廓由吻部至眼部平直，其后位置外凸。鱼的头上没有隆肉。鳞片具黑纹，并具黄绿色或灰绿色横纹，眼后两条黑纹，体青绿色或浅绿色，头具橙色和绿色网状细线，尾鳍圆形，有浅黄色边缘，奇鳍密布细斜线。

涉案波纹唇鱼处理意见：

因此批涉案波纹唇鱼均属活体野生波纹唇鱼，建议由海南海警二支队自行寻找适宜波纹唇鱼放生地点尽早将琼琼海渔 01056 船船上全部涉案波纹唇鱼进行放生处理。

头部有隆肉的波纹唇鱼成鱼照片（来自 Randall）

琼琼海渔 01056 船涉案波纹唇鱼幼鱼照片

附件：

1. 琼琼海渔 01056 船波纹唇鱼清点、统计表
2. 海南海警二支队琼琼海渔 01056 号船查获波纹唇鱼部分样本照片

鉴定单位：中国科学院南海海洋研究所

鉴定人：孔晓瑜 时伟

2017 年 8 月 31 日

物种鉴定意见书

Identification services proposal

出建议，获得国家和地方等相关部门的采纳及批示，并在博物馆内进行了中国首个草地贪夜蛾危害玉米的活体科普展项，一年时间就为近 10 万观众提供了展览，深受好评。

南海海洋生物标本馆充分挖掘馆藏标本资源和发挥职员优势，利用浮游动物样品扫描技术，首次建立了南海浮游动物智能识别系统，为海域监测和环境评价提供了快速准确的浮游动物鉴定平台；提供查获走私珍稀海洋鱼类的物种鉴定服务，为相关部门执法提供依据；多次接受广州媒体的采访，针对广东近岸发生的一系列海洋自然灾害现象，普及海洋生态学知识，并阐述近海污染与人类活动的关系。

海洋生物标本馆是我国海洋生物分类学、生物多样性研究和鉴定中心，服务于国内外海洋生物、生态、环境等多学科领域。为多家企业和贸易公司提供海洋生物物种的鉴定。同时，为海关、商检等部门提供海洋生物物种鉴定等方面的专业服务。2014 年以来先后为青岛、北京海关缉私局等提供 50 余批次濒危海洋生物标本的物种鉴定工作，为相关部门提供执法依据。同时，还和江阴市商检局合作，开展外籍船舶压舱水排放生物检测，有效地防止了外来物种的入侵。

从 2016 年开始，标本馆与沈阳市海关动植物检疫部门建立合作，为其提供截获昆虫和蜘蛛等活体样品的鉴定及标本制作服务，为处置有害外来生物的科学决策提供依据。

菌物标本馆依托馆藏标本，借助国际交换借阅网络及海关合作，建立了我国进境植物检疫性真菌数据库及综合鉴定系统和部分重要农作物病原真菌的溯源系统。制定了国家标准 4 项，行业标准 4 项，协助口岸在进境植物上检出疫情 11 次。

2016 年沈阳市出入境检验检疫局截获的日本长戟大兜虫 (左图为雄虫，右图为雌虫)

The live samples of *Dynastes hercules* from Japan which were intercepted by the Shenyang Customs in 2016 (the left picture is male, and the right one is female)

marine fish for the relevant departments to support law enforcement; through the Guangzhou media, popularize marine ecology knowledge and explain the relationship between offshore pollution and human activities in response to a series of marine natural disasters that occurred off the coast of Guangdong in recent years.

The Marine Biological Museum is the research and identification center of marine organism taxonomy and biodiversity in China, serving the multi-disciplinary fields of marine biology, ecology and environment at home and abroad. To provide identification of marine biological species for many enterprises and trading companies. At the same time, it provides professional services for the customs, commodity inspection and other departments in the identification of marine biological species. Since 2014, it has provided species identification of more than 50 batches of endangered marine biological specimens for Qingdao and Beijing Customs Anti-smuggling Bureau, providing law enforcement basis for relevant departments. At the same time, museum cooperated with Jiangyin Commodity Inspection Bureau to carry out biological detection of ballast water discharge from foreign ships, effectively preventing the invasion of alien species.

Since 2016, the cooperation was erected between the Northeast Biological Herbaria and the Animal and Plant Department of Shenyang Customs. The herbaria help the Customs to identify the samples of insect or spiders, also give them some suggestions on treatments of the harmful alien organisms.

"Database of Chinese phytosanitary fungi and the one-stop-shop identification system" was established by the HMAS base on the specimens housed in the Fungarium and the exchanged data through international herbarium network. These specimens, databases, and systems have been well exploited in the developments of various diagnosis protocols for quarantine fungi, some of which have been accepted as national standards (4 protocols) and quarantine standards (4 protocols). These diagnosis protocols have been well adopted by the Chinese quarantine authorities and based on which, 11 successful intercepts of fungal pathogens were reported recently.

检疫真菌参比物质及标准库

Reference materials and standard library of quarantine fungi

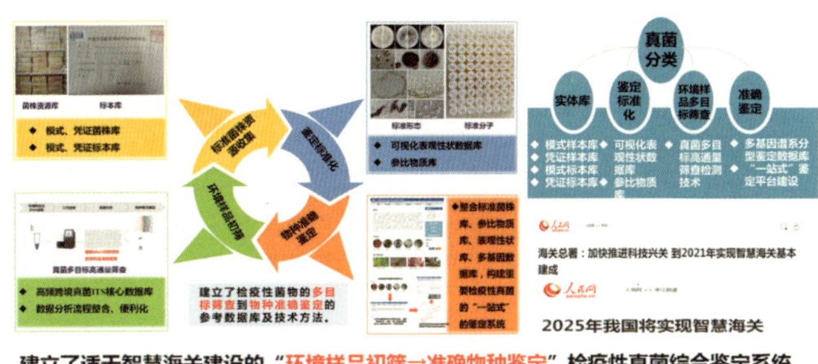

中国进境植物检疫性真菌数据库及综合鉴定系统

Database of Chinese phytosanitary fungi and the one-stop-shop identification system

4.4 Science education

The biological collection is a place to collect the past of natural life and natural history. While providing materials for scientific research, as a non-profit informal education place, the biological collections provide the public with a more intuitive and effective way to understand the natural history with the lowest threshold and bring people a unique spiritual experience. The biological collection is a link connecting the public and scientists in understanding life, and a window for people to understand the past, present and future of life.

With more and more attention paid by the state to science popularization education, it has become an important part of the work of the biological collections to publicize scientific knowledge and ideas through science popularization activities to improve the scientific literacy of the public. In 2014-2019, all museums held more than 2000 kinds of popular science activities, and some collections or museums were opened to the public all year round, with an audience of more than 4.46 million people; it also provides more than 18,000 people with social consulting services, with a total of 190 million people visiting the websites.

The activities held by each museum are various and colorful, forming normalized science popularization publicity,

4.4 科学教育

生物标本馆是收藏自然界生命的过去，收藏自然历史的地方。在为科学研究提供材料的同时，生物标本馆作为一种非营利性的非正式教育场所，以最低的门槛为大众提供了更直观更有效地了解自然历史的途径，给人们带来了独特的精神体验。生物标本馆是将大众和科学家在对生命的理解上连接起来的纽带，是人们了解生命的过去、现在和未来的一扇窗。

随着国家对科普教育越来越重视，通过科普活动来宣传科学知识和科学思想以提高公众科学素养成为生物标本馆工作的重要组成部分。2014-2019 年各馆共举办 2000 余场各类科普活动，部分标本馆 / 博物馆还常年对外开放，受众达 446 万余人次；并提供社会咨询服务 1.8 万余人次，各馆网站访问量总计达 1.9 亿人次。

科普服务情况

Science popularization services

	2014	2015	2016	2017	2018	2019	总计 Total
参观人数（万人）Number of visitors (×10,000)	80.7	56.7	76.2	76.6	75.4	81.0	446.6
科普活动 Number of Popular science activities	242	256	286	484	314	460	2042
科普著作 Number of Popular science works	19	17	13	13	10	12	84
科普文章 Number of Popular science article	77	47	83	104	211	223	745

各馆举办的活动丰富多样，异彩纷呈，形成了常态化科普宣传、规模化科普活动、品牌化科普产品和系列化科普成果。此外，还发表科普著作 80 余部，科普文章 700 余篇，支持拍摄科普纪录片、宣传片 8 部，野外考察实录 500 余篇等。这些成果将整个学科知识成体系地呈现出来，有效地提升了公众的科学素养。部分科研工作也得到了社会的关注和报道。由于工作出色，各馆获得了国家级或地方级各类“科普教育基地”等称号，以及各种集体或个人的奖励，得到了社会各界的高度认可。

1）标本展示和科普活动

对馆藏标本进行展览是各生物标本馆进行科普宣传的最基本手段。展示的内容除了来自中国各个地方以及国外的千姿百态的生命形式外，还有那些蕴含着重要的科学真理和科学家出色的研究成果的生物标本。各馆也围绕这些藏品及其背后的一个个精彩的科学故事开展了多种多样的科普活动，以尽可能容易理解的方式向公众传达着自然生命对人类发展至关重要的理念。

菌物标本馆创建了我国第一个运用多媒体技术展示真菌知识和研究成果的科普展览——“真菌与人类”科普展厅。自 2008 年开展以来，每年接待各地参观者 4000 余人，已成为国内重要的菌物学科普基地。标本馆完成科技部支撑计划和国家自然科学基金资助的菌物学科普项目研究工作，开发制作了冬虫夏草生长多媒体模拟展示系统和我国第一张菌物科普多媒体光盘。此外还与中央电视台科教频道合作完成“真菌的诱惑”和“给‘仙草’正名”2 集科普纪录片的拍摄工作，于 2015 年 9 月，在中央电视台科教频道《走近科学》栏目中播出。

为传播科学知识、弘扬科学精神、提高公众科学素养，自 2010 年起，中国科学院成都生物研究所连续 9 年举办“公众科学日”活动，以标本馆作为活动主要场地，吸引了各界群众和大中小学生前来参观，在

large-scale science popularization activities, brand science popularization products and series of science popularization achievements. In addition, it has published more than 80 popular science works, more than 700 popular science articles, supported the shooting of 8 popular science documentaries, promotional films, and more than 500 field investigation records. These achievements present the whole subject knowledge systematically and effectively improve the scientific literacy of the public. Part of the research work has also been concerned and reported by society. Because of their outstanding work, the museums have won the titles of "Popular Science Education Base" at the national or local level, as well as various collective or individual awards, which have been highly recognized by all sectors of society.

1) Specimen display and popular science activities

The exhibition of collected specimens is the most basic means for the popularization of science. In addition to the diverse life forms from all parts of China and abroad, there are also biological specimens containing important scientific truth and outstanding research results of scientists. The biological collections also carry out a variety of science popularization activities around these collections and the wonderful science stories behind them and convey the idea that natural life is essential to human development to the public in a way that is as easy to understand as possible.

The Fungarium established the first mycological exhibition in China, "Fungi and Humankind", for citizens using multimedia technology. "Fungi and Humankind" exhibition annually has welcomed more than 4,000 visitors since 2008, which became an important popularization base of mycology in China. Supported by the Ministry of Science and Technology of China Support Program and National Natural Science Foundation of China, the Fungarium have developed and produced the multimedium simulation display system of growth of the fascinating fungus *Ophiocordyceps sinensis* and the first fungus science popularization multimedia CD in China. Cooperated with China Central Television Station (CCTV), the Fungarium produced two documentary television films, i.e. "Allure of Fungi" and "Rectifying the Name of 'Lingzhi'". These two films were shown on the program "Approaching Science" of CCTV in September 2015.

In order to propagate scientific knowledge, introduce the scientific virtues, and improve public literacy, since 2010, Chengdu Institute of Biology has held the "Public Science Day" annually for 9 years consecutively, attracting varieties of publics including students of all ages, which influenced on the public positively and honored by the society. In 2019, the Herpetological Museum participated in the 15th "Public Science Day" of the Chengdu Institute of Biology and provided online live streaming, which attracted more than 70,000 audiences at the day, and about 500 people attended the offline event. Comparing with the previous events, the 15th "Public Science Day" focused on the demands and diversity of the public; with the online live streaming, the museum made the demonstrations "visible and tangible" so that the public can have a closer view to the event and biological science.

The Herbarium of Chengdu Institute of Biology has held a series of public science activities such as "Common plant identification", "How to make a herbarium specimen", "Investigation and control on the alien invasive plants in Chengdu" for primary and middle school students and social people since 2010, which had attracted much public attention and received high praise.

In Southern China, Marine Biodiversity Collections of South China Sea is an important public education base for marine sciences. Various forms of science popularization activities have been held during the Public Science Day of

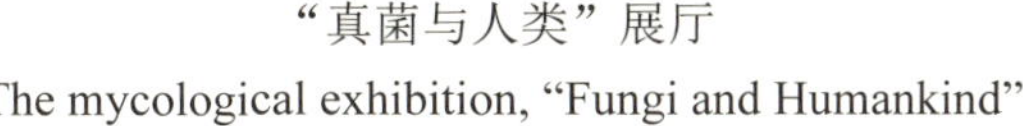

"真菌与人类"展厅
The mycological exhibition, "Fungi and Humankind"

冬虫夏草生长多媒体模拟展示系统
The multi-medium exhibition of growth of *Ophiocordyceps sinensis*

《给“仙草”正名》科普纪录片

Documentary telefilm, “Rectifying the name of ‘Lingzhi’”

社会上引起了广泛的影响和赞誉。2019年两栖爬行动物博物馆参与举办第15届成都生物研究所公众科学日，开展现场直播活动，当日点击量超过7万，约500余人参加现场活动。与前几届活动相比，2019年公众科学日聚焦结合公众和社会的需求，注重参观群体的广泛性，结合线上直播平台，让美丽的动植物“看得到摸得着”，让公众能够更加近距离地了解生物知识。

成都生物研究所植物标本馆也开展了各种丰富的科普活动。自2010年起，连续九年举办公众科学日活动，针对中小学生和社会人士开展了“常见植物识别”、“如何植物标本制作”和“成都市外来入侵植物调查和防控”等专题科普活动，在社会上引起了广泛的影响和赞誉。

南海海洋生物标本馆是华南地区宣传海洋知识的重要科普基地。标本馆在“中国科学院公众科学日”举办形式多样的科普宣传活动；标本馆于2017年被中国海洋学会评为“全国海洋科普教育基地”；作为广州市科普教育基地，标本馆自2014年开始与广州市科协合作，参与承办了以“畅游科学知识殿堂，提升科学文化素质”为主题的“广州科普游自由行”和“科普一日游”活动，同时向公众免费开放科普基地，每年科普受众累计超过1万人次。

海洋生物标本馆依靠丰富的海洋生物标本资源，广泛开展海洋生物科学知识宣传工作，让海洋生物科普走向大众。馆内海洋生物科普展厅、珍贵海洋生物标本展厅等设施免费（预约）对公众开放，每年接待近万人次的学生和市民前来参观。此外，海洋生物标本馆还是中国海洋大学、厦门大学、上海海洋大学、临沂大学等近十所学校学生的实习基地，每年接待近2000名师生来馆参观实习。

沈阳应用生态研究所东北生物标本馆是教育部授予的“全国中小学生研学实践教育基地”，同时也是辽宁省、沈阳市的科普教育基地，近年来该馆充分依托研学科普基地平台，以标本展示、科普讲座、互动实验、室外任务及野外活动等多层次的活动方式，针对小学、初高中、大学生、社会公众等不同受众群体开展专

2019年“公众科学日”来访的学生和家长

Participants of the “Public Science Day” in 2019

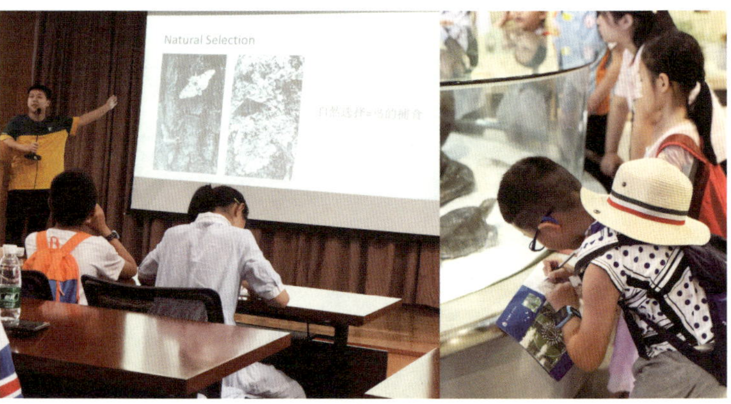

讲解员为来访学生做“自然选择”科普讲座（左）、科普馆内认真做笔记的来访学生（右）

Public science lectures entitled “natural selection” (left), and a student took notes carefully (right)

CAS. MBCSC has been nominated as the "National Marine Science Popularization Education Base" by the Chinese Society for Oceanography in 2017. Guangzhou Public Education Day and the Sciences Tour free Line sponsored by Guangzhou Association of science and technology, held on Sunday in the middle of each month. As the Guangzhou Science Popularization Education Base, MBCSC hosted the "Guangzhou Popular science independent travel of the marine science public education station" activity with the theme of "Changing the Temple of Scientific Knowledge and Improving the Quality of Science and Culture" since 2014, and the visitors exceed 10,000 each year.

标本馆专题科普活动现场

Science communication activities of the Herbarium

Relying on the rich resources, the Marine Biological Museum has carried out extensive publicity of marine biological scientific knowledge, so as to make the popularization of marine biological science popular. The Marine Life Science Exhibition Hall and Precious Marine Life Specimen Exhibition Hall and other facilities are open to the public free of charge (with appointment), receiving nearly 10,000 students and citizens to visit each year. In addition, the museum is also an internship base for students from nearly ten schools, including Ocean University of China, Xiamen University, Shanghai Ocean University and Linyi University. It receives nearly 2,000 teachers and students to visit the museum for internships every year.

标本馆开放日

MBCSCS open days

The Northeast Biological Herbaria of Institute of Applied Ecology is the "National Primary and Secondary School Students' Research and Practice Education Base" awarded by the Ministry of Education, as well as the popular science education base of Liaoning Province and Shenyang City. In recent years, the herbaria carry out multi-level activities such as specimen display, popular science lectures, interactive experiments, outdoor tasks, and field activities for students and public groups. The herbaria timely release the advance notice of scientific activities and reviews through the official account of WeChat. Every year, it receives thousands of primary and secondary school students and the public who come from Liaoning and other provinces, and even many foreign countries to carry out science popularization and scientific practice in the scope of natural ecology, and received warm responses from students, parents, teachers and the public. It has been reported by provincial and municipal TV stations and media for many times. The Northeast Biological Herbaria has become an important base of popular science in the field of natural ecology in Liaoning Province.

小学生参观海洋生物标本馆

Primary school students visit the Marine Biological Museum

As National Popularization Base of Science & Technology and Scientific Research and Popularization base of Qinghai-Tibet Plateau Biology of Qinghai Province, the exhibition hall of Qinghai-Tibet Plateau Museum of Biology opened

东北生物标本馆科普活动
Popular science activities of the Northeast Biological Herbaria

业讲解和活动，并通过标本馆网站和微信公众号及时发布科普活动预告及活动回顾。每年接待辽宁省及全国各地，乃至多个国家前来进行科普和研学的数千名大中小学生和社会公众，在自然生态领域开展了卓有成效的科普活动，收到了学生、家长、教师和社会公众的热烈反响，被省市电视台及媒体多次报道，成为辽宁省在自然生态领域科学普及的重要场所。

作为全国科普教育基地、青海省青藏高原生物学科研科普基地，青藏高原生物标本馆科普展厅常年不定期面向社会各界，特别是中、小学生免费开放，开展讲座介绍青藏高原特有动植物，同时开展学生野外生物多样性调查等研学考察活动及植物标本采集、植物标本制作等实践体验活动。

南京古生物博物馆每年推出两个主题特展，一个偏重介绍古生物学、地层学某个专题的科普内容，另一个偏重研究当年度最新的科研成果；开办“院士论坛”“达尔文大讲堂”等系列科普讲座；连续 12 年举办“化石鉴赏大会”；连续 11 年举行“奥妙的地球与奇特的生命少儿绘画、征文大赛”；举办多种形式的科普活动，如举办了 60 余期的“博物馆奇妙夜”夜宿活动，组织“寻找化石之旅”“化石小猎人”“雨花石探秘”等地质冬、夏令营及地质科普旅游等。

南京古生物博物馆开馆 15 年以来，担负起将研究所古生物学和地层学专业及相关学科领域知识向全社会普及的重任，在科学传播领域做出了自己应有的贡献。为更好提高人民群众科学文化素质，保障人民群众基本的科学文化权益，南京古生物博物馆对社会免费开放。目前，南京古生物博物馆是中国科学技术协会和中国古生物学会的“全国科普教育基地”，中国自然科学博物馆协会“团体会员单位”，江苏省科协、科技厅、教育厅联合颁发的“江苏省科普教育基地”，江苏省科协、教育厅联合颁发的“江苏省青少年科普教育基地”，南京市科学技术协会的“南京市科普教育示范基地”等，同时也是江苏省科普场馆协会、江苏省科普作家协会、江苏省地质协会的副理事长单位，中国古生物学会科普工作委员会依托单位。

科普特展
Popular science exhibition

水生生物博物馆每年举办 10-20 次科普活动，主要形式为参观场馆和科普讲座（“长江大保护要保护什么”、“留住江豚的微笑”、“长江之鱼”等）相结合，每次活动受众 150-300 人。开发 online 博物馆，将标本数字化、场馆数字化。依托新媒体，如 APP、微信公众号；结合各种技术，如 720 度全景视频、VR 虚拟场馆等，博物馆实现全年开放，深度互动、及时更新、资源共享。并通过科普文章、“科学大讲堂”系列电视专题片、网络授课等科普产品和形式，让更多公众了解生态环境和水生生物，积极保护水生生物资源与环境。

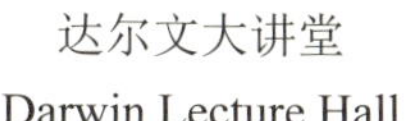

达尔文大讲堂
Darwin Lecture Hall

高端院士论坛
Academician Forum

“奥妙的地球与奇特的生命”少儿绘画、征文大赛
“Mysterious earth and strange life” children’s painting, and essay competition

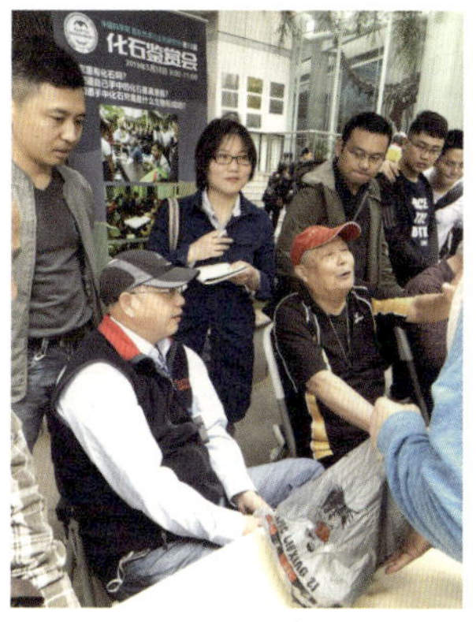

化石鉴赏大会
Fossil Appreciation Conference

寻找化石之旅
Fossil Hunting Tour

古生物模型制作
Paleontology model making

non-scheduled freely to all sections of Society especially middle and primary School student. Lectures of the popularization of life science were held in different schools to introduce the knowledge of the endemic flora and fauna of Qinghai-Tibet Plateau. The practical activities such as the survey of biodiversity, collection and preparation of plant specimen were carried out frequently for students of different levels.

The Nanjing Museum of Palaeontology regularly launches two special thematic exhibitions each year, one focusing on the introduction of paleontology and stratigraphy of a particular topic of popular science content, the other focusing on the research of the latest scientific achievements of the year. A series of popular science lectures, such as “Academician Forum” and “Darwin Lecture Hall”; organizing the “Fossil Appreciation Conference” for 12 consecutive years; held the “Mysterious earth and strange life Children’s painting, and essay competition” for 11 consecutive years. Various forms of popular science activities were held, such as more than 60 sessions of “Night In Museum” activities, “Fossil Hunting Tour”, “Yuhua Stone Exploration”, and other geological winter, summer camps and geological popular science tourism.

Since its opening of 15 years, the Nanjing Museum of Palaeontology has been responsible for popularizing the knowledge of paleontology, stratigraphy and related disciplines to the whole society, and has made its due contribution to

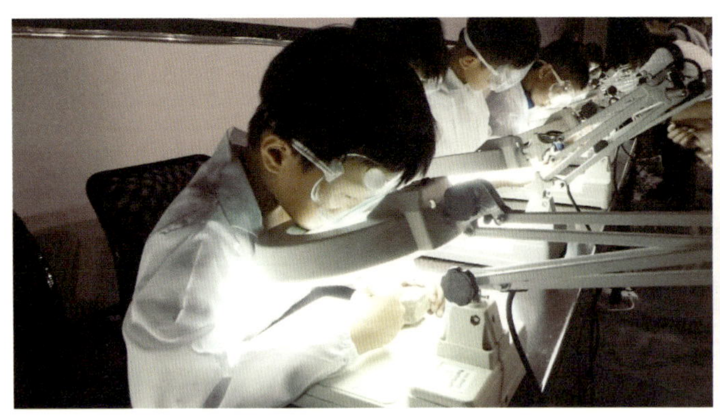

化石修复
Fossil restoration

参观实验室
Visit the laboratory

2）科普形式的创新

为了能引起公众的兴趣，提供独特的体验，各馆从各种人们喜闻乐见的方式入手，让大众参与进来，真正体验科学探索的乐趣，对科普的形式进行了各种创新。

沉下来：深入基层办科普

科普工作需要耐心，需要沉下来，在工作中不断总结经验，从内容上不断思考、探索和创新，走出一条有自身特色的道路。经过近几年的发展，各馆已不再仅仅通过展示标本来进行科普宣传，而是主动走出去，深入学校、社区和基层，将科学知识传播给那些不容易获取的群体。

昆明植物研究所标本馆将科普对象精准定位于学校和学生，尝试将针对学校和学生的科普与教育整合，创造出更多新内容，科普工作者亲自到各级学校中，通过科普讲座、野外实践、标本采集等活动激发学生的科学兴趣。不断深化与各政府科学教育职能部门的合作，为科普群体找到精确的定位。

昆明植物研究所标本馆将植物分类学知识与日本宫崎骏动漫相结合，在第十五届全国公众科学日中“探寻宫崎骏动漫里的植物世界”受到了公众的热烈欢迎，融讲解、互动、展示和竞猜为一体的科普活动取得了极大的反响。活动内容也被人民日报、新华社等国内外多方主流媒体报道。

为配合全国科技活动周活动，新疆生态与地理研究所标本馆连续承办了九届新疆维吾尔自治区“科技活动周”启动仪式暨分场的“科普宣讲在基层”科技活动周活动。通过“请进来，走出去”的方式积极开展形式多样、贴近民众、通俗易懂、参与性强的科学传播活动。足迹遍布新疆的博尔塔拉蒙古自治州、哈密市、塔城地区、阿勒泰地区、和田地区、喀什地区和阿克苏地区的 29 个县市，为 83 所中小学以及政府部门等举办科普报告 461 场，展出科普展板近 500 块，直接听众达 90 000 余人，免费发放多语种科普读物 50 000 余本，吸引 10 万余社会公众参与，受到 60 余家新闻媒体直接报道和转载报道。“科普宣讲在基层”已成为标本馆的一项重要的科普活动品牌。

近年来，海洋生物标本馆的科普工作，由传统的“请进来”向“请进来与走出去”并举转变。每年有多名专家到各地中、小学进行海洋生物科普讲座。同时，与中国科技馆、山东博物馆等强强联合，举办海洋生物科普展，让难得一见的珍贵海洋生物标本走出标本馆，与广大观众零距离接触。仅与青岛水族馆联合举办的“走向深海大洋”深海科学考察成果展，在展览的三年间，参观人数超过 700 万人次。

庐山植物园标本馆通过选取植物标本进校园的方式，展示部分重要和特色类群标本，结合专业技术人员科普报告的形式，面向中小学生开展植物学知识传播，进行植物保护和环境教育，让孩子们从小就能意

the field of science communication. In order to better improve the scientific and cultural quality of the people and protect their basic scientific and cultural rights and interests, the Nanjing Museum of Palaeontology will be opened to the public with free admission for permanent exhibitions. Nanjing Museum of Palaeontology is the "The National Popular Science Education Base" awarded by the China Association for Science and Technology and the Palaeontological Society of China. And it is a member of the Chinese Museum Association of Natural Science, the "Popular Science Education Base in Jiangsu Province" awarded by the Jiangsu Province Association for Science And Technology and Department of Science and Education of Jiangsu Province.

科学讲堂
Popular Science Class

Combined in the form of visiting venues and lecturing on popular science topics including *What Should We Protect in the Protection of Yangtze River*, *Keep the Smile of the Finless Porpoise*, and *The Yangtze River Fishes*, 10-20 popular science activities are held annually by MHBS. There are approximately 150 to 300 visitors in each activity. MHBS also develops online museums, specimen digitalization, venue digitalization, APP, WeChat official account, 720-degree panoramic video, and VR virtual venue. MHBS is open all year round, updates the latest news of science, and hopes to share more resources with the public. In addition, through science education and communication articles, television programs including *Science Lecture Hall*, online lectures, and other popular science products and forms, MHBS lets visitors learn more about the ecological environment and aquatic life, which is a benefit for them to actively protect resources and environments of aquatic organisms.

2) Innovation of popular science forms

In order to arouse the public's interest and provide a unique experience, each museum starts from various ways that people like to see, lets the public participate in, truly experiences the fun of scientific exploration, and innovates various forms of science popularization.

Sink down: go deep into grassroots to do science popularization

The work of popular science needs patience, needs to sink down, constantly summarizes the experience in work, constantly considers, explores and innovates from the content, and walks out a road with its own characteristics. After the development in recent years, the collections no longer only display specimens to carry out science popularization propaganda, but take the initiative to go out, go deep into schools, communities and grassroots, and spread scientific knowledge to those groups who are not easy to obtain.

KUN accurately positioning the objects of science popularization in schools and students, trying to integrate science and education with schools and students to create more new content. Science popularization workers come to schools personally to spread scientific knowledge. Students' interest in science is stimulated through science lectures, field practice, specimen collection and other activities. Constantly deepening cooperation with various government science and education functional departments has found a precise positioning for the science popularization group.

KUN combined the knowledge of plant taxonomy with Hayao Miyazaki's animation, and "Explore the plant world in Miyazaki's anime" was warmly welcomed by the public. The popularization of science activities, including interpretation, interaction, display, and quiz, had a great response. The content of the event has also been reported by mainstream media-*People's Daily*, Xinhua News Agency, etc.

In order to cooperate with the National Science and Technology Activity Week, the Specimen Museum of Xinjiang Institute of Ecology and Geography undertook the launching ceremony of 9 "Science and Technology Activity Week" of Xinjiang Uygur Autonomous Region and the branch science and technology activity week activity of "Popular Science Propaganda at the Grass-roots Level". Through the way of "Invite in and Go out", actively carry out scientific communication activities of various forms, close to the people, easy to understand and strong participation. Covering 29 counties and cities in Bortala Mongolia Autonomous Prefecture, Hami city, Tacheng region, Altay region, Hetian region, Kashi region and Aksu region, it has held 461 science popularization reports for 83 primary and secondary schools

宫崎骏主题科普活动

Hayao Miyazaki theme science popularization activities

参加“SELF格致论道”公益讲坛（第42期）

海盐县教育局邀请马文章中小学科学老师开展科普培训讲座

昆植标本馆科普进校园——走进海之韵

参加第五届昆明科博会的科普大讲坛活动

参加SELF x Kids讲坛

参加2019野生植物种质资源采集保存技术高级培训班

走进学校与科普讲坛

Science popularization forum

识到保护植物就是保护地球，就是保护全人类。

一起玩：公众参与科考和科研

随着人们对生态的关注，国家对环境问题的重视，以及全社会对教育的反思，近年来掀起了一股“自然教育”的热潮。部分场馆组织了大量沉浸式的科学体验活动，让公众能够参与进来，与科学家一起进行

深入农村校园科普

Science popularization in rural campus

and government departments, exhibited nearly 500 science popularization boards, with a direct audience of more than 90,000 people, distributed more than 50,000 free multilingual science popularization books, and attracted more than 100,000 people With the participation of the public, it has been directly reported and reprinted by more than 60 news media. “Popular Science Propaganda at the Grass-roots Level” has become an important popular science activity brand of the XJBI.

“科普宣讲在基层”

“Popular Science Propaganda at the Grass-roots Level”

In recent years, the popular science work of the Marine Biological Museum has changed from the traditional “Invite in” to “Invite in and Go out”. Every year, many experts go to primary and secondary schools to give lectures on marine biology. At the same time, it has jointly held a science popularization exhibition of marine biology with China Science and Technology Museum and Shandong Museum, which let rare and precious marine biological specimens come out of the museum and make joint contact with the audience at zero distance. During the three years of the exhibition, more than 7 million people visited the “Towards the Deep Ocean” deep-sea scientific research achievement exhibition jointly held with Qingdao Aquarium.

免费发放多语种科普读物

Free distribution of multilingual popular science books

展出的部分珍稀深海生物标本
Some rare deep-sea biological specimens on display

标本进校园
Specimens come into the campus

野外科考，在探索大自然的过程中获得知识和乐趣，受到了广大学生及家长的欢迎。

2002 年 7 月 19-21 日，赵尔宓院士带领 23 名中小学生赴峨眉山进行 3 天野外科考，影响深远，其中有 3 名学生日后进入了两栖爬行动物学研究领域。2019 年 7 月 29-31 日，标本馆配合中国科学院成都生物研究所知识管理中心，结合“不忘初心、牢记使命”主题教育活动，为新中国成立、中国科学院建院 70 周年献礼，成都生物研究所两栖爬行动物标本馆举办了“重走院士路 · 重上峨眉山”野外科考活动，9 名同学在此次活动中重走了 17 年前的同样行程和路线。值得一提的是，17 年前参与活动的初中生蒋珂（标本馆工作人员），正是本次带队老师。当年的学生今天的带队老师，科学始终有人延续。

活动中，队员们在夜间深入树林观察各种蛙类，在溪流里寻找龙洞山溪鲵的踪影，聆听仙琴蛙的奇妙叫声，学习蛙类的鉴定方法，通过外形观察及肩带骨骼解剖来探究“蛙”和“蟾”的不同，等等。活动反响强烈，学生、家长以及学校对活动表达了赞誉。多家媒体全程报道本次活动，通过视频和微博图文直播，新浪微博“重走院士路”话题阅读量接近 50 万，新华社和新浪网的现场报道也引发传播热潮，活动影响深远。参与活动的 9 名学生中，贺彦搏同学对两栖爬行动物产生兴趣，在蒋珂的指导下，2019 年底在“成都市第 35 届青少年科技创新大赛”中获论文一等奖与科协主席奖。

上海昆虫博物馆每年举办大量科普活动，品牌活动主要有夏令营、互动体验冬令营和生物多样性调研。

（1）“小小法布尔”夏令营 暑期均会组织对生命科学有兴趣的同学前往浙江天目山国家级自然保护区、安徽金寨自然保护区、浙江景宁畲族自治区、浙江安吉、江苏盐城等地开展为期 4 天左右的夏令营活动。3 年来，共计 300 余名同学参与其中。孩子们在博物馆老师的陪伴下，一起亲近自然，一起探索昆虫世界的神奇。

（2）互动体验冬令营 互动体验冬令营以“昆虫与化学”为主题，已连续举办了五期，受众约 300 人次，是博物馆开展“身边的科普”的一次跨学科的尝试，为孩子们更好地认识科学、参与科学活动体验提供了一个平台，内容涉及昆虫种类、昆虫生境、昆虫与环境、昆虫与化学、化学与生活等。

（3）生物多样性调研 昆虫博物馆近 3 年来组织了举办了 38 场“生物多样性调研”活动，共计 1120 余人次参与其中。活动通过观察、采集、记录、绘画、实验、分析数据等亲身体验形式，带同学们走进公园、走进湿地、走进保护区，去了解不同生物的生境、习性以及相互之间的关系，让大家学习科学的生物调查方法，提高科学的数据整理和分析能力，学习科学家的思维方式，引导学生了解昆虫、了解自然，关爱生命，关注自然。

2018 年 5 月，国家动物博物馆承办了全国科技活动周“科学之夜”活动，受到了公众的一致好评。基

In order to encourage the primary and middle school students to understand the importance of biodiversity, i.e., botanical knowledge dissemination, plant protection and environmental education, the team of Herbarium of Lushan Botanical Garden researchers demonstrate them with some important and characteristic groups of specimens.

Play together: public participation in scientific investigation and research

With people's attention to ecology, the state's attention to environmental issues, and the whole society's reflection on education, a wave of "Natural Education" has been set off in recent years. Some collections have organized many immersive scientific experience activities, so that the public can participate in the field scientific investigation together with scientists and get knowledge and fun in the process of exploring nature, which is welcomed by students and their parents.

On July 19 to 21, 2002, Prof. ZHAO Ermi led 23 elementary and middle school students to Mt. Emei for a three-day field investigation, which inspired several passionate youths. It has far-reaching influence; among the 23 students, three of them later entered the research field of herpetology. On July 29 to 31, a team with 9 vital youngbloods (students) reinvestigated Mt. Emei, led by JIANG Ke (staff of the museum), who was one of the inspired youths back in 2002, had participated in the event "Trace the Academician, Trace the Mountain", which belongs to an educational program "Stay true to your heart and your mission" hosted by the Herpetological Museum and the Knowledge Management Center of the Chengdu Institute of Biology. The program started as an appreciating tribute to the 70th anniversary of the country and its founding. Retraced the route that Prof. Zhao once led, JIANG Ke was once a student, and the student had become the instructor seventeen years later. Science had always left us legacies.

During the investigation, the students went deep into the woods at night to observe various frog species, looked for trails left by mountain salamanders, listened to mysterious calls of Emei music frogs, learned the taxonomic methods of frogs, and explored the distinctions between "frogs" and "toads" morphologically and anatomically...The positive reflections from students, parents, and schools had priced the event, Many media fully reported the event, especially Weibo, posted videography and articles that recorded the event, nearly 500,000 people have read this topic. The live streams by Xinhua News Agency and Sina also created propagation booms on the internet. The event had a solid influence on society. Among the 9 students who participated in the activity, HE Yanbo was interested in amphibians and reptiles. Under the guidance of JIANG Ke, HE Yanbo won the first prize of paper and the chairman prize of Association for Science and Technology in the 35th "Chengdu Youth Science and Technology Innovation Competition" at the end of 2019.

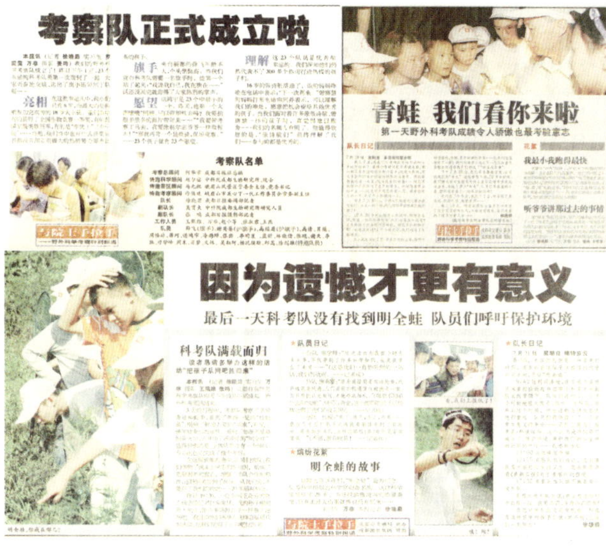

考察队正式成立啦

青蛙 我们看你来啦

因为遗憾才更有意义

最后一天科考队没有找到明全蛙 队员们呼吁保护环境

2002 年赵尔宓院士带领中小学生科考 3 天的新闻报道
News report on the three days for field investigation of students led by Prof. ZHAO Ermi in 2002

带队老师和队员们在峨眉山下合影
Photograph of instructor and participants in Mt. Emei

在安徽天堂寨开展“小小法布尔”夏令营活动
“Little Fabre” summer camp in Anhui Paradise Village

“博物馆之夜”活动
“Museum Night” activity

黄浦江郊野公园研学活动
Research and learning activities in Huangpu River Country Park

“科学之夜”开幕式
Opening ceremony of “Science Night”

于 2018 年度的良好工作，中国科学院动物研究所获得了 2019 年度全国科技活动周“科学之夜”主场活动的承办权，国家动物博物馆依旧肩负了其中的若干项重要工作。在 2019 年的“科学之夜”活动中，中国科学院动物研究所在充分展示中国科学院自身科研创新和高质量发展成就的同时，针对公众对科技日益多样化的需求，推出了十余项充满科学性、体验性和互动性的趣味亲子科学嘉年华活动。除了丰富多样的互动性活动之外，展示馆一层的濒危动物展厅和鸟类展厅、二层的昆虫展厅和蝴蝶展厅，以及标本馆中的昆虫标本分馆、鸟类标本分馆和兽类标本分馆同时向公众开放，展示了丰富的馆藏、悠久的历史，讲解动物标本珍贵的价值，宣传我国多样的动物资源，传达人与动物和谐相处的理念。

走出去：国外巡展与合作交流

借助馆藏标本资源，部分场馆还走出国门，与国外同行开展合作交流，共同举办展示活动。

化石是不可再生的自然资源，是全人类共同的自然文化遗产，也是探究地球演变历史、揭示生物起源

Shanghai Entomological Museum holds a large number of popular science activities every year, including Summer camp, Interactive experience winter camp and Biodiversity research.

(1) "Little Fabre" summer camp. Every summer, Shanghai insect museum organizes students who are interested in life science to go to Zhejiang Tianmushan National Nature Reserve, Anhui Jinzhai nature reserve, Zhejiang Jingning She autonomous region, Zhejiang Anji, Jiangsu Yancheng and other places to carry out a four-day scientific summer camp. Over the past three years, more than 300 students have participated in this program. The children, accompanied by the teachers of the museum, are close to nature and explore the magic of the insect world together.

(2) Interactive experience winter camp. The interactive experience winter camp, with the theme of "insects and chemistry", has been held five times with an audience of about 300 people. It is an interdisciplinary attempt by the museum to carry out "popular science around us". It provides a platform for children to better understand and participate in scientific activities, including insect species, insect habitat, insects and environment, and insects and chemistry, chemistry and life, etc.

(3) Biodiversity research. In the past three years, the museum has organized 38 "biodiversity research" activities, involving more than 1120 people in total. Through observation, collection, recording, painting, experiment, analysis of data and other forms of personal experience, the activity takes students into parks, wetlands and reserves, to understand the habitats, habits and relationships of different creatures, so that everyone can learn scientific methods of biological investigation, improve scientific data processing and analysis ability, learn the way of thinking of scientists, and guide students to understand insects, nature, life and nature.

In May 2018, the National Zoological Museum of China hosted the "Science Night" activity of the National Science and Technology Week, which was highly praised by the public. Based on the good work foundation in 2018, the Institute of Zoology has won the right to host the "Science Night" main event of the National Science and Technology Week in 2019, and NZMC still undertakes several important tasks. In the "Science Night" activity in 2019, the Institute of Zoology fully demonstrated the scientific research innovation and high-quality development achievements of the Chinese Academy of Sciences, and launched more than ten interesting parent-child science Carnival activities full of scientificity, experience and interaction in response to the increasingly diversified needs of the public for science and technology. In addition to a variety of interactive activities, The exhibition hall of endangered animals and birds on the first floor, insects and butterflies on the second floor, insect specimen branch, bird specimen branch and animal specimen branch are open to the public at the same time to display rich collections and long history, explain the precious value of animal specimens, publicize the diverse animal resources in China, and convey the principle of harmonious coexistence between human and animals.

观众参观标本馆

Visitors visit the collection

Going out: overseas itinerant exhibition and cooperation

With the help of the collection of specimen resources, some collections have gone abroad to carry out cooperation and exchange with foreign counterparts and jointly hold exhibition activities.

Fossils are non-renewable natural resources, and a common natural and cultural heritage of mankind. They are the important carriers to explore the evolutionary history of the earth, reveal the origin and evolution of biology, and improve public scientific literacy. As an international well-known fossil collection, the Collection Center of Institute of Vertebrate Paleontology and Paleoanthropology is also responsible for popular science education. Collection Center always insists on carrying out scientific education activities through various forms and strengthening the promotion of scientific research achievements, which not only helps to publicize the true meaning of natural development and evolution, but also helps to promote the concept of biodiversity conservation deeply rooted in people's minds. The center always adheres to the dissemination of scientific ideas to the public through multiple channels, popularization of biological evolution knowledge,

“雪山下的远古世界——青藏高原古生物科考成果展”
“Exhibition of the Paleontological Explorations into the Qinghai-Tibet Plateau”

“中国龙——从撼地巨人到飞羽精灵”特展
The exhibition of “Dinosaurs of China: Ground Shakers to Feathered Flyers” in UK

和演化奥秘，提高公众科学素养的重要载体。作为国际知名的化石收藏机构，古脊椎动物与古人类研究所标本中心也肩负着科普教育的社会责任，始终坚持通过多种形式开展科学教育活动，加强科研成果的推广，既有助于宣传自然发展演变真谛，也有利于促进生物多样性保护理念深入人心；始终坚持通过多元化的途径向公众传播科学理念，普及生物演化知识，激发公众追根溯源的探索精神。借助中国古动物馆的平台，中心每年都会精选大量的标本参与国内外古生物专题巡展活动，如 2019 年在中国古动物馆举办的“雪山下的远古世界——青藏高原古生物科考成果展”、2017 年在英国诺丁汉举办的“中国龙——从撼地巨人到飞羽精灵”特展。同时中心也积极参与相关科普读物的编写，如《听化石的故事》(2018)、《挖掘者手记：一位古生物学者的荒野之旅》(2014)。

基于中国科学家的科学成果，昆明动物博物馆团队也将科普知识传播到法国、英国、日本、马来西亚、马达加斯加等国，在我国少数民族地区、扶贫点、边境口岸、自然保护区、学校、社区、公园、机场、科技场馆等地推广，取得良好的社会效益。此外，科普团队积极开展 STEAM 科普创作，积极与高校、艺术家、软件公司、出版集团等合作开发科普漫画、科普视频、科普游戏、科普绘画、文创衍生品等，2015-2019 年共推出科普展览 12 个、科普教育活动 100 余场，参与观众 200 余万人次，深受社会好评。由于工作出色，昆明动物博物馆获得了全国科普教育基地、云南省精品教育基地、云南省生态文明科普教育基地、中国科学院巾帼建功集体、云南省科学技术奖（科技进步奖）三等奖、中国科学院科学优秀微视频等省部级奖项十余项；获得全国先进科普工作者、中国科协科学使者等个人奖项百余项。

3）专业培训

在向普通大众传播科学知识的同时，各馆还积极配合有关单位，为全国各地的生物相关从业人员提供专业的培训服务，开办了各类培训班。举办培训班是各馆进行高端科学普及的一种有效手段，主要面向知识有所欠缺的一线工作人员，目的是向他们传授一些基础的、实用的科学知识和技能。2014-2019 年各馆共举办各类培训班近 50 期。由于效果显著，培训班规模逐年扩大，参与人数不断攀升，形成了品牌效应，有些甚至影响到了第三世界国家，成为各馆和所在研究所重要的科学普及形式。

and stimulation people's exploration spirit of tracing to the source. With the help of the Paleozoological Museum of China (PMC), the center selects a large number of specimens to participate in special exhibitions of paleontology in China and abroad every year. For example, "Exhibition of the Paleontological Explorations into the Qinghai-Tibet Plateau" in PMC in 2019, the exhibition of "Dinosaurs of China: Ground Shakers to Feathered Flyers" in UK in 2017. At the same time, the center also actively participated in the compilation of relevant popular science books, such as *Listening to the Stories of Fossils* (published in 2018) and *Digger's Notes* (published in 2014).

与日本琵琶湖博物馆合作交流

Cooperation and exchange with Biwa-ko Lake Museum of Japan

The KNHMZ team has shared knowledge with France, the United Kingdom, Japan, Malaysia, Madagascar and other countries based on the scientific achievements of Chinese scientists, and also has achieved good social benefits in China's ethnic minority areas, poverty alleviation points, border crossings, nature reserves, schools, communities, parks, airports, science and technology venues and other places. The science popularization team has actively developed STEAM(Science, Technology, Engineering, Art and Mathematics) popular science content. In recent years, the KNHMZ has actively cooperated with universities, artists, software companies and publishing groups to develop popular science comics, micro-videos, games, posters, and cultural and creative products. From 2015 to 2019, 12 popular science exhibitions and more than 100 popular science education activities have been launched, with about 2,000,000 audiences having been visited, which has been well-received by the community. Because of the excellent work, KNHMZ has won more than ten provincial and ministerial awards, such as the National Popular Science Education Base, Excellent Education Base of Yunnan Province, Popular Science Education Base of Ecological Civilization of Yunnan Province, Women's Merit Group of the CAS, Third Prize of the Yunnan Science and Technology Award (Science and Technology Progress Award) and the CAS' excellent micro-video award; and also more than 100 individual awards, such as National Advanced Science Popularization Workers and Science Emissaries of the Chinese Association of Science and Technology.

3) Professional training

While disseminating scientific knowledge to the general public, the collections also actively cooperate with relevant units to provide professional training services for biological related practitioners all over the country and set up various training courses. Holding training courses is an effective means for collections to popularize high-end science, mainly for front-line staff who lack knowledge, with the purpose of teaching them some basic and practical scientific knowledge and skills. In 2014-2019, nearly 50 training courses were held. Because of the remarkable effect, the scale of the training course has been expanded year by year, the number of participants has been increasing, forming a brand effect, some of which have even affected the third world countries, and become an important form of scientific popularization for various museums and research institutes.

Since its opening in 2009, the exhibition hall of the NZMC has gradually carried out a series of science popularization activities at different levels and for different audiences, and has initially formed its own concept, brand, image and culture in science popularization education and social publicity. In terms of popular science activities, in addition to the regular open exhibitions, there are also a series of theme activities such as "China Animal Specimen Competition" (four sessions), "National Science Venues (Animal Science) Popular Science Training Course" (eight sessions), "Month of Birds Loving" (10 sessions), "Popular Science Lecture Hall" (139 sessions), "Wonderful Night of Museum" (nearly 200 sessions), and a series of field ecological investigation camping activities, and held a variety of special exhibitions, temporary exhibitions, such as "Human Kinship - Primate Diversity and Origin of Human", "Wild China - China's Endangered Animal Protection Image", "Saving China's Tiger to Resist Tiger's Products", "Commemorating the 150th anniversary of the publication of *On the Origin of Species* and the 200th Anniversary of Darwin's Birth" and so on.

Since 2014, NZMC has held "Science Popularization Training Course for Natural Science museums of China" every year. Relying on the platform of NZMC, the training course invited domestic and foreign science popularization experts and scholars, combined with the experience gained by NZMC in conducting science popularization activities for many years, carried out science popularization training to the staff from different industries and units, such as natural science

科普培训班
Science Popularization Training Course

国家动物博物馆展示馆自 2009 年开馆以来，逐步开展了不同层次、面向不同受众群体的系列科普活动，在科普教育和社会宣传方面初步形成了自己的理念、品牌、形象和文化。除了常规对外开放展陈外，还举办了“中国动物标本大赛”(4 届)、“全国自然科学类场馆(动物学科)科普培训班”(8 届)、“爱鸟月”(10 届)、“科普讲堂”（139 期)、“博物馆奇妙夜”（近 200 场）等主题活动和一系列野外生态考察营活动，并举办了各种专题展览、临时展览，如“人类亲缘——灵长类多样性与人类起源”、“野性中国——中国濒危动物保护影像”、“拯救中国虎 抵制虎制品”、“纪念《物种起源》发表 150 周年暨达尔文诞辰 200 周年展”等。

国家动物博物馆自 2014 年起，每年举办“全国自然科学类场馆科普培训班”。培训班依托国家动物博物馆平台，邀请国内外科普专家和学者，结合国家动物博物馆多年来开展科普活动取得的经验，向来自国内不同行业和单位的全国自然科学类场馆、社会科普机构、保护区工作人员开展科普培训，收到了良好的培训效果，得到了社会各界好评。此外，自 2012 年起，国家动物博物馆还承办了由中国科学院动物研究所、中国动物学会、国际动物学会主办的历届“中国动物标本大赛”，已承办 4 届。大赛组织全国主要的动物标本制作厂家和个人将作品送至国家动物博物馆展示馆展陈，并邀请国内外标本制作专家、动物学家等组成评委团，对参赛作品进行评审，评出不同等级奖项。在大赛期间，国家动物博物馆还邀请国内外专家围绕动物标本制作与保藏、野生动物保护法规、动物多样性研究等主题做专题报告。“中国动物标本大赛”的举办，推动了国内动物标本行业的沟通交流和工艺水平的提高，加快了与国际接轨，也将使标本制作更好地服务于动物教学、研究、科普教育以及标本收藏工作，推动我国标本行业的发展。

此外，为了进一步推进国际合作项目“农业有害昆虫的监测预警和绿色防控技术联合研究”的实施，提升各参加单位相关工作人员的理论水平和操作技能，增进国内外参加单位间的了解与合作，国家动物博物馆于 2019 年 6 月举办了“现代农业害虫监测与防控技术国际培训班”。共有 10 名来自哈萨克斯坦、吉尔吉斯斯坦、乌兹别克斯坦、塔吉克斯坦、蒙古国的青年科研骨干和 5 名国内学员参加了此次培训班。通过理论与实践相结合的授课方式，使学员们掌握了农业害虫监测与防控技术，并对我国在害虫监测和防控方面取得的进展产生了浓厚的兴趣。国外学员纷纷表示要将学习到的理论知识以及研究方法带回自己的国家进行推广。该培训班达到了良好的效果，加强了我院与“一带一路”沿线国家在农业有害生物防控和国门生物安全方面的合作基础。

2013 年，华南植物园标本馆率先在中国科学院系统内开展植物标本采集与鉴定培训，截至 2019 年，已经成功举办 8 期培训班，400 多名主要来自华南地区的学员参加了培训。7 年来，我们的培训班从初级班到高级班，再到精品培训项目；培训对象由最初的主要针对自然保护区的技术骨干，后来逐步扩大到院校

中国动物标本大赛
Animal Specimen Competition of China

museums, science popularization institutions and natural reserves in China, and received good training results, which was praised by all walks of life. In addition, since 2012, NZMC has also hosted four sessions of the "Animal Specimen Competition of China", which were sponsored by the Institute of Zoology, the Chinese Society of Zoology and the International Society of Zoology. The competition organized major animal specimen manufacturers and individuals in the country to send the works to the exhibition hall of NZMC for exhibition, and invited domestic and foreign specimen-making experts, zoologists and other members to form a jury to evaluate the entries and awards. During the competition, NZMC also invited experts at home and abroad to make special reports on animal specimen production and preservation, wildlife protection laws and regulations, animal diversity research and other topics. The holding of the "Animal Specimen Competition of China" has promoted the communication and exchange of the domestic animal specimen industry and the improvement of the technology level, accelerated the integration with the international community, and will make specimen production better serve the animal teaching, research, popular science education and specimen collection, and promote the development of the specimen industry in China.

Besides, in order to further promote the implementation of the international cooperation project "joint research on monitoring and early warning of agricultural pests and green control technology", improve the theoretical level and operational skills of relevant staff of all participating units, and enhance the understanding and cooperation between domestic and foreign participating units, the NZMC held the "International Training Course on Modern Agricultural Pest Monitoring and Control Technology" in June 2019. A total of 10 young scientific research backbones from Kazakhstan, Kyrgyzstan, Uzbekistan, Tajikistan and Mongolia and 5 Chinese students participated in the training course. Through the combination of theory and practice, the students have mastered the technology of agricultural pest monitoring and control, and have a strong interest in the progress of pest monitoring and control in China. Foreign students have said they want to bring the theoretical knowledge and research methods back to their own countries for promotion. The training class has achieved good results, and it has strengthened the cooperation foundation between CAS and other B&R countries in agricultural pest control and national biosafety.

In 2013, Herbarium of South China Botanical Garden took the lead in carrying out plant specimen collection and identification training within CAS. By 2019, the herbarium has successfully held eight training courses, and more than

现代农业害虫监测与防控技术国际培训班

International Training Course of Advanced Technologies in Agricultural Pest Monitoring and Control

2019 年培训班合影

Group Photo of the Training Course in 2019

师生，乃至植物爱好者；招生范围从广东省到华南地区，再到邻近省，下一步将要扩大到全国范围；培训内容也由最初的植物分类知识培训，到后来科普能力培训的增加。经过多年的发展，“标本采集与鉴定培训班”已成为该馆的一张名片。

植物研究所标本馆每年定期举办植物分类研究高级培训班。培训班为中国科学院植物科学继续教育基地项目，培训一批植物分类与系统进化学方面的高级专业人才，至今已连续举办 5 届，赢得了大量专业学者和技术人员的认可，极大地推动了中国植物研究的发展。

沈阳应用生态研究所东北生物标本馆除了在 2014-2017 年为沈阳药科大学和沈阳农业大学的项目调查人员开展药用植物标本鉴定培训外，2018 年开始与农业部全国农技推广中心合作，每年对来自 20 多个第三世界国家的学者和学员进行技术培训，学习生物资源开发和利用技术。标本馆科研人员专门组织开发关于生物多样性和资源的系列课程。这些技术培训及参访提升了这些学员对生物多样性保护和生物资源的可持续利用的认识，为进一步开拓沈阳应用生态研究所和这些国家的科技合作提供了重要的交流渠道，同时也宣传了研究所和标本馆在生物资源领域的科研进展。

庐山植物园标本馆依托庐山优越的区位优势，以及庐山植物园科普教育基地和高等院校教学实习基地，通过专题报告和野外考察的形式对高等院校的学生进行植物学专业知识教育，培养学生植物识别能力和兴趣，每年吸引全国 20 多所高校的近万名学生参与。

植物分类研究高级培训班
Advanced training course on plant taxonomy

400 students mainly from southern China participated in the training. Over the past seven years, the training courses have developed step by step: 1) Upgrading from primary course to advanced course, and finally to the top-quality training project; 2) Trainees are mainly for the technical staff of nature reserves at first, then gradually expanded to college teachers and graduate students, and even the nature lovers; 3) The enrollment range is expanded from Guangdong Province to South China region, and then to neighboring provinces, and to the whole country at last; 4) The content of the training was initially only to teach the knowledge of collection and classification, and then increased the training of science popularization ability. After years of development the training course has become a business card of the IBSC.

PE has organized training courses on plant taxonomy and phylogeny in recent years. The training course is a project of continuing education of CAS. Until now, the herbarium had held five sessions, and many senior professional researchers and technicians were trained. The training course has received high praise in the field of plant study and promote the development of Chinese plant research greatly.

In addition to the medical plant specimen identification training for the project investigators of Shenyang Pharmaceutical University and Shenyang Agricultural University from 2014 to 2017, the Northeast Biological Herbaria of Institute of Applied Ecology began to cooperate with the National Agricultural Technology Promotion Center of the Ministry of Agriculture in 2018, providing technical training for scholars and students from more than 20 third world countries every year, learning about the development and utilization of biological resources technology. Researchers of the herbaria specially organize and develop a series of courses on biodiversity and resources. These technical training and visits have improved their understanding of biodiversity conservation and sustainable utilization of biological resources, provided an important communication channel for further exploring the scientific and technological cooperation between the Institute of Applied Ecology and these countries, and also publicized the research progress of the Institute and the herbaria in the field of biological resources.

第三世界国家学员参加生物多样性和资源的系列课程
Participants from third world countries take part in a series of courses on biodiversity and resources

植物考察和识别培训
Training of field plant survey and plant identification

4）科普著作和文章

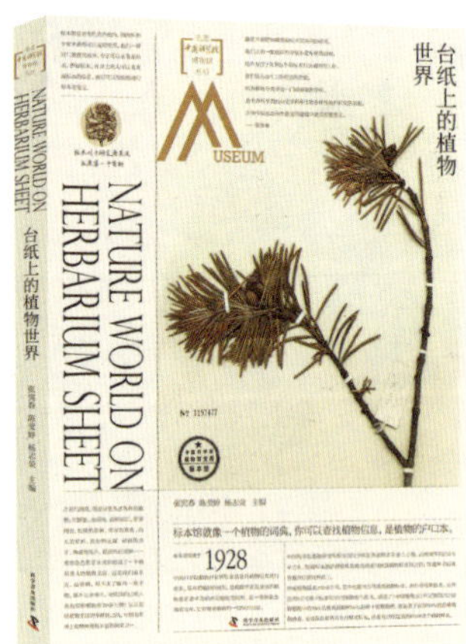

科普图书《台纸上的植物世界》获 2018 年 9 月中国好书奖

Popular science book *Natural World on Herbarium Sheet* won China good book award in September 2018

“中国野生动物生态保护 · 国家动物博物馆精品研究”丛书之《动物与人》

Animals and Humans of the series of “Ecological Protection of China's Wildlife - Boutique Research of National Zoological Museum of China”

在不断发展科普事业的过程中，产生了一系列科普成果，其中最主要、传播范围最广的是各类科普图书和文章。2014-2019 年各馆发表科普著作 80 余部，科普文章 700 余篇，以及野外考察实录 500 余篇。这成为社会大众了解自然生命和科研人员与标本管理工作的最主要的窗口。部分书籍还获得了国家和地方的奖励。

植物研究所标本馆除积极参与全国科普活动周、科学院公众开放日及其他对外科普接待工作外，还出版《台纸上的植物世界》、《西藏野生花卉》、《中国常见植物野外识别手册》、《欧洲园林花卉图鉴》等科普著作。PE 公众号已累计发布植物领域新闻、科普文章 30 余篇，累计阅读量 20 000 多次。

在 2015 年，由中国科学院动物研究所暨国家动物博物馆的科研人员编写的“中国野生动物生态保护 · 国家动物博物馆精品研究”丛书出版，该丛书共分 9 册，全面展现了我国野生动物研究与保护的主要成就、科研成果，图文并茂，融科学性、知识性、趣味性于一体。面世后即受到广大科普爱好者的极大欢迎。

成都生物研究所两栖爬行动物标本馆有着优良的科普工作传统。赵尔宓院士自 20 世纪 80 年代起，就发表了大量关于两栖爬行动物的科普文章，参与编写或审校了部分科普书籍，如《邮票里的动物世界》、《蛇类》等。近年来，标本馆团队利用业余时间译著科普书籍 2 部，即《蛙类博物馆》和《爬行动物》。此外，还在《博

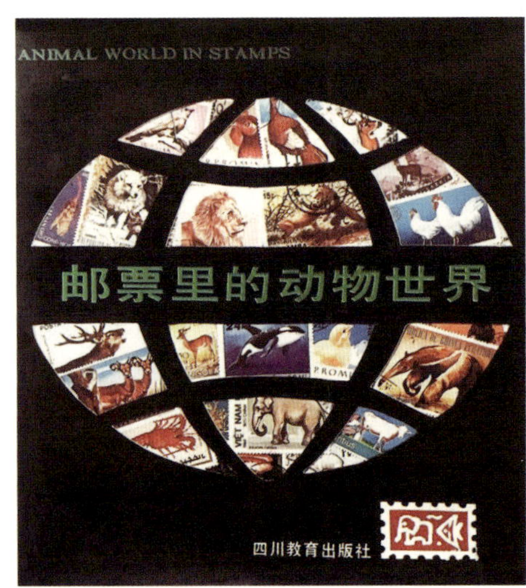

赵尔宓院士参编的科普书籍《邮票里的动物世界》封面

Popular science book edited by Prof. ZHAO Ermi *Animal World in Stamps*

标本馆人员翻译的科普书籍《蛙类博物馆》封面

Popular science book *The Book of Frogs* translated by staff of the museum

标本馆人员审订的科普书籍《爬行动物》

Popular science book *The Reptiles* reviewed by staff of the museum

Relying on the superior geographical advantages of Lushan, as well as the popular science education base of Lushan Botanical Garden and the teaching practice base of colleges an aniversities, the LBG carries out botanical professional knowledge education for students in form of special reports and field trips, cultivates students plant identification ability and interest, and attracts nearly 10,000 students from more than 20 colleges and universities across the country every year.

4) Popular science works and articles

In the process of continuous development of science popularization, a series of science popularization achievements have been produced, among which the most important and widely spread are all kinds of science popularization books and articles. In 2014-2019, each collection published more than 80 popular science works, more than 700 popular science articles, and more than 500 field investigation records. This becomes the most important window for the public to understand natural life and the work of researchers and specimen management. Some books are also awarded by the national and local governments.

PE actively participates in "Science Popularization Week", "Public Open Day" of CAS, and other popular science works. PE also published many books, including *Natural World on Herbarium Sheet*, *Wild Flowers in Tibet* and *Field guide to wild plants of China*, *Flowers of European Garden*, etc. PE's public media in WeChat has released more than 30 news, science articles, and received more than 20,000 reading.

In 2015, the series of "Ecological Protection of China's Wildlife - Boutique Research of National Zoological Museum of China" compiled by the researchers of the Institute of Zoology and the NZMC was published. The series is divided into nine volumes, which comprehensively show the main achievements and scientific research achievements of research and protection of China's wildlife, with rich pictures and texts, integrating science, knowledge and interest. After its appearance, it has been greatly welcomed by the majority of science popularization fans.

科普文章《树蛙，攀援于树的精灵》
Popular science article entitled "Tree frogs, sprites in trees"

There is a fine tradition of popular science work in the Herpetological Museum of Chengdu Institute of Biology. Since the 1980s, Prof. ZHAO Ermi has published a large number of popular science articles on amphibians and reptiles, and participated in the compilation or review of some popular science books, such as *Animal World in Stamps*, *Snakes*, etc. In recent years, the museum staff translated and published two popular science books, *The Book of Frogs* and *Reptiles*. In addition, the museum had also published more than 20 popular science articles in

科普文章《横纹树蛙——墨脱雨林湖泊的精灵》
Popular science article entitled "Medog Flying Frog, sprites from rainforests of Medog"

科普文章《热泉蛇影》
Popular science article entitled "Snakes living in hot-spring"

国家科学技术进步奖

证 书

为表彰国家科学技术进步奖获得者，特颁发此证书。

项目名称：远古的悸动——生命起源与进化

奖励等级：二等

获 奖 者：冯伟民

证书号：2014-J-204-2-01-R01

科普图书获得国家科技进步奖二等奖

Science popularization book won the Second Prize of National Science and Technology Progress

科普图书

Science popularization books

书签：寒武霸主

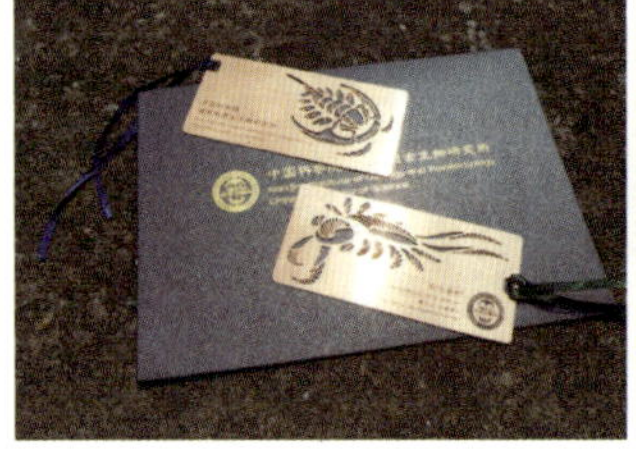

T恤：
志留纪
石炭纪

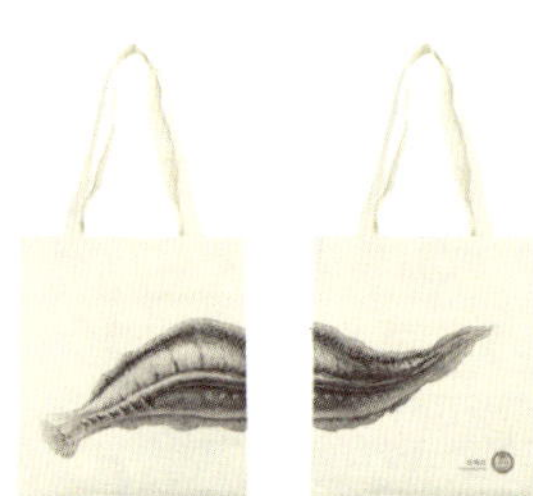

手提袋：云南虫　　手提袋：奇虾

科普文创产品

Popular science and cultural products

物》、《森林与人类》、《大自然》和《动物世界》等杂志上发表科普文章20余篇，网络在线发表科普文章若干。《蛙类博物馆》于2018年7月由北京大学出版社出版，共计45万字，介绍了全世界具代表性的600种蛙类（约占蛙类总数10%），包括外形、色斑、体型、生境、生活史、濒危等级、相近种比较等信息，尤其是每种蛙类都配有与其实际尺寸相同的彩图，能使人直观感受每种蛙类的大小。该书不仅是科学性与艺术性结合的蛙类高级科普读物，也为蛙类的多样性与进化、行为学、保护生物学等研究提供了重要的参考信息。

南京古生物博物馆工作人员积极创作、编写科普图书与科普文章，收获颇丰，远古系列丛书中的《远古的悸动—— 生命起源与进化》获得了2014年度国家科学技术进步奖二等奖。博物馆还积极开发科普文创产品，如结合博物馆展陈特色，开发了5个场馆内容的多媒体触摸软件，恐龙魔方、澄江多点触摸、奇妙涂鸦—— 寒武纪虫虫秀、与恐龙合影等多个大型互动软件，开发设计了3D古生物模型和骨骼制作、澄江生物群扑克牌、澄江生物钥匙扣、樱花书签等文创产品；与全市多家中小学合作，推广科普教育课程。利用传统媒体、网站、新媒体等多种形式向全社会传播科学新知，扩大自身的影响力。

水生生物博物馆出版了具有影响力的科普读物：《瞬间——用镜头留住长江濒危动物》；《青少年环境教育系列读本》（9册，共约63万字，发行数量超过百万册，约有100万学生学习使用），该系列读本获2018年度湖北省科技进步奖二等奖；《小学课本里的奥秘》（1、2册），共计40万字，首次发行1万册。

新媒体科普文章

New media articles of science popularization

magazines, including *Nature History*, *Forest & Humankind*, *China Nature*, *Wildlife*, as well as several online articles. The museum staff had translated and published, 450,000 words popular science book *The Book of Frogs*, which introduces 600 of the most significant frog species (about 10% of the frog species of the world) around the world, including information about morphs, coloration, size, habitations, life history, conservation status, and the distinctions between similar species, the actual size and colorations of every frog species was also provided for a better sense when reading. This book is not only a combination of scientific and artistic but also a fine resource for information of frog diversity, evolution, behavior, and conservation status.

The Nanjing Museum of Palaeontology also runs a series of educational and public engagement programs. These include, for example, the written books on paleontology by staffs, "the Pulse of Ancient Life" received the Second Prize of National Science and Technology Progress Award in 2014; In addition, the museum also actively develops popular science Cultural and creative products, cooperates with many primary and secondary school in the city to promote popular science education courses, and uses traditional media websites, new media.

MHBS published influential science education books: *Moment - Save the Endangered Animals of the Yangtze River with Lens*; *Series of Environmental Education for Youth*. This series have 9 volumes and about 630,000 words in total; its circulation is more than one million copies; and about 1 million students learned this series. This series won the Second Prize of the 2018 Hubei Science and Technology Prize. *The Mysteries in Primary School Textbooks*: This series have 2 volumes and 1,000,000 words in total; it was first released in 10,000 copies.

科学技术奖励证书

获奖项目：青少年环境教育系列科普读本

奖励类别：科技进步奖

获奖等级：贰等奖

获奖单位：中国科学院水生生物研究所

证书编号：2018J-214-2-086-055-D01

二〇一八年十二月

"青少年环境教育系列科普读本"获 2018 年湖北省科技进步奖二等奖（排名第一）

Series of Environmental Education for Youth won the Second Prize of the 2018 Hubei Science and Technology Prize（Rank 1st）

4.5 Application of new technology

The function of the specimen lies in all kinds of information it contains. The use of specimens is actually the process of information extraction, preservation and processing. In past, limited by technology, we can only obtain and use simple macro morphological information. With the development of molecular biology and micro CT technology, the use of specimens has been deepened to the micro level.

With the progress of science and technology, the collections are also exploring new methods, new ideas and new technologies to serve specimen collection, management, as well as information acquisition and utilization, which greatly mobilize the enthusiasm of scientific research and management personnel, and improve the efficiency of information collection and specimen management. Some collections also extend it to science popularization, which has aroused public

4.5 新技术的应用

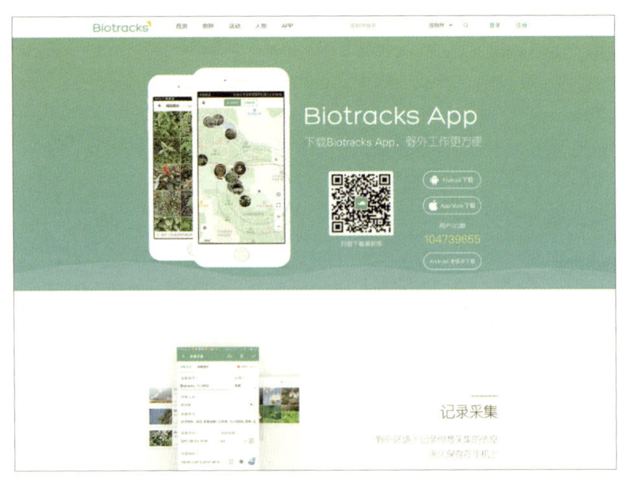

Biotracks 网页
Webpage of Biotracks

标本的作用，本质上在于其蕴含的各种信息。对标本的利用实际上就是信息提取、保存和加工的过程。以往受技术所限，只能对其简单的宏观形态信息进行获取和利用。而分子生物学、显微 CT 等技术的发展促使对标本的利用深入到了微观层面。

随着科技的进步，各馆也在极探索用新方法、新理念、新技术，服务于标本采集、管理，以及信息的获取和利用，极大地调动了科研和管理人员的积极性，提高了信息收集和标本管理的效率。部分场馆还将其延伸到科学普及，引发了公众对生物标本的兴趣，带来了全新的体验。

1）信息的智能收集

昆明植物研究所标本馆开发的 Biotracks 是一套野外科考数字辅助系统，包括相互连接的软件、网站和手机 APP。该系统在国内率先实现了野外与室内工作的全程信息化，实现了“一部手机搞科考”。

Biotracks APP 是国内首个具有普适意义的可以完全替代传统纸质记录本的野外科考数据及时记录应用，它可以帮助个体用户在野外无网络的环境下快速记录诸如采集编号、采集日期、地理坐标、海拔等标本采集信息。同时，它还可以通过 LBS（基于位置的服务，Location Based Services）获得周边物种、路线、地理空间等信息。

由于良好的交互设计，APP 一经推出便大受欢迎，已被国内上百所高校和科研院所的用户使用，上线一年便获得了 1812 位种子用户，富集了上万个物种近百万张数字照片，其中近半数具备详细的地理位置信息，是目前国内成长速度最快的野外生物多样性信息平台。这些数据覆盖了中国 32 个省区市、137 个州市、

Biotracks APP 界面
Interface of Biotracks APP

interest in biological specimens and brought a new experience.

1) Intelligent collection of information

"Biotracks" is a set of digital assistant system for field scientific examination which developed by KUN, including interconnected software, website and mobile APPs. The system takes the lead in realizing the whole process information of field and indoor work in China and realizes "Only use one mobile phone to carry out a scientific investigation".

Biotracks APP is the first field scientific data recording application with universal significance in China that can completely replace the traditional paper record book. It can help individual users quickly record specimen collection information such as collection number, collection date, geographic coordinates, altitude and so on in the field without a network. At the same time, it can also obtain surrounding species, routes, geographical space and other information by LBS (Location Based Services).

As a result of good interaction design, this APP has become popular once it is launched, and has been used by hundreds of universities and scientific research institutes. In the first year of its launch, it has obtained 1,812 seed users, enriched nearly one million digital photos of tens of thousands of species, nearly half of which have detailed geographic location information, and is the fastest-growing field biodiversity information platform in China. These data cover 32 provinces, 137 cities, 320 counties and regions in China, covering 371 families, 2,820 genera and 12,220 species. All these data, except for sensitive data such as endangered, resource species and protected species, can be accessed and queried through http://www.biotracks.cn.

The new APP to be released will also introduce AI species identification, multi-group record, field sample record, code scanning to form a team, project cloud collaboration and other highly attractive functions.

In addition, KUN also carried out a series of promotional activities for the Biotracks APP. On the "National Science Day" on September 15, 2019, the herbarium is responsible for the "Ecology" section. For the public popularization of vegetation change and biodiversity knowledge. Mr. Xu Zhoufeng of the herbarium introduced the use of Biotracks software to the public, and lead the participants to practice in the botanical garden.

In addition, Biotracks also cooperates with Institute of Vertebrate Paleontology and Paleoanthropology of CAS to develop Geotracks APP suitable for users in the field of geology and paleontology, and cooperates with Tencent in all aspects of species identification, scientific public welfare, scientific content, etc., which will greatly expand the user scale and influence of Biotracks and its related platforms. In the future, Biotracks and its related platform system will hopefully grow into the largest and most authoritative natural observation platform in China.

为公众介绍 Biotracks 软件的使用方法

Introduce the usage of Biotracks software to the public

320 个县区，涵盖类群 371 科、2820 属、12 220 个物种。所有的这些数据，除了濒危、资源物种、保护物种等敏感数据，其余均可通过 http://www.biotracks.cn 访问和查询。

即将发布的新版 APP 还将引入 AI 物种识别、多类群记录、野外样地记录、扫码组队、项目云协同等高吸引力功能。

标本馆同时对 Biotracks APP 开展了一系列推广活动。例如，在 2019 年 9 月 15 日全国科普日活动，标本馆负责其中的“生态”板块，为民众科普植被变化及生物多样性知识。标本馆徐洲峰为公众系统介绍 Biotracks 软件的使用方法，并带领大家在植物园中进行实际操作。

此外，Biotracks 还与中国科学院古脊椎动物与古人类研究所合作共同开发适合地质与古生物领域用户的 Geotracks APP，并与腾讯携手在物种识别、科学公益、科学内容等方面展开全方位的合作，这些举措都将极大地扩展 Biotracks 及其相关平台的用户规模和影响力。未来 Biotracks 及其相关平台体系将很有希望成长为中国最大、最权威的自然观察平台。

2）物种的智能识别

“花伴侣”APP
“Flower Companion” APP

植物研究所标本馆在科技部“国家标本资源共享平台”和中国科学院标本数字化项目等支持下，已完成 211 万份标本数字化工作，收集彩色图片 293 万张，目前所有标本数字信息资料已上交战略生物资源服务网络（BRS）；所有新拍摄照片均已提交到“中国植物图像库”。

在此基础上，标本馆利用深度学习技术开发了植物识别软件——“花伴侣”，目前已推出手机端 APP。利用该 APP，只需要拍摄植物的花、果、叶等特征部位，即可快速识别植物，能识别中国野生及栽培植物 3000 属，近 5000 种，几乎涵盖中国植物大部分科属和身边所有常见花草树木。

“花伴侣”APP 主要有以下 4 个功能。

- 识别：只需拍照、选取照片或者从相册分享到花伴侣，即可快速识别；
- 分类：物种科属按照最新分子系统学成果，附有常用俗名，点击名称可进入植物百科；
- 记录：自动保存识别历史，方便后续查看，也可以向左滑动后删除记录；
- 分享：微信好友、朋友圈、QQ、QQ 空间、微博等。

该应用尤其适合园艺工作者、植物爱好者、大中小学生及学生家长，无论在街头、公园或者郊外游览，只需一拍照片就可以认识植物，让人们在亲近大自然的同时，了解身边的植物和花卉。该应用极大地帮助了普通民众识别常见植物，促进了植物学知识的普及。此外，还开发了可识别 1.5 万种植物的“花伴侣”专业版，未来将会为大众提供更多的专业服务。

南海海洋生物标本馆也充分发挥其浮游动物分类专家的优势地位，利用馆藏浮游动物标本，运用浮游动物样品扫描技术已经获得了大量的数据，取得了多种浮游动物种类的多角度标本照片，主要包括 13 个门类，被囊类、水螅水母、管水母、栉水母、枝角类、桡足类、端足类、糠虾、磷虾、介形类、等足类、毛颚类、多毛类等。在此基础上，首次建立了南海浮游动物图像识别种类专家库和浮游动物智能识别系统，囊括了南海浮游动物种类 393 种。通过图像别软件与拟分析样品进行对比，根据专家识别库中已有的浮游动物种

2) Intelligent identification of species

With the support of the "National Specimen Information Infrastructure (NSII)" of the Ministry of Science and Technology and the specimen digitization project of the CAS, the PE has completed the digitization of 2.11 million specimens and collected 2.93 million color pictures. At present, all digital information materials of specimens have been submitted to the Biological Resources Service network (BRS); all new photos have been submitted to the "Plant Photo Bank of China (PPBC)".

On this basis, the herbarium has developed the plant identification software "Flower Companion" by using deep learning technology, and now has launched the mobile APP. With this APP, everyone can quickly identify plants by photographing the flower, fruit, leaf and other characteristic parts of plants. It can identify 3,000 genera and nearly 5,000 species of wild and cultivated plants in China, almost covering most families and genera of Chinese plants and all common flowers and trees around.

The "Flower Companion" APP has four main functions:

- Identification: just take photos, select photos or share them from the album to the "Flower Companion" to quickly identify;
- Classification: according to the latest achievements of molecular systematics, species families and genera are attached with common names; click the names to enter the plant encyclopedia;
- Records: automatically save the identification history for subsequent viewing, or delete the records after sliding to the left;
- Sharing: share to WeChat friends, circle of friends, QQ, QQ space, microblog, etc.

This application is especially suitable for horticulturists, plant lovers, primary and secondary school students and students' parents. No matter they are on the street, in the park or in the countryside, they can know the plants just by taking a picture, so that people can know the plants and flowers around when they are close to the nature. This application greatly helps ordinary people to identify common plants and promotes the popularization of Botany knowledge. In addition, a professional version of "Flower Companion" has been developed, which can identify 15,000 kinds of plants, and will provide more professional services for the public in the future.

The Marine Biodiversity Collections of South China Sea also gives full play to its dominant position as an expert in zooplankton classification. A large number of data have been obtained by using zooplankton specimen collection and zooplankton sample scanning technology, and multi angle specimen photos of a variety of zooplankton species have been obtained, mainly including 13 categories, including cysts, Hydra jellyfish, tube jellyfish, ctenophore jellyfish, Cladocera, copepod, Amphipoda, mysis, krill, Ostracoda, isopod, Chaetognatha, Polychaeta, etc. On the basis of the above analysis, an expert database of zooplankton image recognition and an intelligent zooplankton recognition system were established for the first time, including 393 species of zooplankton in the South China Sea. Through the comparison between the image software and the samples to be analyzed, the long-time series of zooplankton can be continuously collected, observed and quickly identified according to the existing characteristics of zooplankton in the expert identification database. Through the intelligent identification system of zooplankton, the time of zooplankton identification is greatly saved, the accuracy of zooplankton identification and counting is improved, the identification method of zooplankton sample is simplified, and a fast and accurate identification and analysis platform for marine monitoring and environmental evaluation is provided.

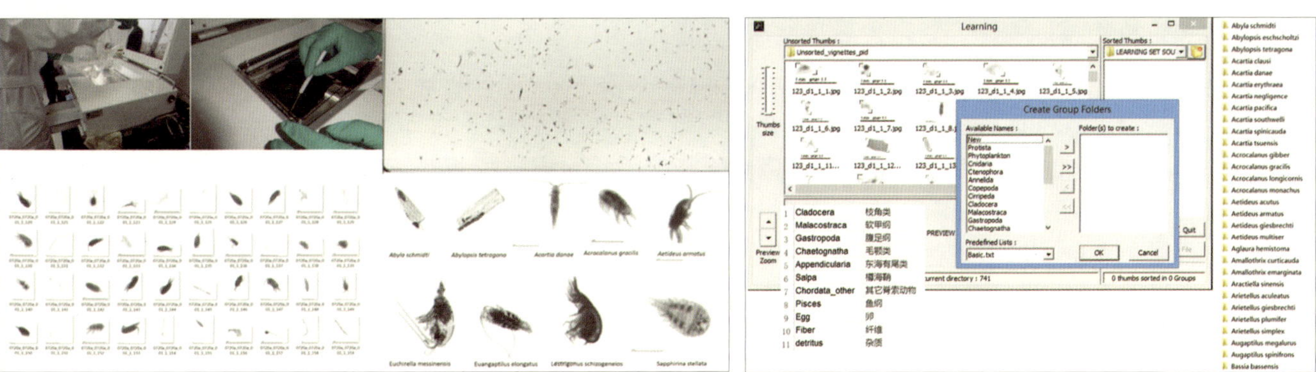

浮游动物智能识别系统

Intelligent zooplankton recognition system

类特征实现对浮游动物的长时间序列连续采集观测和快速鉴定。通过浮游动物智能识别系统大大节省了浮游动物鉴定的时长，同时提高了浮游动物鉴定计数的准确度，简化了浮游动物样品鉴定方法，为海域监测和环境评价提供了快速准确的浮游动物鉴定分析平台。

此外，上海昆虫博物馆还构建“智能鉴别技术应用”，保障第十届花博会举办期间的生物安全。2021 年，第十届中国花博会将在上海崇明举办，生态安全研究工作是花博会运营维安的重要保障和组成部分，一旦松懈，不但会对花博会展期的花卉安全造成严重危害，还会给崇明岛的花卉产业带来严重损失，乃至影响整个岛屿生态系统的稳定。上海昆虫博物馆参与了上海市科委的项目《花博会重要害虫智能鉴别技术研发》，将通过对花博会展区及外围生境害虫开展本底调查、分类研究、基础名录构建和智能鉴别技术应用，为花博会展期本土害虫暴发的预警和控制提供基础服务，为花博会携入害虫的应急防控提供技术支撑，同时通过后期阻断技术研究，减小外来害虫在崇明岛长期定殖危害的风险，在保障花博会举办期间生物安全的基础上，充分维护花博会后崇明岛的生态安全。

3）标本的智能化管理

Kingdonia 数字标本馆系统
Kingdonia digital herbarium system

以往的标本管理工作主要由管理人员通过手工操作完成，费时费力且容易出错。为了克服这些缺点，昆明植物研究所标本馆开发了 Kingdonia 数字标本馆系统。该系统有效地打通了标本管理的各个环节，极大地提高了标本管理效率。

Kingdonia 后台系统深入到标本馆的日常工作之中，并重塑了标本馆的标本数字化和标本管理流程，它不仅支持基于影像的标本数据转录，还可以直接调取普通用户在 Biotracks 上记录的野外采集信息和 AI 物种识别结果，并自动完成这些信息与馆藏标本影像的匹配，从而极大地提高了标本数字化的效率和品质。

Kingdonia 系统上线三年以来，已经将标本馆主要的日常工作整合入信息系统，并使其成为标本馆连接内部各个成员工作的平台，该系统运行至今，平均每天要处理上千次内外用户的访问请求，并使得标本馆从标本采集、标本入库到数据生产、数据服务等各个环节融为一体。

Kingdonia 平台正逐步成长为一套数字标本馆解决方案系统，目前已被中国科学院华南植物园标本馆、西双版纳热带植物园标本馆、新疆生态与地理研究所标本馆采用。该系统不仅可以应用于植物标本馆，还可以根据其他用户的需求快速生成个性化的数字馆藏系统，昆明植物研究所标本馆目前已经与昆明动物博物馆合作，通过 Kingdonia 系统定制动物标本馆的馆藏信息化体系，共同推动 Kingdonia 动物版的开发与普及。未来，Kingdonia 还将进一步开放合作，与各标本馆一起推动中国生物战略资源向更加高效、可控和灵活的体系发展。

此外，植物研究所标本馆还研发了标本伴侣 iSpecimen，实现了万种植物腊叶标本的图像识别，同时开始了对植物花粉和种子图像的采集工作。

In addition, the Shanghai Entomological Museum also built an "Intelligent Identification Technology Application" to ensure biological safety during the 10th Flower Expo. In 2021, the 10th China Flower Expo will be held in Chongming, Shanghai. The ecological security research work is an important guarantee and component of the operation and safety of the Expo. Once relaxed, it will not only cause serious harm to the flower safety during the exhibition period of the Expo, but also bring serious losses to the flower industry of Chongming Island, and even affect the stability of the island's ecosystem. The Shanghai Entomological Museum has participated in the project "Research and Development of Intelligent Identification Technology for Important Pests in Flower Expo" of the Shanghai Science and Technology Committee. Through the background investigation, classification research, basic directory construction and application of intelligent identification technology for pests in Flower Fair area and surrounding habitats, it will provide basic services for the early warning and control of local pest outbreaks during the Flower Expo period, and provide emergency services for pests brought into Flower Expo. With the prevention and controlling technical support, at the same time, through the later blocking technology research, reduces the risk of long-term colonization of foreign pests in Chongming Island, and fully maintains the ecological security of Chongming Island after the Flower Expo on the basis of ensuring the biosafety during the flower fair.

3) Intelligent management of specimens

In the past, the management of specimens was mainly done by the managers by hand, which was time-consuming, laborious and error prone. In order to overcome these shortcomings, KUN has developed the Kingdonia digital herbarium system. The system has effectively opened up all links of specimen management and greatly improved the efficiency of specimen management.

Kingdonia background system goes deep into the daily work of the herbarium and reshapes the process of specimen digitization and specimen management. It not only supports the transcription of specimen data based on image, but also can directly retrieve the field collection information and AI species identification results recorded by ordinary users on Biotracks, and automatically complete the matching of this information with the specimen image of the collection. So, it can greatly improve the efficiency and quality of specimen digitization.

Since the Kingdonia system was launched three years ago, it has integrated the main daily work of the herbarium into the information system and made it a platform for herbarium to connect the work of its members. Up to now, the system has to process thousands of access requests from internal and external users every day, and make the herbarium from specimen collection, specimen storage to data production, data service and other aspects inosculate as a whole.

Kingdonia platform is gradually growing into a digital herbarium solution system, which has been adopted by the Herbarium of South China Botanical Garden, the Herbarium of Xishuangbanna Tropical Botanical Garden, and the Specimen Museum of Xinjiang Institute of Ecology and Geography. The system can not only be applied to the herbarium, but also can quickly generate a personalized digital collection system according to the needs of other users. At present, the Herbarium of Kunming Institute of botany has cooperated with the Kunming Natural History Museum of Zoology to customize the collection information system of the animal collection through Kingdonia system, so as to jointly promote the development and popularization of Kingdonia animal edition. In the future, Kingdonia will further open up and cooperate with various collections to promote the development of China's biological strategic resources to a more efficient, controllable and flexible system.

"iSpecim

Besides, PE also developed "iSpecimen"App, a software for the management of herbarium, which realized the image recognition of ten thousand kinds of plant dehydrated specimens, and started the collection of plant pollen and seed images at the same time.

5

生物标本馆的未来发展趋势

Future development trend of biological collections

目前世界一流的生物标本收藏机构有 10 余家，均位于欧美发达国家。例如，位于美国的史密森学会（Smithsonian Institution），成立于 1846 年，拥有 16 家博物馆，保藏着许多珍贵标本，其馆藏总量已达到惊人的 1.45 亿件，是世界最大的博物馆体系，同时也是美国公共教育、国民服务以及艺术、科学和历史等领域的研究中心；位于欧洲的英国自然历史博物馆（Natural History Museum）原为 1753 年创建的不列颠博物馆的一部分，目前馆藏量超过 8000 万件，是欧洲最大的自然历史博物馆。而其他几个位于美国的自然历史博物馆，如美国自然历史博物馆（American Museum of Natural History）、菲尔德自然历史博物馆（Field Museum of Natural History）、加州科学院金博尔自然历史博物馆（Kimball Natural History Museum），以及位于欧洲的法国国家自然历史博物馆（Muséum National d'Histoire Naturelle）、德国自然历史博物馆（Museum für Naturkunde）等，馆藏量均在 3000 万件以上。

这些三千万级的收藏机构大多历史悠久，多数已有近 300 年的历史，标本藏量多，代表性广，各有特色；拥有一批著名的分类学家和稳定的科研队伍，便于长期从事分类学研究，出版了大量覆盖全球的专科、专属分类学和系统学专著，大多有自己发行的正式学术刊物，并已在世界范围产生极大影响；同时有一支稳定的高水平技术管理队伍，有固定的、较为充裕的研究和运行经费，保持与国内外有关机构进行标本交换，并经常有目的地组织队伍进行标本采集；标本管理的科学性强，已建立起相应的标本数据库，尤其大都对模式标本建立了信息数据库。

与发达国家相比，中国科学院生物标本馆在标本收藏、管理及可持续利用方面还有差距，如标本量、物种数和代表性仍有许多空白，还不能全面反映中国的战略生物资源的实况和优势，分类研究力量日趋薄弱，对面向国家重大需求、为解决重大科研问题等提供技术支撑方面的作用仍然没有充分发挥，科普工作需要加强，现代化管理水平也有待提高。工委会成立后，中国科学院生物标本馆建设事业奋起直追，经过十年的发展，已有了明显的进步。馆藏量达到百万级别的标本馆已有 5 个，其中国家动物博物馆馆藏量已达 800 万以上，是亚洲最大的生物标本馆。

在“一带一路”倡议等不断推进的背景下，中国科学院生物标本馆正面临着前所未有的机遇和挑战。未来生物标本馆必须不断增加馆藏并拓展收藏范围、深入挖掘资源价值、加强资源的整合与利用，并根据国家需求适时调整运行模式，才能更好地服务于科研、国家和社会。

因此，中国科学院生物标本馆未来将会有以下几个发展趋势。

At present, there are more than 10 world-class biological specimen collection institutions, all located in developed countries in Europe and America. For example, the Smithsonian Institution in the United States, founded in 1846, has 16 museums and many precious specimens. Its collection has reached an amazing 145 million pieces. It is the largest museum system in the world, as well as a research center in the fields of public education, national service, art, science and history in the United States; the Natural History Museum in Europe, was originally part of the British Museum founded in 1753, with a collection of more than 80 million. It is the largest natural history museum in Europe. Other natural history museums in the United States, such as the American Museum of Natural History, the Field Museum of Natural History, the Kimball Natural History Museum of the California Academy of Sciences, and the French National Museum of Natural History (Muséum National d'Histoire Naturelle), and the German Museum of Natural History (Museum für Naturkunde), etc. all have collections of more than 30 million.

Most of these 30 million level collection institutions have a long history of nearly 300 years. They have a lot of specimen collections, wide representativeness and unique features. They have a group of famous taxonomists and stable scientific research teams, which are convenient for long-term taxonomic research. They have published a large number of specialized, exclusive taxonomic and systematic monographs covering the whole world. Most of them have their own official academic journals at the same time and have a stable and high-level technical management team with fixed and abundant research and operation funds, maintains specimen exchange with relevant institutions at home and abroad, and often organizes teams to collect specimens at a destination. Their management is scientific, and corresponding specimen databases have been established, especially for the type specimens.

Compared with developed countries, there is still a gap in specimen collection, management and sustainable utilization in the biological collections of CAS. For example, there are still many gaps in specimen quantity, species number and representativeness, which can not fully reflect the reality and advantages of China's strategic biological resources. The strength of classified research is becoming weaker and weaker, and its role in providing technical support for major national needs and solving major scientific research problems has not been brought into full play. The role of science popularization needs to be strengthened, and the level of modern management needs to be improved. After the establishment of the Working Committee, the construction of the biological collections of CAS has made great progress after ten years of development. There are five collections with a collection of more than one million, among which the National Zoological Museum of China has a collection of more than 8 million, making it the largest biological collection in Asia.

Under the background of the B&R Initiative and so on, the biological collections of CAS are facing unprecedented opportunities and challenges. In order to serve the scientific research, the country and the society better, it is necessary for the future biological collections to continuously increase their collection and expand their collection scope, deeply tap the value of resources, strengthen the integration and utilization of resources, and adjust its operation mode according to the needs of the country.

Therefore, there will be the following development trends in the future of Biological Collections of CAS.

5.1 Continue to collect biological specimen resources and international cooperation to increase the number and scope of the collections

Since its birth, biological collections have been the main contributor to the development of basic life science. It is one of its missions to continuously collect and accumulate specimen resources. Each collection will continue to increase specimen resources as an important work to carry out. With the support of relevant resource investigation projects, it will go to domestic hot spots, weak or blank areas to collect specimens, and constantly increase the number and scope of collections.

Under the premise of national strategic cooperation, all collections are steadily promoting scientific research exchange and cooperation, collecting specimen resources of relevant countries and regions. In the future, closer ties will be established with various countries along the B&R. Systematic biological resources investigation and cooperation will be organized. Based on this, the feasibility of the "B&R Alliance for Biological Collection" can be explored. At present, the "B&R Alliance for Biological Resources and Sustainable Utilization" is being actively promoted and will make more contributions to the B&R Initiative in the future.

5.2 Promote the digitization of specimen information and dig the value of resources

Biological specimens, as well as their related taxonomic and systematic studies, have also lasted for quite a long

5.1 持续开展生物标本资源收集和国际合作，增加馆藏数量和范围

生物标本馆，自其诞生以来就是基础生命科学发展的主要贡献者，不断收集和积累标本资源是其使命之一。各馆将不断增加标本资源作为一项重要工作开展，在有关资源调查类项目的支持下，赴国内热点地区、采集薄弱或空白地区采集标本，不断增加馆藏数量和范围。

在国家战略合作的前提下，各标本馆正在稳步推进科研交流与合作，收集相关国家和地区的标本资源，尤其是未来将与“一带一路”沿线国家建立更密切的联系，组织系统的生物资源考察和合作，并在此基础上探索构建“一带一路”标本馆联盟的可行性。目前正在积极推进“‘一带一路’生物资源与可持续利用联盟”建设，未来将为国家“一带一路”倡议做出更多贡献。

5.2 推进标本信息数字化，深入挖掘资源价值

生物标本，以及与其相关的分类学和系统学研究，也已经持续了相当长的时间。随着各国对生物资源的重视和保护，以及生物分类热潮的退去，未来将不再可能有大规模的标本采集活动。取而代之的是已有标本馆之间的交流与融合，资源共享和深入挖掘利用将是未来标本馆发展的趋势之一。如何发挥已有标本资源的价值，是我们必须要考虑的问题。在这种形势下，拥有生物标本的机构应该以“了解地球生命”为使命，这将有助于在日益重视生态环境的背景下增加对生物标本的使用，从而使社会对这些标本、相关研究以及标本所在机构的赞赏、鼓励和支持也随之增加。

在一个对生物多样性信息比以往需求更大的时代，生物标本馆保存的数亿份植物、真菌和动物标本正是提供这样的信息、数据和知识的信息源，能够揭示其他数据来源无法观察到的模式，有着推动全球变化生物学领域发展的巨大潜在价值。生物标本作为生物多样性信息的重要载体，已被越来越多的国家和机构重视，各国正积极开展生物标本数字化工作，中国也已从2003年开始，在科学技术部国家科技基础条件平台项目的支持下，对生物标本开展了大规模的数字化建设工作，取得了巨大的进步。在“国家标本资源库”建立之后，未来将进一步推动数字化工作，向全面建设地球生命数据保藏中心迈进，从而为国家、科研和社会发挥更有力的支撑作用。

5.3 加强对已有资源的整合，统一管理

2018年9月，动物研究所国家动物博物馆组织院内7家研究所标本馆和院外6家高校标本馆联合申报国家科技资源共享服务平台“国家动物标本资源库”，同时植物研究所标本馆也组织全国16家植物标本馆联合申报“国家植物标本资源库”。2019年6月5日，“国家动物标本资源库”和“国家植物标本资源库”顺利获得科技部和财政部批准。

两个国家平台旨在通过整合中国动植物标本资源，依据相关标准、规范开展动植物标本的收集、整理、制作、保藏、研究等工作，对其进行数字化建设，以此推进中国生物标本资源保藏、管理、建设水平。同时以实物资源、数字化资源和科研资源为依托，通过实体馆、门户网站等途径面向社会进行资源共享，实现生物标本资源在科学研究、国家建设和科学普及等方面的服务功能。

国家平台的建立，对于在时间和空间上大尺度的研究非常有利，因为在大多数情况下，只有大型标本馆（馆藏量大于100万号）才能提供全世界不同地域、不同时期收集的标本，并从生物地理和全球气候变化入手，结合新一代基因测序技术，对标本的研究可能在不久的将来提供令人振奋的前所未有的时空视角，从而揭示更多自然奥秘，并为国家战略决策提供指导。

time. With the attention and protection of biological resources and the decline of the upsurge of biological classification, there will be no large-scale specimen collection activities in the future. Instead, the exchange and integration between the existing collections, resource sharing and in-depth exploration and utilization will be one of the future development trends. How to give full play to the value of existing specimen resources is a problem we must consider. In this situation, organizations with biological specimens should take "understanding life on earth" as their mission, which will help to increase the use of biological specimens in the context of increasing emphasis on the ecological environment, so that the community's appreciation, encouragement and support for these specimens, related research and specimen organizations will also increase.

In an era of greater demand for biodiversity information than ever before, the hundreds of millions of plant, fungus and animal specimens preserved in the biological collections are exactly the information sources that provide such information, data and knowledge, can reveal the patterns that cannot be observed by other data sources, and has great potential value to promate the development of the field of global change biology. As an important carrier of biodiversity information, more and more countries and institutions have attached importance to the biological specimens. Countries are actively carrying out the digital work of biological specimens. China has also carried out large-scale digital construction of biological specimens since 2003, with the support of the National Science and Technology Conditions Platform project of the Ministry of Science and Technology and made great progress. After the establishment of the National Specimen Resource Bank, it will further promote the digitalization work and move forward to the comprehensive construction of the earth life data preservation center, to play a more powerful supporting role for the country, scientific research and society.

5.3 Strengthen the integration and unified management of existing resources

In September 2018, the National Zoological Museum of China of the Institute of Zoology organized 7 CAS collections and 6 university's collections to jointly apply the national science and technology resource sharing service platform: "National Animal Collection Resource Center". Meanwhile, the Herbarium of the Institute of Botany also organized 16 herbariums to jointly apply the "National Plant Specimen Resource Center". On June 5, 2019, the "National Animal Collection Resource Center" and the "National Plant Specimen Resource Center" were approved by the Ministry of Science and Technology and the Ministry of Finance.

The two national platforms aim to promote the level of conservation, management and construction of biological specimen resources in China by integrating the resources of animal and plant specimens in China, carrying out the collection, sorting, production, preservation and research of animal and plant specimens in accordance with relevant standards and specifications, and carrying out digital construction of them. At the same time, relying on the resources of specimens, digitalized information and scientific research, resources sharing is carried out for the society through collections, portals and other ways, so as to realize the service function of biological specimen resources in scientific research, national construction and scientific popularization.

The establishment of the national platform is very beneficial to the large-scale research in time and space, because in most cases, only large-scale collections (with a collection of more than 1 million) can provide specimens collected in different regions and periods of the world. Starting from biogeography and global climate change and combining with the new generation of gene sequencing technology, the research on specimens may reveal more natural mysteries and provide guidance for national strategic decision-making in the near future.

After the establishment of the national platform, with the support of special funds, it is expected that the collections will also have a greater improvement in operation management, resource collection and preservation, scientific research support and social services, talent team construction, popular science activities and social benefits.

5.4 The change of operation mode of biological collections

In order to cooperate with the construction of the national specimen resources platform, the operation mode of biological collections will also be changed. The platform will adopt the principle of "Equality, Mutual benefit and Achievement sharing", and build separately and share resources collectively. The units which the platform rely on and co-construction units will form a network structure. Relying units are fully responsible for the operation of the platform, making construction and development plans, rules and regulations, standards and specifications, information collection, review, release, and portal website management and construction, assessing the work of co-construction units, building and operating online service systems, organizing and carrying out resource sharing services, and implementing platform funds

国家平台建立之后，在专项资金的支持下，预期未来各馆在运行管理、资源收集与保藏、科研支撑与社会服务、人才队伍建设、科普活动普惠社会大众等方面也会有较大的提升。

5.4 生物标本馆运行模式的改变

为配合国家标本资源平台建设，其运行模式也将发生变化。平台将采取“平等、互利、成果共享”原则，以分别建设、集中共享的方式进行资源共建、共享。依托单位与共建单位组成网络状结构，依托单位全面负责平台运行，制定建设发展规划、规章制度和标准规范，负责信息汇交、审核、发布，以及门户网站管理建设，对共建单位进行工作考核，负责在线服务系统建设和运行，组织开展资源共享服务，实施平台经费使用。共建单位遵守平台管理规章制度，接受工作考核和绩效评价，根据建设规范和标准开展各项工作，提供资源共享服务。平台实物标本资源归各单位所有和保藏，信息数据均须汇交平台管理使用，所有权和使用权归共建单位及平台共有。根据平台运行与管理需求，管理人员也将进行相应调整。特别是要设立数据分析挖掘人员和共享服务人员，将极大地促进标本资源的利用与共享。

可以相信，未来“国家动物标本资源库”和“国家植物标本资源库”的建设将充分展示我国作为资源大国的优势和科学发展的实力，推动对生物多样性的认知与长期监测，支撑学科的交叉融合与创新，促进生物标本资源面向社会开放共享，提高资源的利用效率。

usage. The co-construction unit shall abide by the platform management rules and regulations, accept the work assessment and performance evaluation, carry out various works according to the construction specifications and standards, and provide resource sharing services. The physical specimen resources of the platform shall be owned and preserved by all units, the information data shall be collected and handed over to the platform for management and use, and the ownership and use rights shall be jointly owned by the co-construction unit and the platform. According to the operation and management requirements of the platform, the manager will adjust accordingly. In particular, the establishment of data analysis and mining personnel and sharing service personnel will greatly promote the use and sharing of specimen resources.

It can be believed that in the future, the construction of the "National Animal Collection Resource Center" and the "National Plant Specimen Resource Center" will fully demonstrate the advantages of a large resource country and the strength of scientific development, promote the cognition and long-term monitoring of biodiversity, support the cross integration and innovation of disciplines, promote the opening and sharing of biological specimen resources to the society, and improve the utilization efficiency of resources.

主要参考文献

References

Chen Y Y. 2019. Preface to the topic “Protection and Utilization of Strategic Biological Resources”. Bulletin of the Chinese Academy of Sciences, 34(12): 1343-1344. (in Chinese) [陈宜瑜 . 2019.“战略生物资源的保护与利用”专题序言 . 中国科学院院刊 , 34(12): 1343-1344.]

Emily K M, Davies T J, Daru B H, et al. 2019. Biological collections for understanding biodiversity in the Anthropocene. Philosophical Transactions of the Royal Society B, 374(1763): 1-9.

Graham H P, Ehrlich P R. 2010. Biological collections and ecological/environmental research: A review, some observations and a look to the future. Biological Reviews, 85(2): 247-266.

He P, Chen J, Qiao G X. 2019. Current situation and future of biological collections of Chinese Academy of Sciences. Bulletin of the Chinese Academy of Sciences, 34(12): 1359-1370. (in Chinese) [贺鹏 , 陈军 , 乔格侠 . 2019. 中国科学院生物标本馆(博物馆)的现状与未来 . 中国科学院院刊 , 34(12): 1359-1370.]

Johnson N F. 2006. Biodiversity informatics. Annual Review of Entomology, 52: 421-438.

Kamenski P A, Sazonov A E, Fedyanin A A, et al. 2016. Biological collections: Chasing the ideal. Acta Naturae, 8(2): 6-9.

Lavoie C. 2013. Biological collections in an ever changing world: Herbaria as tools for biogeographical and environmental studies. Perspectives in Plant Ecology, Evolution and Systematics, 15(1): 68-76.

Miller S E. 2001. A Smithsonian jewel: Biological collections. Science, 293(5534): 1433.

Mwebaze P, Bennett J. 2015. Valuing access to biological collections with contingent valuation and cost-benefit analysis. Journal of Environmental Economics and Policy, 4(3): 238-258.

Office of the First National Movable Cultural Relics Census of the National Cultural Heritage Administration. 2016. Report on Special Survey of the First National Movable Cultural Relics. Beijing: Cultural Relics Press. (in Chinese) [国家文物局第一次全国可移动文物普查工作办公室 . 2016. 第一次全国可移动文物普查专项调查报告 . 北京 : 文物出版社 .]

Robbirt K M, Davy A J, Hutchings M J, et al. 2011. Validation of biological collections as a source of phenological data for use in climate change studies: A case study with the orchid *Ophrys sphegodes*. Journal of Ecology, 99(1): 235-241.

Sa R, Hong D Y. 2014. Brief introduction to the compilation and research project of *Flora of Pan-Himalayan*. Bulletin of Biology, 49 (1): 1-3. (in Chinese) [萨仁 , 洪德元 . 2014.《泛喜马拉雅植物志》编研项目简介 . 生物学通报 , 49(1): 1-3.]

Suarez A V, Tsutsui N D. 2004. The value of museum collections for research and society. BioScience, 54(1): 66-74.

Wilson E O. 2000. A global biodiversity map. Science, 289(5488): 2279.

Xiao C, Luo H R, Chen T M, et al. 2017. Progress and analysis about present situation of National Specimen Information Infrastructure. E-science Technology & Application, 8(4): 6-12. (in Chinese) [肖翠 , 雒海瑞 , 陈铁梅 , 等 . 2017. 国家标本资源共享平台数字化进展与现状分析 . 科研信息化技术与应用 , 8(4): 6-12.]

Yeates D K, Zwick A, Mikheyev A S. 2016. Museums are biobanks: Unlocking the genetic potential of the three billion specimens in the world's biological collections. Current Opinion in Insect Science, 18: 83-88.

Zeng Y, Zhou J. 2019. Strengthening effective protection and sustainable utilization of strategic biological resources in China. Bulletin of the Chinese Academy of Sciences, 34(12): 1345-1350. (in Chinese) [曾艳 , 周桔 . 2019. 加强我国战略生物资源有效保护与可持续利用 . 中国科学院院刊 , 34(12): 1345-1350.]

Zhang L L, Chen J, Qiao G X. 2016. Status and prospect of biological specimen museums in China. World Environment, (Supplement): 88-90. (in Chinese) [张莉莉 , 陈军 , 乔格侠 . 2016. 我国生物标本馆现状与展望 . 世界环境 , (增刊): 88-90.]

Zhang L L, Li D L, Zhou L, et al. 2017. Construction and prospect of animal specimen resource sharing platform in China. E-science Technology & Application, 8(4): 32-35. (in Chinese) [张莉莉 , 李大立 , 周丽 , 等 . 2017. 我国动物标本资源共享平台建设与展望 . 科研信息化技术与应用 , 8(4): 32-35.]